Joseph E. Hodgins

# The Veterinary Science

The anatomy, diseases and treatment of domestic animals also containing a full description of medicines and receipts. Thirteenth Edition

Joseph E. Hodgins

**The Veterinary Science**
*The anatomy, diseases and treatment of domestic animals also containing a full description of medicines and receipts. Thirteenth Edition*

ISBN/EAN: 9783337236564

Printed in Europe, USA, Canada, Australia, Japan

Cover: Foto ©berggeist007 / pixelio.de

THE ANATOMY, DISEASES AND TREATMENT

OF

# Domestic Animals

ALSO CONTAINING

A FULL DESCRIPTION OF MEDICINES AND RECEIPTS.

BY

J. E. HODGINS, V. S.,
*(Honorary Graduate of the Ontario Veterinary College)*

AND

T. H. HASKETT,
*(Secretary of the Veterinary Science Company)*

*THIRTEENTH EDITION.*

LONDON, CANADA:
THE VETERINARY SCIENCE COMPANY,
1897.

# PREFACE.

AT THE request of some of the leading stockowners and prominent farmers of Canada and the United States, and believing that a work of this kind has long been wanted by those interested in domestic animals, the authors have undertaken the publication of THE VETERINARY SCIENCE in the simplest English language. They have endeavored to make it a profitable and interesting study for stockowners and their sons, who, by carefully consulting the pages of this book, will find it to contain information of untold value.

The first fifty-eight pages contain the practical anatomy or conformation of the horse, and should be carefully studied.

The diseases and treatment of the horse follow these, and Chapter I. lays the plan of how to examine a sick horse, and as the reader proceeds it will be found to be simple in the extreme. Mistakes in doctoring can hardly be made, as the causes, symptoms and treatment of the diseases are so fully explained that the stockowner has but to follow the directions laid down in this book.

The comparative anatomy, diseases and treatment of cattle, beginning at page 243, will likewise be found to be as simple and satisfactorily explained as that of the horse.

The diseases and treatment of sheep, found between pages 316 and 337 of the book, are fully dealt with, and it will be found by persons who read and study this part of the book that sheep are as easy to doctor as any of the other animals. The same may be said of the diseases and treatment of pigs, dogs and poultry, found at page 337 and following pages.

The medicines and receipts constitute Part V. of the book, and are a very important part of the study; they are fully and practically explained. The receipts for making liniments, lotions, powders, blisters, etc., are very valuable to everybody, especially those interested in stock, which it may now be said is what the farmer looks to for a livelihood.

The index, as you will see, is a prominent feature of the book, being an index of symptoms as well as of diseases. This makes it very easy for anyone to find out what their animal is suffering from, as they can look for the symptom the animal is showing in the index, which directs them to the right disease.

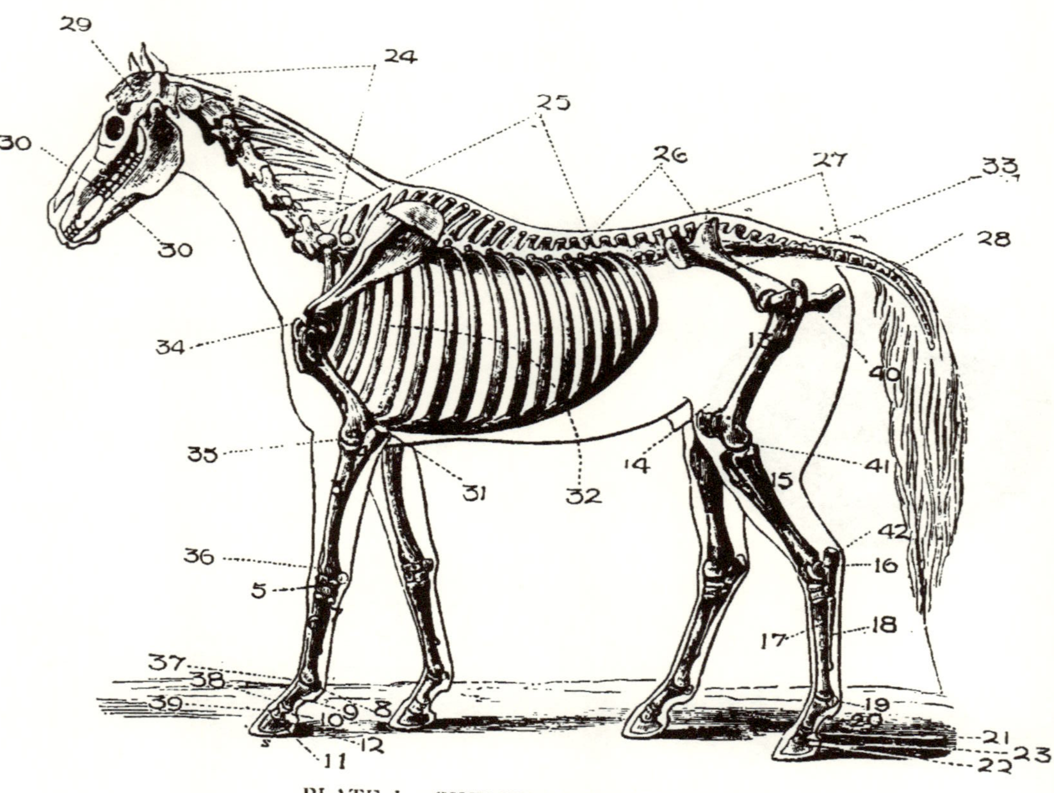


PLATE I. SKELETON OF THE HORSE.

# EXPLANATION OF PLATE I.

## SKELETON OF THE HORSE.

This cut gives the names of all the joints and bones in the body, also the number of bones in each section. Names marked according to numbers.

1. Scapula, or shoulder blade.
2. Humerus, or shoulder bone.
3. Radius, or bone of fore-arm.
4. Ulna, or bone of fore-arm.
5. Carpus, or knee (contains 8 bones).
6. Metacarpal, or large shin bone.
7. Small Metacarpals, or splint bones (2 in number) 1 on each side.
8. Sesamoids, or fetlock bones, 2 small bones at the back of the joint.
9. Os Suffraginis, or large pastern bone.
10. Os Coronæ, or small pastern bone.
11. Os Pedis, or foot bone.
12. Os Naviculare, or shettle bone, situated behind the coffin-joint, and is the seat of the coffin-joint lameness.
13. Femur, or hip bone.
14. Patella, or stifle bone.
15. Tibia and Fibula, or thigh bones, the Fibula being a small bone running down just outside the Tibia.
16. Tarsus, or hock joint (contains 6 small bones).
17. Metatarsal, or large shin bone.
18. Metatarsals, or small shin bones (being 2 in number).
19, 20, 21, 22 and 23 receive the same names in the hind leg as 8, 9, 10, 11 and 12 in the front leg, being the same from the fetlock down.
24. Cervical region, or neck bone (being 7 in number).
25. Dorsal region, or back bones (being 18 in number) to correspond with the 18 pairs of ribs.
26. Lumbar region, or small of the back bones (being 6 in number).
27. Sacral region, or rump bones (being 5 in number).
28. Coccygeal region, or tail bones (being 18 in number).
29. Cranium bones, or bones that protect the brain.
30. Upper and lower jaw bones.
31. Sternum, or breast bone.
32. Ribs, 18 pairs.
33. Pelvis, or hip bones.
34. Shoulder joint.
35. Elbow joint.
36. Carpus, or knee joint.
37. Fetlock joint.
38. Pastern joint.
39. Coffin-joint, which is situated within the hoof.
40. Hip joint.
41. Stifle joint.
42. Tarsus, or hock joint. The joints receive the same name from the hock down in the hind leg, as from knee down in the front leg.

# THE VETERINARY SCIENCE

## PART I.

## ANATOMY OF THE HORSE.

### CHAPTER I.

### BONE.

BONES are hard, yellowish-white, insensitive objects which form the skeleton of animals. Living bone is bluish pink, insensitive and elastic. On exposure to air it becomes diseased and blackened, and is very sensitive and painful.

Bones are made up of two kinds of matter—animal, which makes the bone tough and elastic, and earthy, which makes it hard and brittle. In young animals the animal matter forms one-half of the substance, which afterwards diminishes to one-third as the animal advances in age. This is why we account for old animals' bones being easier to fracture and harder to mend.

#### COVERING OF BONE.

Bones are covered by a tough, fibrous, inelastic membrane called periosteum, which can be seen on examining a bone of an animal which has just died. The only exception we have to this is at the joints where one bone articulates with another, and where a tendon or muscle plays over a bone; here we find its place taken by articular cartilage.

#### CONTENTS OF BONE.

We find in the extremities or near the ends of bones, red marrow, while in the shaft we find white marrow.

#### CLASSES OF BONE.

Bones are classed as long, flat and irregular. Long bones are found in the extremities or legs, and serve as levers for travelling and pillars to support the body. Always remember, that long bones are divided into a shaft and two extremities. Flat bones are found where vital organs need protection. Example: Shoulder blade and ribs, to protect the heart and lungs; pelvic or hip bones, to protect the bladder, rectum and other urinary and genital organs; also the cranial bones, to protect the vital organ called the brain. Irregular bones are found in the spinal column and in the joints, such as the knee or carpus hock or tarsus, where great strength is required.

### HEAD BONES.

In speaking of the different points of importance in connection with the head bones of the horse, in making a cross or longitudinal section of the head, you will notice it is full of cavities or sinuses. The uses of these are to lighten the head and also to warm the air as it passes into the nostrils on its road down to the lungs; you will also find it is full of foramen or small holes for the nerves from the brain and various blood vessels to pass through to the organs situated in the head, such as the tongue, lips, and the various glands in and around the head.

Then we have the cavities in which the eyes are situated, one on each side of the head, called the orbital fossa. In examining this fossa you will find a small opening or foramen, through which the optic nerve passes in coming from the brain to the eye. This is the nerve of sight. Then the most important part of all to consider is the cranial cavity in which that very important organ is situated called the brain, which controls all the various functions and movements of the body.

Then we have another point, viz.: the situation of the drum of the ear. It is situated in the hardest bone we find in the whole skeleton, called the petrosal. The nerve that gives the function called hearing comes from the brain down to the petrosal bone and enters by a small foramen or hole into the drum of the ear to give hearing. This nerve is called the auditory nerve.

### CERVICAL OR NECK BONES.

In considering these bones, we find seven in number. The first and second bones proceeding from the head receive special names. The first one is called the atlas, from which the head is suspended and attached to; it somewhat resembles the body of a bird with wings out. The next bone receives the name of dentata. This is the bone which allows the head to turn in any direction, hence it is sometimes called the axis or pivot of the neck. Between these two bones, on the upper side, is the only place where the spinal cord is not covered with bone—a spot about the size of a twenty-five cent piece. Here is where the operation of pithing is performed in destroying the horse. It is done by running a sharp instrument, such as a knife, fairly in the centre of the neck about four inches behind the ears, and passing in this hole through the coverings of the spine into that very vital organ called the spinal cord, causing death instantaneously. The next four bones receive no name, and are about the same in size and length. The last, or seventh bone is only about one-half the length of the preceding ones and receives no special name.

### DORSAL OR BACK-BONES.

In considering these bones we find eighteen in number. The chief points of interest about them are the height of the spines on the upper part of the bones. These large spines form the withers of the horse, as you will notice in the skeleton. On either side of these bones we find the ribs attached, eighteen pairs corresponding with the number of bones in this region.

### LUMBAR OR SMALL-OF-BACK BONES.

In considering these bones we find six in number, and they are situated immediately above the kidneys.

### SACRAL OR RUMP BONES.

In considering these bones we find five in number in the young horse, which become united and form one in the adult. These are situated between the upper hip bones and help to form the rump of the horse. Beneath these bones the bladder is situated.

### COCCYGEAL OR TAIL BONES.

In considering these bones we find eighteen in number. The principal points of interest in these are, they become smaller as they proceed towards the point of the tail.

### HIP OR PELVIC BONES.

These are three in number, viz.: Ilium, ischium, and pubis. The largest is the ilium, passing upwards and forwards, the outer part of it forming the pin of the hip or outer angle. The back or posterior part of this bone forms a third of the articular surface of the round joint and also helps to form the upper wall of the pelvic or hip cavity. The ischium passes back from the round joint and helps to form the floor of the pelvic cavity, and can be felt in a horse that is poor in flesh projecting out below the tail on each side. The pubis, the smallest of the three bones, is situated immediately in front of the ischium bone and forms the floor of the pelvic cavity.

### STERNUM OR BREAST BONE.

This bone is small and short in the horse and is situated on the lower side of the chest cavity. The principal points of this bone are that it is the softest one in the horse, and the first eight pairs of ribs are attached to it on either side. This bone, in the horse, resembles a small, flat boat.

### RIBS.

In considering the ribs we find eighteen pairs. They form the lateral wall of the thorax or chest cavity and afford protection to the heart and lungs. They are attached above to the dorsal region of the back bone. Below, the first eight pair are attached to the sternum, or breast bone; these are called the true ribs of the horse; the next ten pair are not attached to the sternum below, the lower ends being free, and are continued downwards and forwards by a small piece of cartilage or gristle, and are just slightly attached to the breast bone; these are called the false ribs. Also remember that, starting with the first rib, they get longer until we get to the ninth rib; afterwards they get shorter until the last rib is reached, it being only a few inches long.

### NUMBER OF BONES IN SKELETON OF HORSE.

In the head of the colt is found thirty-eight bones, not including the teeth, but in after life we find a great many bones become attached to each other by a process called ossification.

| | |
|---|---|
| Head | 38 |
| Spinal column | 54 |
| Pelvic cavity or hips, three pair or six single | 6 |
| Ribs, eighteen pair | 36 |
| Breast | 1 |
| Both front legs | 42 |
| Both hind legs | 36 |
| Ear, bones very small and unimportant | 3 |
| Total, not including the teeth | 216 |

The teeth are dealt with separately in another section of this book.

## CHAPTER II.

# CONSIDERATION OF JOINTS.

THE various structures which enter into the formation of joints are the extremities of bone, ligaments, cartilage, and sometimes we have muscles which help to support the joint. All articular joints are supplied with a synovial membrane, which secretes the synovia or joint oil for lubricating the joint.

Cartilage, or what is familiarly called gristle, are of two kinds, viz.: articular and fibrous. The fibrous is not of much importance and does not need much consideration ; it is found in various parts of the body. But the articular cartilage, which covers over the ends of bones where they form a joint, is of more importance.

Ligaments are strong fibrous bands found principally around joints. They are of two kinds, capsular and binding. A capsular ligament is a thin membrane, attached around the end of each bone, which helps to form a joint, and is, as it were, air-tight. The uses of these ligaments are to keep the joint oil from escaping, and partly helps to hold the bone together. On the inside lining of this ligament we find the synovia membrane which secretes the joint oil. Binding ligaments are generally found on each side of of the joint, and are very strong bands of fibrous tissue. The uses of these ligaments are to hold the bones of the joint firmly together.

### THE IMPORTANT POINTS OF JOINTS IN THE LEGS.

**Shoulder Joint.**—This joint is formed by the shoulder blade, or scapula, and humerus or shoulder bone. This is a ball-and-socket joint, and has a strong capsular ligament, and is held to its place also by several large muscles, the most important one being the flexor brachii, which passes down over the shoulder joint through a pulley-like process on the bone, and is held down by a binding ligament which is supplied by a synovial membrane, or sack. This is the seat of what is called shoulder joint lameness. The motion of this joint is outwards or inwards, backwards or forwards.

**Elbow Joint** is formed by the lower part of the humerus and the upper part of the radius and ulna. It has a capsular ligament and binding ligaments, one on the inside and one on the outside. The action of this joint is just forwards and backwards, or flection and extension, but has no side motion.

**Carpus, or Knee.**—This is a very important joint, constructed of eight bones, having two rows, four in each, held together by strong, short ligaments. Thus we have three separate articular surfaces in this joint. The top side of the upper row articulates with the bone of fore arm; this surface gives the most motion to the knee. Another articulation between the two rows of bones gives slight motion to the knee, but not so much as the upper articular surface. Between the lower part of the lower row of bones and the upper part of metacarpal, or shin bones is another articular surface which gives slight motion to the knee. This joint has a large capsular ligament, and has binding ligaments inside and outside. The action of this joint is the same as elbow joint.

**Fetlock Joint.**—This is formed above by the lower part of the shin, or metacarpal bone, and below by the upper part of the large pastern bone. At the back of this joint we have two small bones called the sesamoid bones. This joint has a capsular and two lateral binding ligaments. The motion of this joint is backwards and forwards, same as knee.

**Pastern Joint.**—This joint is situated just above the hoof, and is formed by the lower end of the large pastern bone and the upper end of the small pastern bone. It has a capsular and lateral binding ligaments, same as fetlock joint. It also has same motion as fetlock joint, and is sometimes the seat of what is termed a high ringbone.

**Coffin Joint.**—This joint is situated within the hoof of the horse. It is formed above by the lower end of the small pastern bone, and below by the upper part of the foot bone, or os pedis. Immediately behind this joint, and articulating with the other two bones, we have the navicular, or shuttle bone—it gets its name from its likeness to the shuttle of a sewing machine. This bone is the seat of navicular disease, or otherwise called coffin-joint lameness, and also is affected with what is sometimes called a low-down ringbone.

### JOINTS OF HIND LEG.

**Hip Joint.**—This is a ball-and-socket joint, similar to shoulder joint. It has a capsular ligament and what is called a round ligament, in the joint, holding the head of the bone in the socket, which can be seen plainly on examining the joint. This is an important ligament, as it often becomes strained, which is the seat of hip joint lameness. It is also held together by the heavy muscles of the hip. Its motion is similar to that of shoulder joint.

**Stifle Joint.**—This joint is formed above by the lower end ot the femur, or hip bone, and the upper end of the tibia, or thigh bone. These two bones in front form a pulley-like surface on which the patella, or stifle bone is situated. This bone, when the joint is in motion, glides up and down over the pulley-like surface. It sometimes becomes displaced, and this is termed dislocation of the stifle bone. This is an important point to notice about this joint. It has a capsular and lateral, or binding ligaments; also three very important ligaments, called the straight ligaments, which holds the stifle bone to its place as it plays upon the pulley-like process of this joint. The motion is backwards and forwards, same as elbow joint.

**Hock or Tarsus Joint.**—This joint contains six bones. The two upper bones, one of which is a pulley-like bone placed in front, and the other one placed behind, forms that part of the hock which is called the point of the hock to which the muscles of the gamb are attached, which can be easily seen or felt. The upper surface of these two bones articulate with the lower extremity of the tibia, or thigh bone, and forms a true articulation of the hock joint. This part is what gives most of the motion to the joint. Below those we find three other small, flat bones, placed one upon the other, having articular surfaces between each one. These are called the cuneiform bones of the hock. Immediately behind these three small bones we have what is termed the cuboid bone. This bone also articulates with the cuneiform bones, helping to form the articular surface of the hock. The lower articular surfaces help to give a small amount of motion to the joint. This joint is the seat of the disease termed spavin.

**Fetlock, Pastern and Coffin Joints** are similar to those in front leg.

In speaking of ligaments, there is another very important one to mention which is called the ligamentum nucha, which supports the head when the muscles are at rest. This extends from the pole of the head backwards under the mane and is attached to the spine of the dorsal vertebræ or, commonly called, the withers of the horse. This ligament is chiefly made up of fibro elastic tissue, and will stretch much the same as a piece of elastic.

## CHAPTER III.

# THE MUSCULAR SYSTEM

### COMMONLY CALLED FLESH.

MUSCLES are the chief organs of motion, each one being separated from the other by a thin, delicate membrane made up of connective tissue, which forms a sheath for the muscle. A muscle is divided into two parts, viz.: muscular and tendinous. The muscular part is the larger of the two and forms the larger portion of it, and is sometimes called the belly of the muscle. It is made of muscular tissue, which has a reddish meaty color. At both ends of the muscle we find a tendinous part, or the hard, white portion of the muscle where it becomes attached to the bone. Every muscle is attached to two or more places on different bones, and, upon their contraction, moves the joints of the body. They are well supplied with nerves, which give strength and feeling, and also well supplied with small blood vessels, from which the muscle is fed. Muscles are found in separate groups, all of which have different actions to perform. There are two kinds of muscles—voluntary and involuntary. The voluntary muscles are under the control of the will of the animal; example—the muscles of the legs, hip, back, head, and neck. The involuntary muscles are beyond control of the animal, and will act even though the animal were asleep—such as the heart, the large muscular curtain which separates the chest cavity from the abdominal cavity, which is one of the great muscles of breathing; also the muscles around the chest which assist in breathing.

**Voluntary Muscles** are in groups. The first one we call attention to, after removing the skin, is a thin muscle almost entirely covering the body, and is sometimes accidentally removed if not careful in skinning the animal. The action of this muscle is to shake the skin when flies or something bother the horse. At the head we find a group of muscles which assist in chewing, or masticating, the food. Around the throat is another set of muscles, sometimes called the muscles of the gullet, or pharynx, which assist in swallowing. The neck muscles are divided into two groups, one on each side. The action of these are to raise and lower the head, also to turn the neck and head from side to side. The muscles of the back are generally divided into two groups, one above the spinal column and the other below. The muscles above the spine assist the animal in running, jumping and rearing. The muscles below the spine are sometimes called the lumbar, or psoæ, situated below the lumbar bones, or the bones of the small of the back. Below these muscles are the kidneys. The action of these muscles is to assist the animal in getting up. These muscles are important, for when paralyzed the horse cannot use his hindquarters in trying to get up.

**The Gluteal Muscles,** or the muscles of the hip, are very large, filling in around the hip bones. The action of these is much the same as those of the back, as they assist in jumping, running, rearing, and in raising the hind leg.

**The Abdominal Muscles,** or belly muscles, are four large, flat muscles on each side of the abdomen, or belly. The outer edge of these muscles are attached to the outer ends of the false ribs, also to the processes of the lumbar bones and the outer angles of the hip bones. They unite below to what is called the linea alba, which is a hard, white, fibrous cord, and is attached in front to the back part of the breast-bone, passing back in the centre of the belly, and is attached to the front of the pubic bones, or what is called the lower bones of the hip cavity. About ten inches from where it is attached here, passing forward, we find a small slit, or hole, which is called the navel, or umbilical opening, where the naval vessels pass in and out during the fœtus life, or before the colt is foaled. This is a point of importance to note, for sometimes at the time of birth this opening does not close and allows the bowels to come down and form what is known as navel, or umbilical rupture.

Before finishing the description of this group of muscles we will mention a very large, important ligament which is found spread all over the abdomen, or belly, of the horse. It is of a yellowish color and about one-eighth of an inch thick, attached in front to the back of the breast bone and to the hip bones behind. This is the first structure seen after removing the skin from the belly. This ligament gives great support to the organs contained in the abdominal cavity. The action, or uses, of the abdominal muscles are to support the organs contained in the abdominal cavity, to flex the back-bone and assist in the passing of the fæces, or manure. In the mare these muscles assist her in foaling, or parturition.

**The Coccygeal, or Muscles of the Tail.**—In these we find four important ones—one situated on the upper side of the tail when it is straight out, its action is to raise the tail; two are situated one on each side of the tail, they have the power of drawing the tail to either side; the last one we mention is situated under the tail and is the smallest one of the four, its action is to draw the tail down.

**The Shoulder Muscles** are very large and powerful ones. There are only three of great importance, viz.: The two situated on the outside of the shoulder blade; these are important as they are the seat of the disease called sweeny. The other important one is that which passes down over the shoulder joint through the groove or pulley-like surface on the shoulder bone. This is a long, powerful muscle, attached above to the lower end of the shoulder blade, down through the groove mentioned to the upper part of the bone to the fore arm, or radius, at the front side. Its chief point of importance rests in its action in raising the front leg, where it passes through the pulley-like surface mentioned, when it becomes injured or diseased; it is also the seat of shoulder joint lameness.

**Muscles of Front Leg,** from shoulder down, are divided into two separate kinds, viz.: the extensor and flexor muscles.

The extensor muscles are the ones which bring the leg forward. These muscles, above, are attached to the bones around the elbow joint, passing down in front of the arm bones. About three inches above the knee they become changed into the tendinous part of these muscles, or what is called the cords of the leg. Some of them are attached to the knee joint, while others pass over the front of the joint and are held down to their place by a band or ligament, forming a loop, as it were, for this muscle to glide in when the leg is in action. Each one of these loops, where the muscles pass through, are supplied with a synovial membrane to secrete the synovia, or oil, which lubricate it during action, the same as in the joint. This is a point of importance, as sometimes, on account of injury or strain of this part of the joint where the muscle plays through, we find a small puffy enlargement containing oil secreted by the synovial membrane. This disease is what is called a bursal enlargement.

The flexor muscles are the ones situated at the back part of the leg, attached above to the back part of the elbow joint, passing downwards at the back part of the leg. About two or three inches above the back part of the knee joint they become tendinous, and from there down to the back part of the foot bone, where two of the principal muscles are attached; these form what is known as the back tendons, or cords, of the leg. Some of them become attached to the back part of the knee, same as the muscles on the front part of the leg, while the other two principal tendons pass through a loop formed by ligaments, the same as those mentioned in the front part of the knee. In tracing these tendons down from the knee to the fetlock, they pass through another large loop or sheath formed at the back of the fetlock, where some of the fibres are attached, while others continue down at the back part of the pastern bones, and are attached to the os pedis, or foot bone. These tendons are important as they are known, when they are strained, as the strain of the back tendons. The action of these muscles is to flex the leg, bend the knee, pastern joints and fetlock.

**Muscles of Hind Leg.**—These are also divided into two groups, extensor and flexor.

The extensor muscles are situated in front of the hind leg; attached, above, around the stifle joint, passing downwards in front of the thigh bone, one being attached to the front part of the hock, while the other passes through sheaths, or loops, which is supplied by a synovial membrane, formed by ligaments, to hold the muscles firm in front while the leg is in action. In tracing them down, in front of the shin bone to the fetlock, we find them passing through loops, or sheaths, continuing down in front of the pastern bones to where they are attached. The action of these is to bring the leg forward.

The flexor muscles of the hind leg are attached, above, around the back part of the stifle joint. In tracing them down it is found they become tendinous. Two of the principal ones pass down to that part of the hock joint, which sticks up behind, known as the cap. These form what is called the gamb of the leg, and are partly attached at the point of the hock, the other part passing down to the fetlock joint through a loop, or sheath, along the back part of the pastern bones, and are attached to the foot bone. This muscle, from the hock down, forms one of the back tendons of the hind leg. Another important muscle is found passing down underneath the ones already mentioned, through a loop, or sheath, at the back part of the hock, where it is supplied with a synovia sack. This is a point of importance, for when it becomes strained it is the seat of what is called thoroughpin. It then passes down the back part of the shin bone beneath the other tendon already mentioned, through the loop at the fetlock to the back part of the foot bone, where it is attached. The action of these muscles are to flex or bend the fetlock and raise the hock joint in travelling.

**Involuntary Muscles,** or muscles which are not under the control of the will. The first we call attention to are the muscles of breathing, or respiration; they are a group situated around the chest in such a way as to enlarge the chest cavity and draw the air into the lungs—this action is called inspiration—while others in acting decreases the size of the chest cavity and forces the air out of the lungs, which is called expiration. The diaphragm is a muscular curtain which separates the chest from the abdominal cavity, and also assists greatly in drawing the air in, when it contracts; this muscle also assists in passing manure, and in the mare foaling. It separates the heart and lungs from the bowels, liver and stomach. Everyone interested should examine this muscle, which can be seen by opening any dead animal.

There is one muscle which is both voluntary and involuntary. It is situated in the penis, surrounding the urethra, or the tube, which carries the urine from the bladder to the penis in the male animal. Its action is voluntary while the animal is passing urine, or water. It acts involuntary during sexual intercourse, forcing the semen down through the penis. There is what is known by the name of fat situated between the muscles.

---

## CHAPTER IV.

# THE NERVOUS SYSTEM.

THIS system is a very important set of organs which give motion and feeling to the body, and the different senses, such as seeing, hearing, smelling and tasting. The two principal parts of the nervous system are the brain and spinal cord. The brain, being the centre of the whole nervous system, is situated in the cranial cavity, surrounded by three delicate membranes, the outer one being attached to the inner wall of the cranial cavity. The brain contains several important nerves called the cranial nerves, which are given off from the brain and passed down through the various foramen or openings in the head to supply the different organs situated there, such as the optic nerve, which passes down to the eye, giving the sense of sight. The auditory nerve passes down to the drum of the ear to give the sense of hearing. The aulfactory nerves, which give the sense of smell, are situated in the mucus membrane lining the nose. The nerves passing down to the tongue give the sense of taste. Other nerves pass down to the lips, teeth, mouth and face, giving motion and feeling to the parts mentioned; others pass down to the gullet or pharynx, giving it the power of swallowing. In passing from the brain along the spinal cord, which is situated in the opening of the bones of the back, there are numerous small nerves given off to supply the muscles of the neck, giving the neck motion and feeling. About opposite the shoulder blade, or withers, the spinal cord gives off a large bunch of nerves, part of which supplies the heart and lungs with nervous power. This is a point of importance, for if the spinal cord becomes injured in front of these nerves it causes immediate death. The other part of this bunch of nerves supply the shoulder, chest, and muscles of the front legs. Passing backwards along the spinal cord is found the sympathetic system of nerves, which go to supply the bowels, stomach, liver, kidneys, and other organs situated in the abdominal cavity. Continuing backwards along the spinal cord, to about opposite the hip bones, we find another lot of nerves, one of which goes to supply the rectum, or back bowel; this gives the power of passing manure. Others go to the womb and bladder; these assist in urinating by contracting the bladder. Other nerves pass to the small organs situated in the pelvic cavity; some of these nerves pass down to the hind legs, supplying them with nervous power. The balance of the nerves of the spinal cord go to supply the tail.

Nerves have the appearance of bunches of white thread held together by connective tissue.

## CHAPTER V.

# CIRCULATORY SYSTEM.

THIS is an important system to understand on account of it being the means by which the various parts of the body are fed or nourished. The principal points to consider are the heart, arteries, capillary vessels, veins, and the very important fluid they carry, called the blood. Considering the heart, we find this the main organ of circulation; it weighs about six and one-half pounds in the average horse; it acts as a force pump to force the blood through the vessels already named. It is made of strong muscular tissue, which acts involuntary, and is situated between the lungs, which are divided by what is known as the mediastinum, which is a division between the lungs and is made up of two folds, the heart being between these. The bottom end, or apex, of the heart is downwards and rests just above the breast-bone; the base, or upper part, being directed upwards and to the left side, the left lung having a hollow on its inside for the heart to work in. There is a covering or sack around the heart which helps to protect and support it in its place, attached above to back-bone and below to the bones of the sternum, or breast-bone. This sack is made up of fibrous tissue and is of a whitish appearance; inside it is smooth, and has numerous small glands which secrete an oily substance called serous fluid, which lubricates the outer wall of the heart and the inner wall of the sack so that in action it does not irritate the walls. The cavity in the heart is divided into two parts, the right and left sides; each one of these parts are again divided into an upper cavity called auricle and a lower cavity called ventricle; thus we have the right and left ventricle and right and left auricle. The right auricle communicates with the right ventricle through an opening in the septum, or partition in the right side of the heart. This opening is guarded by a valve to keep the blood from flowing back into the auricle. The left auricle communicates with the left ventricle, same as on the right side. The right side of the heart is sometimes called the venous; this side only deals with the impure blood. The left side is sometimes called arterial side, and only deals with pure blood; this side of the heart is very much stronger and thicker than the right side.

In tracing the blood through the heart, commencing at the right auricle, we find the two large veins of the body, called the anterior and posterior vena cavas, which empty the impure blood of the body into the right auricle of the heart. It is there guarded by two small valves at the mouth of each vein, while the right auricle contracts, forcing the blood down through the hole in the septum into the right ventricle. It is there guarded by a valve to keep the blood from flowing back, while the right ventricle contracts and forces the blood up into the pulmonary artery, which passes only two or three inches above the heart and divides into two branches, one to the right lung and the other to the left lung. These again divide into other small arteries, which flow into what is known. as the capillary network. This network is situated around the air cells of the lungs, where, by a process, the blood gives off carbonic acid gas, which is breathed out along with the air. The blood takes in the oxygen from the pure air, which changes the color of the blood from a black red to a bright red. This is a point of importance to note as it shows that stables should be well ventilated so that animals can have plenty of pure air. The blood now flows into the pulmonary veins, which carry the pure blood back to the heart to be emptied into the left auricle. Here these veins are guarded with valves so as not to allow the blood to flow back while the left auricle contracts to force the blood down through the opening mentioned before into the left auricle. This opening is also guarded by valves so as not to allow the blood to flow back while the left ventricle contracts, with great force, to drive the blood up into the common aorta, the largest artery in the body, which passes above the heart two or three inches, and, just below the backbone, breaks into branches, one passing forward and supplying the parts of the body in front of the heart, the other branch passing backward under the spinal bones, supplying the parts of the body behind the heart.

The branch which goes forward runs under the spine just a few inches, where it breaks into other branches, some going to supply the shoulder and front leg; the other branches being two large ones, one passing on each side of the neck under the jugular vein, which are called carotid arteries. These give off small branches, as they pass up the side of the neck, to feed the muscles and other parts thereof, while just below the butt of the ear, this artery breaks into three large branches, which go to supply the brain and different parts of the head. In considering the branch which runs backwards from the heart, we find it a very long, large artery, passing just below the spine, between the kidneys, breaking up about six inches behind them into four large branches—two on the left and two on the right side. One on the left goes to supply the left hip and organs in the pelvic or hip cavity, while the other passes down the left leg to supply it with blood; one on the right side helps to supply the right hip and pelvic cavity, while the other passes down the right leg. This large branch, in passing back along the spine, gives off small branches—one to the liver, one to the spleen, one to the stomach, and branches to the large and small bowels, and one to each kidney.

Arteries are the vessels which carry the blood away from the heart to the different parts of the body. They always carry the pure blood of the body, which is a bright red color. When the left ventrical contracts it causes a wave, as it were, to pass all down through the arteries. This is an important point in connection with the pulse of a horse. The walls of the arteries are made up of elastic tissue, and after death are always lying open, and, also, you never find and blood in them after death, the reason is because they contract and force the blood all out before they loose the power of contracting.

**How to Tell When an Artery is Cut.**—The blood comes out in spurts every time the heart beats, and is of a bright red color. Arteries are always found deep-seated and well protected with muscles and bone; as, for instance, the large arteries of the legs always pass down on the inside of the leg very close to the bone, and on account of this we very rarely have large arteries injured. Towards the end of the arteries they are found to break up into very small ones which run into the capillary network of the body. These are numerous very small vessels about $\frac{1}{3000}$ of an inch in diameter. Their walls are very thin and cannot be distinguished except under a microscope, and are found in all parts of the body. As the blood passes slowly through these small vessels, the nourishment is absorbed from the blood through the very thin walls to supply the tissues of the body. When the blood passes through this capillary network it again enters into larger vessels called the veins, which carry it on its road back to the heart. In starting at the head to trace the blood back to the heart, we find it carried from the head by two very important vessels called the jugular veins; these are important on account of being the veins which are used to bleed the animal from. The veins which run up the inside of the front leg, carrying the blood back from the leg, unite with the jugular veins, forming what is known as anterior vena cava, which empties into the right auricle of the heart. In tracing the blood back from the hind leg, a large vein is found on the inside of the leg, passing up under the hip, where it unites with the veins of the hip region, forming what is known as the posterior vena cava. As it passes forward it takes in veins from the various organs such as the stomach, liver, kidneys, spleen, and small bowels, and finally empties into the right auricle of the heart. This is the complete circulation of the blood.

Blood is that which carries nourishment to all parts of the body, and also carries away all the waste material of the body, where it is thrown off in the form of urine, which is secreted from the blood by the kidneys. We find the waste material also thrown off in sweat or perspiration through the skin, and also by the lungs. The blood varies in color in the different parts of the body—in the arteries it is a bright red, while in the veins it is a dark red color.

---

CHAPTER VI.

## LYMPHATIC SYSTEM.

This is also known as the absorbent system, and is in connection with the blood vessels, and is made up of very fine minute tubes and glands, which convey from the tissues of the body a clear fluid known as lymph, and pours it into the blood of the veins as it is on its way back to the heart. These glands are found all through the body; for instance, a large group of these are found inside the thigh or stifle joint of the horse, also another large group inside the shoulder. These are important points to note as they sometimes become inflamed and the leg swollen ; then they are the seat of the disease called weed in the leg, or lymphangitis.

## CHAPTER VII.

# DIGESTIVE ORGANS.

THE whole digestive track from the mouth to the anus, which is situated just below the tail, is sometimes called the alimentary canal. The chief parts are the mouth, pharynx, or gullet, œsophagus, or the tube which leads from the gullet to the stomach and the intestines, or bowels. The mouth is an oval cavity at the commencement of the alimentary or digestive canal. In front of the mouth are the lips, one above and one below; at the sides are the cheeks; it is lined by what is known as the mucous membrane, which has several small openings in it from the glands, which are situated around the mouth, through which the saliva is poured. On the upper part of the mouth the mucous membrane is thrown into ridges, or folds, and are from eighteen to twenty in number. This is a point of importance in connection with bleeding a horse with lampers. Always remember it is not safe to bleed back of the third bar because there is a large artery which runs down through the roof of the mouth and enters up through a hole in the bone just before it reaches this bar. The tongue, which has the chief nerves of the sense of taste, is situated in the mouth; this organ also has a very important part to perform in masticating the food and mixing it with saliva. The teeth, which also take a very active part in the masticating of food, are mentioned later on in the book under the heading of "Teeth."

**Salivary Glands.**—These are the glands which secrete the saliva that is poured into the mouth while the animal is eating. There are only three pairs of much importance. One large pair, one on each side of the throat below the ears, filling up the space between the jaw bone and the neck; this pair has tubes passing around and under the lower jaw and up into the cheek muscles, entering the mouth opposite the fourth molar tooth; these tubes are about as large as straws and convey the saliva from the gland into the mouth. The next pair are situated under the pair first mentioned; their tubes enter into the bottom part of the mouth. The third pair are situated under the tongue, one on each side; they pour their secretion into the mouth by several small openings near the front under the tongue, which can be seen by examining closely. This is a very important fluid in connection with the digesting of the food.

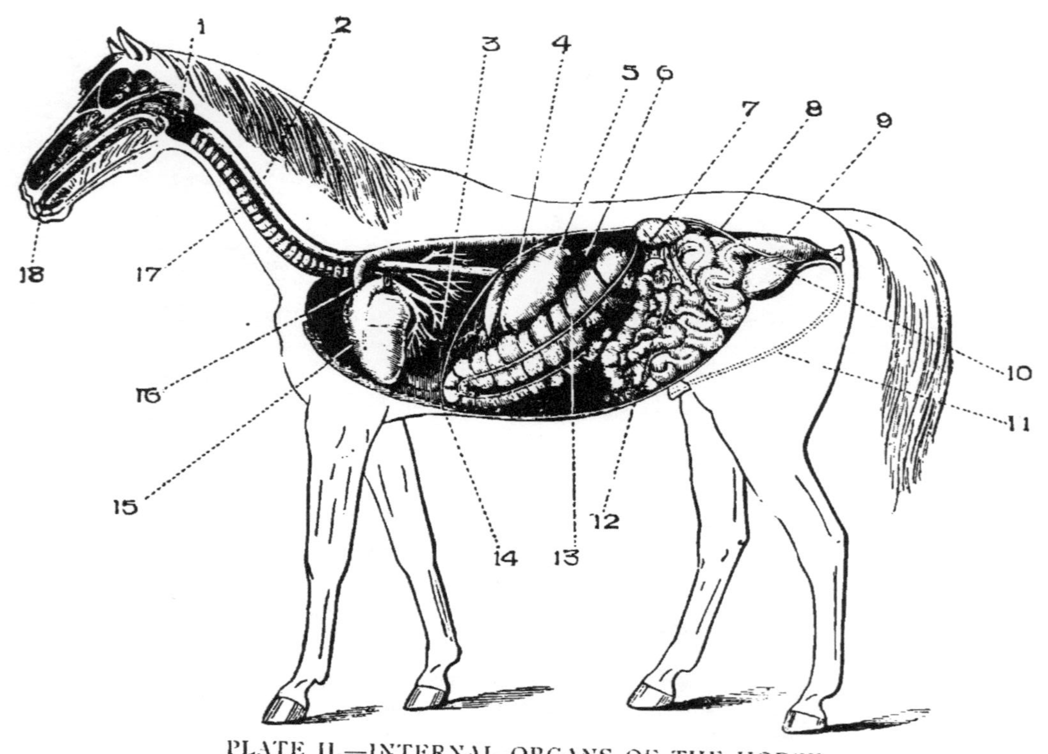


PLATE II.—INTERNAL ORGANS OF THE HORSE

## EXPLANATION OF PLATE II.

### INTERNAL ORGANS OF THE HORSE.

This cut shows the position of each organ in the body.

1. Pharynx, or throat.
2. Œsophagus, or the tube which carries the food to the stomach.
3. Right lung.
4. Spleen.
5. Stomach.
6. Liver.
7. Kidney.
8. Ureter. This is the tube which carries the water or urine from the kidneys to the bladder. There is one tube to each kidney.
9. Rectum, or back bowel.
10. Bladder.
11. Urethra.
12. Small bowels.
13. Large bowels.
14. Diaphragm, or curtain which separates the stomach and bowels from the lungs and heart.
15. Heart.
16. Common Aorta, which is the largest artery in the body.
17. Trachea, or wind pipe.
18. Mouth.

SPECIAL NOTICE.—Every place where Tablespoonful is mentioned in this book should read: SMALL, or DESERT TABLESPOONFUL, which is equal to two Teaspoonfuls.

**The Gullet** is a cavity situated just behind the mouth. It is chiefly made up of muscles which perform the act of swallowing the food. It is lined with the continuation of the mucous membrane of the mouth.

**The Oesophagus,** or the continuation of the gullet, which is a tube extending from the gullet to the stomach, is used to carry the food to that organ. It is made up of two coats, the muscular and the mucous. The former contains fibres which, when once the food enters the tube, contracts behind it, forcing it along to the stomach. Its lining is a continuation of the mucous membrane of the mouth and gullet. In tracing the tube down the neck from the gullet, it passes down the left side of the neck, entering the thoracic, or chest cavity, between the lungs over the heart through the large muscular curtain known as the diaphragm, entering the stomach an inch or two after passing the diaphragm.

**The Stomach.**—This organ is very small in the horse in comparison with the ox; it only holds about four gallons, and is situated just behind the curtain which separates it from the lungs; it is mostly to the left side. The walls of the stomach are composed of three coats, one on the outside called the serous membrane—this is a name applied to membranes which line closed cavities, such as the abdominal, or belly cavity. The inside lining is a continuation of the mucous membrane lining the organs before mentioned. When the stomach is empty this membrane is thrown into loose folds, and there are also found several small openings through which the gastric juice and pepsine from the glands, situated in the walls of the stomach, enters. These are very important fluids as they assist greatly in digesting the food. The third coat is known as the muscular coat, being made of muscular fibres, and is situated between the two coats already mentioned. Its action is to give the stomach a churning motion, rolling the food around and mixing it with the juices. The stomach is guarded by a valve where the food enters which prevents it from passing back through the gullet. There is also a valve at the opening of the bowels, preventing any coarse, undigested food from entering them. The stomach is held in its place by five large ligaments, and is well supplied by blood vessels and nerves. Digestion of food takes place very quickly in the horse in comparison to other animals, and if, through a change of food or working too soon after eating, the digestion is interfered with it sets up what is known as indigestion, which is a very painful disease in the horse. After the food is acted upon by the juices in the stomach it changes into what is known as chyme, which pass into the bowels.

**The Bowels.**—They are divided into two parts—the large and small. The small bowels are seventy-two feet in length and about one inch in diameter, and are made up of three coats, the same as the stomach. The serous coat on the outside contains small glands which secrete an oily material that lubricates the outside of the bowels, which comes in contact with the inner walls of the belly, so as to prevent friction when the bowels are being jolted around in the belly. The muscular coat, made up of muscular fibres, is situated between the other two coats, the same as in the stomach; its action is to contract the bowels, giving it motion to carry the food on through them. The mucous coat is a continuation of the mucous coat of the stomach. Along this coat are found small glands known as villi and lacteal; these are to absorb the nourishment out of the food as it passes along through the bowels and pours it into the blood. This long bowel is found to be attached on the upper side to what is known as the mesentery, which is attached above to the back-bone, and can be seen in any of the smaller animals upon examination. About six inches from the stomach, in the bowels, are found two openings, one for the hepatic duct, which is a tube used to carry the bile from the liver to be poured in on the food as it passes through the bowels. The other hole is for the duct of the gland known as the pancreas. It secretes a clear fluid known as the pancreatic juice, these act on the food in the first part of the small bowels, changing it into chyle. After this, the action of the rest of the bowels is to absorb the nourishment out of the food as it is passing back. The small bowels and stomach, when in a healthy condition, should be found empty one hour after food has been eaten. The small bowels are situated mostly on the left side just behind the stomach.

Large bowels have three coats the same as the small ones. The first part of the large bowels is known as the blind gut and is about three feet long; this is generally the first thing to fly out in opening a horse's belly. Its use is to act as a reservoir to hold the water and fluid parts of the food; this is where the water and fluid parts are mostly taken up into the system. The next part of the large bowels is known as the large colon; it lays along the floor of the belly, is about nine feet long, and is doubled on itself three times. In this bowel the solid part of the food is found, where it is worked about by the contracting of the bowel and the nourishment taken out. After the food is worked back out of this bowel, and the nourishment taken out of it, it enters into what is known as the floating colon, which is ten feet long and about two inches in diameter, or double the size of the small bowels. It is thrown into folds or pleats, and, as that part of the food that has no nourishment in it passes through, it is worked into balls which pass back and are emptied into the rectum or back gut. This part of the bowels, same as the small intestines, is suspended by a sheet or fold from the back bones, and is situated at the back part of the belly, on the left side, behind the small bowels.

**Rectum or Back Gut.**—It is sometimes known as the straight bowel, which is the last part of the intestines, and is about eighteen inches long. The coats of this bowel are a continuation of the large bowels, but each coat is thicker and heavier. Above this bowel, are the bones of the rump; below it, in the horse, we find the bladder and other small glands. In the mare we find what is known as the womb and the vagina, which is the passage into the womb from the outside, and at the sides we find the bones which help to form the hip cavity. At the back part is found what is known as the anus, which is situated immediately under the tail. The use of the rectum is to hold the balls as they pass back from the floating colon and empty in here. When the rectum gets full and presses on the sides of the wall, it stimulates the nerves, which cause the muscular coat of the rectum to contract, while the muscle of the anus dialates or opens out, causing the fæces or manure to pass out.

**The Glands,** which assist in digestion, are the liver, pancreas and the spleen.

**The Liver** is the largest gland in the body; it weighs from ten to twelve pounds in the horse, and is situated between the stomach and diaphragm or curtain which separates the cavities, and is held to its place by several strong ligaments; it is of a dark brown color, and its use is to secrete the fluid called bile, which is a greenish color and very bitter to the taste. There is no gall bladder in the horse, simply a tube which passes down from the liver to the small bowel emptying six inches behind the stomach. It is along this tube where gall stones are sometimes situated. The liver is largely supplied with blood vessels and nerves.

**The Pancreas** is another very important gland situated behind the liver and in front of the kidneys, lying along near the back-bone. This gland is of a grayish, fatty color and can be found by examining closely in front of the kidneys, its use is to secrete a clear, colorless fluid known as the Pancreatic juice. This is an important fluid in connection with digesting the food and enters the bowel just behind where the tube from the liver enters.

**The Spleen** is situated along the left side of the stomach, is long and flat, being about fifteen inches in length, lying along and closely attached to the stomach. This gland is of a grayish-red color and is soft to handle; it is a ductless gland, that is, it has no secretion passing from it. The uses of this organ are not clearly understood, but is supposed to regulate the temperature of the stomach during digestion, and also acts as a reservoir for the blood, and is sometimes called the burying ground of the red corpuscles of the blood, it is largely supplied with blood vessels and nerves and weighs from two to four pounds in the horse.

## MEMBRANES OF THE BODY.

In speaking of the membranes of the body it is found they are of two kinds, serous membranes and mucous membranes.

**Mucous Membranes** are found lining all open cavities of the body, as for example, the lining all through the digestive organs; also lining the respiratory or breathing organs; again it is found lining the organs in connection with the urine, also the genital organs or the organs which reproduce the young animals.

**Serous Membranes** are membranes which line closed cavities, that is, air tight cavities and do not communicate at all with the air, such as are found between the bowels and the wall of abdomen or belly, it is called the peritoneum; there is also another one found lining the chest cavity and the lungs, this membrane is called pleura. There is also another one found in the brain, but is not of much importance. This membrane is always made up of two coats the inner one being attached to the organs contained in the cavity, the other coat is found closely lining the inside of the cavity in which it is found. The inner side of these membranes always secretes an oily fluid which lubricates the two surfaces of this membrane, and if it was not for this oily fluid, the walls would become sore from the friction caused by the moving of one coat on the other. These membranes are an important point to note as sometimes they become chilled and get inflamed and set up inflammation. Inflammation of the pleura membrane mentioned is called pleurisy, and the membrane in connection with the bowels called the peritoneum which when inflamed is called peritonitis, a very dangerous disease, more so than pleurisy.

### PROCESS OF DIGESTION.

In tracing the food from the mouth to where it nourishes the muscles in different parts of the body, we first find the food taken into the mouth by the use of the lips and front teeth where it is masticated or chewed by means of the tongue and teeth, and while it is being rolled around and chewed the salivary glands keep pouring in the saliva, mixing the food and preparing it for digestion. After it is thoroughly chewed it is rolled into a ball at the back part of the tongue, where by the action of the muscles of the gullet it is grasped and forced down into the tube which, by the action of its muscular fibres, is forced down into the stomach where it is rolled about and becomes thoroughly mixed with the gastric and pepsine juices which act chemically on the food and changes it into what is known as chyme. From the stomach it passes up into the front part of the small bowels where there is the secretion from the liver and pancreas which also act chemically and changes the chyme into what is called chyle. After this the rest of the bowels, by the action of these little villi and lacteals, which are in the coats of the bowels, absorbs the chyle or nourishment from the food, these vessels empty the chyle or nourishment into the veins of the body which is then carried to the heart, where it is forced out from the heart through the arteries down into the capillaries which are all through the body, the nourishment is given through the walls of these small thin vessels and gives life and strength to the body.

## CHAPTER VIII.

# ORGANS OF RESPIRATION.

THESE are commonally known as the organs of breathing, they comprise the nostrils, chambers in the head or nasal chambers, larynx or sometimes called Adam's apple. trachea or windpipe, bronchial tubes and the lungs.

**The Nostrils** are two openings, one on each side of the nose; these are held open by the aid of cartilage or gristle, and muscles. About one and one-half inches up the nostril on the under side is a small opening about the size of a grain of shot, this is where the tube or duct which carries the tears down from the eyes empty into the nose. The nostril is lined with a thin delicate skin which afterwards changes into the mucous membrane as it passes up into the chambers of the head.

**Nasal Chambers or Chambers of the Head.**—These give passage to the air from the nostril into the larynx or Adam's apple. There are two of these chambers divided in the center by a thin partition of cartilage called the septum nasi, these communicate with the sinuses of the head. The horse cannot breathe through the mouth on account of the formation of the throat, and this compels him to always breathe through the nostrils. This is a point which should be remembered.

**Larynx or Adam's Apple.**—It is a box made of cartilage, or gristle, gives passage to the air and is also the organ of voice; it is situated in the floor of the gullet. This box has an opening on its upper side, which is guarded by a valve, this valve is always open except when the animal is swallowing food or water. When the food is being swallowed it passes over the valve which closes the opening while the food passes over it. This is important, for if the valve does not close properly and either food or water drop into the windpipe it causes the animal to have a fit of coughing, and is spoken of as the animal to have the food going the wrong way. On the outside of this box of cartilage is found several small muscles attached which help to hold it in its place, it is lined inside by a continuation of the same membrane, as in the chambers of the head. Along the inside of this box is found what is called the vocal chords which come into play when the animal is whinnying. These chords are not nearly so well marked as in the human being, and if they or the cartilage of the larynx become affected by disease it generally gives rise to the disease called roaring.

**Trachea or Windpipe.**—This is a tube which carries the air down from the larynx to the bronchial tubes in the lungs. This tube is made up of forty or fifty rings of cartilage which are united to each other by strong elastic ligaments which gives the windpipe its flexibility, that is, it can be bent in any direction almost like a piece of elastic. In tracing the windpipe from Adam's apple it is found to enter the chest where it terminates into two small tubes, one going to the right lung and one to the left lung; these are called the bronchial tubes.

**Bronchial Tubes and Air Cells.**—These are made up of the same material as that of the windpipe, but are only about half the size. These tubes after they pass into the substance of the lungs they break up into other small tubes which pass all through the lungs and terminate into what is known as the air cells. These small tubes and air cells are lined inside by a very thin mucous membrane which is a continuation of the membrane lining the other organs already mentioned. Just inside this thin mucous membrane is found the capillary network of the lungs, and while the blood is slowly passing through this network of vessels it gives off to the air in the air cells carbonic acid gas and takes in the oxygen from the pure air while it is in the lungs.

**The Lungs** are the most important organs in connection with breathing, they are spongy, yellowish organs, two in number, one situated on the right side and the other on the left; the right lung is the largest on account of the left one having a hollow in its side for the heart. The lungs are separated by a partition known as the mediastinum, also by the heart which is in the folds of this partition and also by the large blood vessels and œsophagus or tube which passes on its way to the stomach. The lungs are made up of light elastic tissue and are full of air cells and tubes, they are very large while the animal is alive and fills up nearly the whole chest cavity, but after death they collapse and are not nearly so large. Between the lungs and the ribs is found a serous membrane called the pleura or the lining membrane of the chest. It is made up of two folds, one being attached around the outer part of the lungs while the other is attached to the ribs at the side and at the back to the large curtain which separates the lungs from the bowels. The little glands situated in this membrane secretes an oily fluid which serves to lubricate these parts while the lungs are working in the chest so as not to cause friction. When this membrane becomes inflamed from a chill or injury it sets up the disease called pleurisy.

The trachea, or windpipe, bronchial tubes and air cells are sometimes compared to a tree, the windpipe being the trunk while the bronchial tubes and air cells represent the branches and leaves of the tree. The lungs are largely supplied by blood vessels and nerves.

**Breathing** in the horse consists of first drawing the pure air in and then forcing the impure air out. These two acts are performed by the muscles of the chest, part of them in contracting in such a manner as to dilate or enlarge the chest cavity and on account of the space between the lungs and chest being air tight the lungs themselves enlarge and the air rushes in to fill up space. The act of forcing the air out is performed by these muscles which relax while others contract in such a manner as to close the chest cavity and makes it smaller, thus the lungs also become smaller and forces the air out. The act of breathing is performed in a horse in perfect health sixteen times a minute.

---

## CHAPTER IX.

# URINARY SYSTEM.

THIS system consists of the kidneys, ureters, bladder and urethra. The action of these organs is to deal with the urine or what is commonly called the water of the animal, which is a watery fluid secreted by the kidneys. It varies in color, according to the condition of the animal's blood.

**The Kidneys** are two in number—one on the right side and one on the left side, and are situated just below the small of the back—the right one being the furthest ahead. In shape, they are long and narrow, and resemble the liver in color. In cutting one of these kidneys open, it is found to be full of glands and tubes, which secrete the urine from the blood while it is passing through the kidneys. These tubes pass to the centre of the kidneys, where they empty the urine into what is called the pelvis of the kidneys. The glands are largely supplied with blood vessels and nerves. In examining the kidneys, one will generally find a large quantity of fat, which help to hold them to their place. The use of the kidneys are to secrete the urine from the blood, which contains a large amount of what is known as ureaic acid, and if not taken out of the blood by these glands, acts as poison to the system.

**The Ureters** are the tubes which carry the urine down from the pelvis of the kidney to the bladder. They are two in number—one situated on the right side of the pelvic or hip cavity and the other on the left side close to the walls—and they enter one on each side at the upper part of the bladder. They are only about the size of an ordinary straw.

**The Bladder** is situated in the pelvic or hip cavity. When it is full it sometimes stretches out into the abdominal or belly cavity. It consists of a body and neck. The bod·

is the large part, and is placed in front; the neck being at the back part of the bladder. This is where the urine or water passes out of the bladder. The bladder is made up of three coats, somewhat similar to that of the bowels. The serous coat is just a continuation of the serous coat found in the belly cavity lining the bowels. The inside of the bladder is lined with mucous membrane similar to that of the bowels, and when the bladder is empty this is all thrown into folds. Another coat is found, between the two membranes above mentioned, called the muscular coat, and is made up of muscular fibres. Its action is when the bladder is full and presses on the nerves of the coat, these nerves causing the fibres in the coat to contract, thus contracting the bladder, forcing the urine out. The bladder is held to its place by ligaments attached to the wall of the pelvic cavity, and above the bladder is found the rectum. The bladder in the horse rests on the floor of the pelvic cavity. The position of the bladder in the mare differs from that of the horse. Instead of the rectum or back bowel being immediately above it, as it is in the horse, the womb is found just above the bladder or between it and the rectum. The use of the bladder is to act as a reservoir to store up the water until the bladder is full; when it is full it presses on the walls and nerves, giving a peculiar sensation to these parts, and causing the walls of the bladder to contract, forcing the water into a tube which carries it from the body; this tube is called the urethra. The neck of the bladder is simply an opening at the back part of the bladder, and is guarded by a valve which prevents the urine from dribbling out except when the animal is passing its water.

**The Urethra** is the tube which carries the water from the bladder out of the body, and is situated much differently in the mare from that of the horse. In the mare it is very short, passing from the neck of the bladder along below the womb and vagina, which is the passage from the outside into the neck of the womb, it opens up into the underside of this passage about four inches in from the outside. This opening is guarded by a small thin valve, and can be felt by passing the finger along the under side of the passage which leads into the womb. In the horse this tube is a great deal longer than in the mare, it commences at the bladder, passes along below the rectum or back bowel to just below the anus, here this tube bends downwards and forwards passing into the penis of the horse and passes down to the end of the penis, where it terminates. This tube is used to carry the urine from the bladder out of the body and is also used in connection with the genital organs, these are the organs which bring forth the young animals. This tube is lined with a continuation of the membrane of the bladder.

---

## CHAPTER X.

# GENITAL ORGANS.

THESE organs in the horse are those which reproduce the young animals. To bring forth the young there must be two animals, one the horse or male animal, the other the mare or female animal, or in other words, there must be one of each sex, male and female. These organs are different in each sex or in the horse and mare and require to be considered separately.

**Genital Organs in the Horse** are as follows: Scrotum or bag, testicles, spermatic cord, vesiculæ seminales or the pouches which holds the semen of the horse, urethra, penis and the sheath.

**The Scrotum** is the sac or bag which contains the testicles and is situated between the hind legs, and is covered on the outside by a very fine soft skin. Passing up in the center under the sheath and scrotum or bag is a well marked line in the skin called median raphe, this can be plainly seen when the horse is on his back, and is found to be continued up gradually getting fainter until it reaches under the anus. Under the skin is found layers of white fascia or tissue, which can be seen on cutting through the scrotum. There is found to be a partition in the scrotum separating the two testicles. The size of the scrotum is affected very much by the weather, the cold weather contracts the fibres in the scrotum causing it to get very much smaller, while in warm weather the fibres relax causing the scrotum to get very much larger. The use of the scrotum is to contain, support and protect the testicles.

**The Testicles** are two in number, one situated on the left side the other on the right, they are oval in shape, and are attached above to the spermatic cord. Before the animal is born the testicles are situated in the abdominal or belly cavity and attached to the serous membrane which has already been spoken of in connection with this cavity. At or about the time of birth, there is what is known as the descent of the testicles into the scrotum; in their downward course, they pass through a slit or small opening at the back part of the muscles of the belly, where they are attached to the under part of the hip bones. These slits or openings are known as the inguinal rings, these rings can be felt in the horse by pressing the fingers well up into the groins. The descent of the testicle is an important point to be remembered for, if it does not come down into the scrotum he is then called what is known as a rig or ridgeling horse, in this case the testicle is not found in the scrotum. At the front part of the testicle is found a small ridge called the globus major, and at the back of it is found another small ridge called the globus minor. Passing between these two ridges there is another well marked ridge called the epididymis, these points can be easily seen upon examining the testicle after the animal is altered or castrated. The substance of the testicle is made up of small glands and fine tubes, these tubes, as they pass towards the back of the testicle, form into larger tubes and finally unite to form one tube, which is used to carry the semen up the back part of the spermatic cord, which these glands in the testicle secrete.

**Spermatic Cords,** or the cords of the testicles, are attached above to the inguinal rings or openings mentioned before, they are about five or six inches long and have the testicles attached to them below. In each cord is found a small muscle which goes by the name of the spermatic muscle, the rest of the cord is made up of the spermatic artery, veins and nerves. Running up at the back of these cords is found a tube about the size of a straw, which upon examination is found to be hard and has a small opening passing up through the centre where the semen passes up through it. This tube is called the vas deferens. Around the spermatic cords and testicles is a serous membrane, one layer being attached to the testicle and cord, while the other is closely attached around the inside of the scrotum or bag. In this membrane are small glands which secrete an oily fluid to lubricate the parts, so as not to cause friction when they are jolted around in the scrotum, this fluid will be noticed to fly out as soon as the scrotum is cut. This is an important point to remember, because sometimes from a slight injury the glands will secrete a large amount of this fluid mentioned, which causes the scrotum to look large and swollen, this disease is known as hydrocele or water in scrotum or bag.

**Vas Deferens.**—These are the tubes which carry the semen up the back part of the cord through the inguinal rings before mentioned. They then pass backwards and upwards, one on each side, to the upper part of the bladder, where they empty into two small pouches or sacs, called the vesiculæ seminales, which store up the semen as it is secreted by the testicles, and when full present the appearance of a pear.

**Vesiculæ Seminales.** These sacs or pouches are situated at the upper side, over the neck of the bladder, one on each side, and have the tube which carries the semen emptying into it at the front end, while at the back end of them is a small opening in each one that leads out into another small tube which passes backward and empties into the urethra, which has been mentioned before as carrying the water out from the bladder. The use of these sacs or pouches is to store up the semen or seed of the horse. While the horse is performing sexual intercourse, these sacs or pouches contract, forcing the semen through these little tubes mentioned out into the urethra, which is a tube leading down to the penis.

**The Penis** is the main organ connected in sexual intercourse; its substance is formed of what is known as erectile tissue, which, under certain circumstances, becomes enormously distended with blood. Passing up the under side, there is what has already been mentioned, the urethra, or the tube, which carries the water or urine out of the body; and also in the act of intercourse, it carries the semen, thus it is noticed this tube is used for two purposes, as we have already mentioned.

**The Sheath** is a loose process of skin which passes downwards from the scrotum or bag, generally from about four to six inches, according to the size of the animal, and is attached to each side, leaving a hole or opening in the centre through which the penis comes down. The outside of the sheath is covered by a thin, delicate skin, same as that of the scrotum; inside it is lined by a membrane having a lot of small glands, which secrete a thick dark fluid to lubricate this passage. Sometimes this fluid collects in here and has the appearance of tar. This is an important point to remember, for when it collects to a large extent the sheath has to be washed out.

**The Semen** or seed of the horse, when examined under a microscope, is found to contain small objects called spermatozoa, which move around, and when in the womb it there meets the ovum of the female, which is secreted by a gland called the ovary. When these two small objects unite, they form the fœtus, or what might be called the animal in its first stage.

**The Female Genital Organs,** or organs of the mare.—These are very different from those in the horse, and are named as follows: Ovaries, fallopian tubes, or the tubes which carry the ovum from the ovaries to the uterus or womb, uterus or womb, vagina, and the vulva.

**The Ovaries** in the mare represent the testicles in the horse. They are about the size of a pigeon's egg, and resemble it much in shape. They are held to their place by ligaments, and at the back part have a tube leading from them called the fallopian tubes. The use of the ovaries are to secrete the ovum or egg. This is a very minute body, which has to be examined under the microscope, being only $\frac{1}{150}$ of an inch in diameter.

**The Fallopian Tubes** are two canals, one on each side, which pass backwards and upwards, and enter into the front part of the uterus or womb. The use of these small tubes are simply to carry the ovum or egg up from the ovaries and empty it into the womb or uterus.

**The Uterus or Womb** is a muscular sac situated in the hip cavity, bounded above by the rectum, below by the bladder, and on the sides by the walls of the hip cavity. It is divided into what is known as a body and a neck. The body of the womb is very small, only about four to six inches long and a couple of inches in diameter when the animal is not pregnant, and near the front end, at the upper side, there are openings where the ovum enters in. When the animal becomes pregnant, the body of the womb becomes enlarged and passes forward and to the left side of the belly or abdominal cavity, getting larger as the time of pregnancy passes on, until the fœtus, or young, has attained its full size. After the mare has had her young the womb begins to get smaller until it attains its natural size again. The womb is very largely supplied with blood vessels and nerves, especially so when the animal is pregnant, as it takes a large amount of blood to nourish the fœtus, or young animal, before birth. The womb is made up of three coats; the inner one is called mucous membrane, and is found to be in the mare, while pregnant, covered over with numerous small processes about the size of peas, to which the placenta or cleaning of the foal is attached. The muscular coat is next to that of the mucous coat, and lies between the outer and inner coats of the womb. It is made up of muscular fibres, and is strong and thick in the womb, much thicker than it is in the bowels or other organs already mentioned. The use of this coat is to support and protect the fœtus or young while it is being carried in the womb, and at the time of parturition, or what is commonly known as foaling, this coat then comes into use, as it contracts the womb very forcibly on the foal, while the neck of womb lies open, helping to force the foal out of the womb. This is important to note as the contraction of this coat is known as labour pains. Lying outside, and covering around the womb, is found a serous coat, which is a continuation of the serous coat of the bowels. The womb is held to its place by strong ligaments attached to the sides of it, and from there to the hip bones, these are called broad ligaments. At the back part of the womb is found the neck. It consists of an opening, formed by a projection, which is about the size of an egg and has a hard, gritty feeling when the animal is not in season and the neck is closed. The neck of the womb is under control of the muscle around it, and this muscle is under control of the nerves of the womb. When the mare comes in season, this muscle is relaxed to a certain extent, thus allowing the neck to open large enough for the passage of a couple of fingers into it; but upon working around it with the fingers it can be forced large enough for a man's hand to pass into it at this period. If the mare is put to the horse at this time, and becomes pregnant or with foal, the muscle in the neck of the womb contracts, firmly closing it, which remains closed until the time of foaling. When, at the time of foaling, the labour pains come on, the muscle in the neck dilates, allowing the neck of the womb to open large enough for the foal to pass out The neck of the womb can be felt easily by oiling the hand and passing it into the passage to the womb, and it will be noticed that the neck spoken of projects into the passage.

**Vagina and Vulva.**—These two organs together make up the passage which leads into the womb from the outside In the young mare they are separated by a thin curtain, or partition, made up of mucous membrane. This curtain is found about four inches from the outside, and is known as the hymen. This membrane is destroyed, or should be, when the mare is first put to the horse, although we often have it broken down in other ways, and in some cases it will disappear of its own accord. The part of the passage in front of the hymen is called the vagina. This passage, in structure, resembles the womb, but is not so strong. There are numerous glands situated along the inner coat or lining of this passage which secrete a fluid to lubricate it. The principal use of this organ is to guide the penis when the animals are performing sexual intercourse, and also serves at the time of foaling as a passage for the foal to come out through The part of the passage behind the hymen is known as the vulva. It is about four inches long and about two or three inches high, varying according to the size of the mare. In front it is separated from the vagina by the hymen membrane. It resembles the vagina in structure, and also has little glands in its inner coat to secrete a fluid to lubricate the passage. At the back part of the vulva, or around the outside, is what is known as the lips of the vulva, one on each side of the opening. The outside of the lips are covered by a very fine skin, and, just below the skin, they are made up of erectile tissue, which is the same kind of tissue as is found in the penis of the horse. This tissue is found more abundantly in the lips of the vulva of the young mare than in the lips of the vulva of the old mare. The opening between these lips is situated just below the anus, or the opening where the back bowel ends. At the back part of the vulva, on the under side, is an opening, or hole, about large enough to allow a man's finger to pass in ; this hole is where the tube leading from the bladder comes up into the passage and allows the urine, or water, to pass into the vulva, where it runs out of the body. The clitoris is situated on the under side of this passage, just inside the lips, and can be seen in the mare after passing water when she works the vulva. Just below the clitoris are found two or three small glands which secrete the fluid that passes away when the mare is horsing.

**Mammary Glands,** or what is known as the mare's bag, are two glands situated between the thighs, the use of which is to secrete the milk after birth to feed the young animal. In the young mare they are very small, but after the mare is with foal a few months these glands begin to get large, and at foaling time they attain their largest size. These glands are covered outside by a thin, smooth skin. The substance of them are made up of small glands and tubes—the glands secrete the milk from the blood, while the tubes retain or hold the milk until it is drawn away from the bag either by milking or the young animal sucking During the time of suckling the young, the glands are largely supplied with blood, from which the milk is secreted. On the under side of each gland is found the teat, or the part the young animal takes hold of in sucking. The end of the teat is pierced by several small holes, where the milk comes out.

### THE FOETUS, OR YOUNG ANIMAL BEFORE BIRTH.

In considering this we must first speak of the ovum, or egg, which is secreted by the ovary of the mare. Every time she comes in season (which occurs every three weeks during the hot weather) this ovum, or egg, passes down the tubes before mentioned into the womb, where it remains a few days and then dies if she is not put to the horse ; but, if during the time this ovum is in the womb she is put to the horse and one of the little bodies which is found in the semen of the horse comes in contact with it the ovum and this little body unites together, the rest of the semen dies and passes away, while the neck of the womb gradually contracts until it is perfectly tight. These two little bodies begin to grow when united and forms the fœtus, or foal. The three parts connected with the fœtus, are the fœtus, navel string, and cleanings, or placenta The cleaning, or placenta, is the part which is found covering the foal and is attached to the little pea-like elevations on the inside of the womb. This covering is found to be full of small blood vessels which run to one point where they unite to form two larger vessels, known as the navel veins, which carry the blood up through the navel opening of the foal where it passes up to its heart ; by the action of the heart it is forced out all through the body of the foal and returned to the heart and then forced down another artery which passes it down to the navel opening, along the navel cord, into the cleaning or placenta again, where it is distributed through the small blood vessels. As the blood comes down this cord from the foal it is in its impure state, and while it is passing through these small vessels in the cleaning it comes very close to the small blood vessels in the womb. The blood is cleansed and nourished from the blood of its mother by a process similar to that which was spoken of in connection with the lungs. The fœtus, or foal, does not grow so fast the first month as it does later on ; at the age of seventeen weeks the first hair appears on the lips and the tip of the tail; between the thirty-fifth and fortieth week the foal begins to show signs of life, and is completely covered with hair. After this time the foal grows very rapidly and can be seen moving around by watching at the flank. The mare carries her foal eleven months, but in some cases in aged mares they have been known to carry their foal over twelve months, and in rare cases in young mares they lack a few days of eleven months.

**How to Tell When a Mare is With Foal.**—The first thing that is noticed is that she does not come in season at the end of three weeks, and if felt at the flanks she will be noticed to be peevish and cross, and also ugly to other horses. The mare usually feeds and thrives better at this period, and at the end of three or four months she begins to get larger at the flanks, and gradually continues getting larger until foaling time. Mares that are fed on hard feed and worked do not usually get as large as mares fed on rough feed and not worked. At about the fifth or sixth month the foal begins stirring in the womb, which can be seen at the flank ; this is noticed mostly after the mare has had a drink of cold water ; it also can be felt by pressing the hand against the flank on the left side. At about the sixth month in the young mare the mammary glands, or bag, begins to get large, and gradually gets larger until the time of foaling.

**Signs of Foaling.**—The muscles and ligaments gradually become relaxed until there is quite a hollow at each side of the tail. The vulva gets quite large at foaling time and wax usually runs from the teats of the mare a few days before. A few hours before foaling she is noticed to be walking around and acting quite uneasy until the labor pains come on, when her restlessness increases to getting up and down and forcing, until what is known as the water bag comes out and breaks ; the labor pains increase, and she lies down, forcing violently, until the front legs and head of the foal appear, when it soon slips out, and the cleaning generally comes with it. Sometimes the foal comes backwards, which is harder on the mare. If the mouth of the foal is examined immediately it is found to contain what is known as the melt, which looks like a piece of liver.

CHAPTER XI.

# THE SKIN.

THIS is the membrane which covers the body, and consists of two layers called the dermis and epidermis. The epidermis is the outer layer of skin, and is made up of epithelium and protects the under layer from the air and slight injuries, this layer undergoes a continual process of being made up and passing away in dandruff. The dermis or true skin is well supplied with blood and nerves, part of the nerves being the nerves of touch. What is known as the sweat glands are found in this layer. When the skin is injured, the outside layer being knocked off, this part of the skin is very painful. The skin is attached on the inside to the body by a layer of white tissue which is known as the areolar tissue, this being the tissue which is cut through in skinning an animal. The skin varies in thickness on different parts of the body, being thinnest on the under parts.

### HAIR.

There are three kinds of hair on the horse, the common, which covers most of the body, being the finest of the three. The mane and tail, which is coarse and long. Around the muzzle or nose and the lips are found long hairs, usually black and called cat hairs.

On the inside of the front legs, just above the knee, and on the inside of the hind legs, about the hock, are rough, horny spots which are called chestnuts.

---

CHAPTER XII.

# THE HOOF.

THIS is a very important point in anatomy in connection with the lameness of the horse. The hoof of the horse corresponds to the finger nail of the man—it is divided into three distinct parts, the wall, the sole, and the frog.

**The Wall** is the part of the hoof that is seen when the foot is resting flat on the ground ; it is divided into the toe, the quarters, the heels and the bars. The toe forms the front of the hoof, and is the thickest and strongest part of the wall. The quarters are situated at the side of the hoof. The walls are not nearly so thick here as at the toe, but are almost straight up and down. The heels

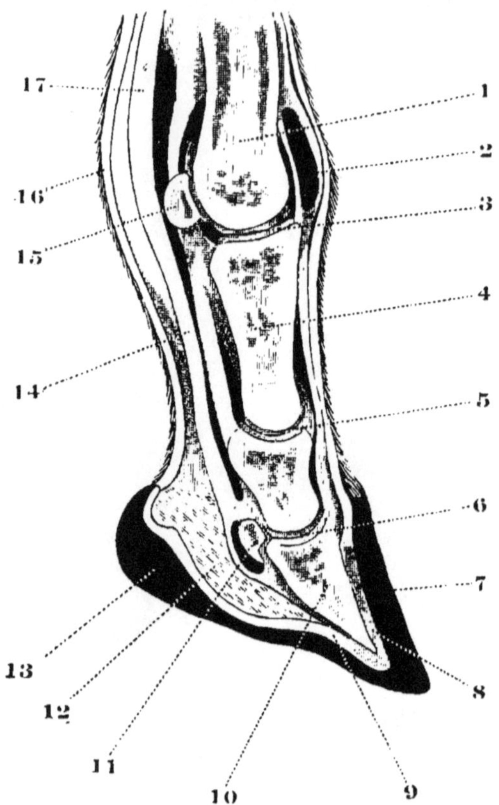


PLATE III.—FOOT OF THE HORSE.

# EXPLANATION OF PLATE III.

## FOOT OF THE HORSE.

This cut represents the foot of a horse sawed from above the fetlock down through the centre of the foot, showing the structure of the foot, and giving the names of each part according to number.

1. Lower end of large metacarpal, or shin bone.
2. Bursa, which secretes the joint oil that lubricates the place where the tendon, or cord, on the front of the leg passes down over the front of the fetlock joint. This is important as it sometimes gets injured and becomes enlarged, it is then called a puffy or bursal enlargement, and is of the same nature as a wind gall.
3. Fetlock joint.
4. Large pastern bone.
5. Pastern joint. This joint is important, for when diseased it is the seat of a high-up ringbone.
6. Coffin joint. This joint is important for when it is diseased it is the seat of a low-down ringbone.
7. Wall of the hoof.
8. Quick of the foot, or sensitive wall.
9. Quick of the foot, or sensitive sole.
10. Os Pedis, or foot bone.
11. Coffin, or navicular bone. This is important for when diseased it is the seat of coffin joint lameness.
12. Fatty Frog.
13. Frog of the foot, or horny frog.
14. Back tendons below fetlock.
15. Fetlock bones (2 in number) one on each side of the joint.
16. Skin.
17. Back tendons above fetlock.

SPECIAL NOTICE.—Every place where Tablespoonful is mentioned in this book should read: SMALL, or DESERT TABLESPOONFUL, which is equal to two Teaspoonfuls.

are situated at the back part of the foot. From the heel is a process of hoof, which looks like a bar, passing forward between the frog and the sole of the foot ; this can be seen plainly by raising up the horse's foot. There is one of these at each side of the frog, and they act as a brace to the heel and the quarters of the wall ; these are called the bars. Covering the outside of the wall is a fine membrane which gives the hoof the polished appearance ; this is called the periople. This can be best seen when a horse's hoof is well washed off, as it is after travelling through wet grass, which gives the hoof a shining appearance. The use of this membrane is to keep the moisture in the hoof and protect it from water. This is a point of importance in connection with shoeing horses, as it is very injurious to file the wall too much. Around the top part of the wall, where it unites with the skin, is found a groove which contains a white band, called the coronary substance, or band. The use of this is to nourish the wall of the hoof, or, in other words, it is from this the wall of the hoof grows. The under part of the wall, or the part which rests on the ground in the unshod animal, or the part to which the shoe is nailed in the shod animal, is called the spread of the foot. On the inside of the wall, attaching it to the bone of the foot called the os pedis, is the part called the quick, or sensitive laminæ. This is a point of importance in connection with driving nails in shoeing, so as not to drive the nail into this membrane or even press on it, for it is very sensitive. When a nail has been driven in so as to injure this membrane it is the common expression, "you have pricked that horse's foot."

**The Sole** is a thick plate of horn which helps to form the under part of the hoof. It is situated between the inner border of the under part of the wall already mentioned and the front of the frog. The under part of the sole is concave, or hollowed out ; the upper part of the sole is attached to the under part of the os pedis bone, or bone of the foot, by a membrane called the quick, or sensitive sole—this membrane is just a continuation of the sensitive laminæ. The outer part of the sole is attached to the inner part of the wall. When pared down a white ring is seen where the sole and the wall is united. At the back part of the sole there is a notch the shape of the letter V ; in this notch the frog is situated. An important point to remember in shoeing is never to let the shoe rest on any part of the sole, and, also it is not well to pare off too much of the barky-looking substance of the sole as it helps to keep the moisture in the foot. When this is taken off it allows the moisture to escape and it becomes dry and contracted.

**The Frog** is the prominent spongy horn found in the V shaped notch in the back of the sole. It is wide at the back, helping to form the heels of the foot, the pointed part in the front is called the apex of the frog. The under part of the frog is triangular in shape and has a hollow in it called the cleft of the frog. There is a hollow at each side of the frog, between it and the bars, called the commissures of the frog. On the upper part is a membrane, known as the sensitive frog, which attaches the frog to the under part of the os pedis, or foot bone. This membrane is simply a continuation of the sensitive sole spoken of in connection with the sole. The back part of the frog is the widest part and spreads out forming the heels.

To get the best idea of the structure of the foot, get a hoof and the bones of the leg as far up as the fetlock, and saw them down through the centre, which will show you the exact conformation. The shape of the hoof differs in animals—large draft horses are apt to have what is termed a flat foot, while in the road horses the chief trouble is what is known as contracted feet and weak heels.

---

## CHAPTER XIII.

# THE EAR.

THE ear of the horse is divided into several parts. The inner part, or drum of the ear, is situated in the hardest bone in the body, called the petrosal. The nerve which passes into the drum of the ear, and gives the sense of hearing, is called the the auditory nerve. From the drum a small opening passes out into the outer part of the ear; this is the portion which is seen on top of the head, and is made up of a membrane known as the cartilage which gives the ear its stiffness. This cartilage is covered by a fine, delicate skin, which is covered on the outside by fine, short hair. Situated on the inner side of the outer ear are numerous long hairs projecting outwards, the use of which is to keep foreign bodies from dropping into the ear. The ear is moved backwards and forwards by small muscles which are attached around it.

## CHAPTER XIV.

# THE EYE.

THE eye is the chief organ of sight, and is situated in the orbital fossa, which was mentioned in the bones of the head. It is chiefly made up of several coats around the outside and in the centre, by the humours of the eye. On the inner side of the coats is a thin membrane called the retina, which contains the branches of the optic nerve, this receives the reflections of objects as they pass through the humours of the eye and they pass along the optic nerve to the brain. The oblong openings seen in the middle of the eye are known as the pupils of the eye. In leading a horse out of a dark stable into the light and watching the pupils of the eye, they will be noticed to get smaller, then on returning it to the stable the pupils will be noticed to dilate or get larger, thus it is seen the pupil does not always remain the same size. The chief use of the pupil is to gauge the sight. At the back part of the eye are several muscles which are attached from around the eye to the bones in the fossa, the use of these muscles are to move the eye and assist in holding it to its place. Around the front part of the eye are two movable curtains, one above and the other below, these are called the eyelids, the use of which are to open and close the eye, and also to potect it from injuries. Around the free border of the eyelids are what is known as the eyelashes, the use of which are to keep foreign substances from falling into the eye. Situated in the inner angle of the eye is what is known as the haw of the eye, this membrane also helps to protect the eye. In the corner of this angle is a small duct or opening, where the tears of the eye pass down through into the nasal tubes, where it is carried down through the bones of the head and emptied into the under part of the nostril or nose. A small gland is situated on the upper part of the eye, secreting the tears which lubricate the eye. The color of the eye is generally brown, but in some cases it is white and receives the name of a walled eye.

CHAPTER XV.

# THE TEETH.

THE teeth are situated in the mouth, in the upper and lower jaws. They are made up of the same tissue as bone, but they contain 10½ per cent. more of the earthy salts, this is what makes them so very hard. Unlike bone, they can stand exposure to air and friction without becoming diseased. The teeth are made up of three hard substances, viz.: dentine, or ivory; enamel; and crusta petrosa. The dentine or ivory is situated in the upper part of the tooth around the pulp or nerve cavity; it is largely supplied with nerves which pass through it from the pulp cavity, and is of a yellowish color. The enamel is the hardest substance of the tooth and covers over the outside of all the exposed part of the tooth, this substance is characterized by its whiteness, and unlike the dentine, there is no blood vessels or nerves in it, and if part of the enamel is broken off it is never replaced again, and the tooth below the part broken off generally becomes decayed. The crusta petrosa is found in the fangs or roots of the teeth and the parts situated below the gum, this substance is the softest part of the tooth. Each tooth is divided into the body or crown, which is the part above the gum, the table of the tooth, which is the part that comes into wear on the top. The neck of the tooth is the part where the gums are attached, and the fangs or roots are the parts situated down in the bone.

The uses of the teeth are to masticate or chew the food, and are also used to tell the age of the animal. There are three kinds of teeth found in the horse, viz.: The incisors or front teeth, which are situated in the front part of the mouth just inside the lips, are twelve in number, six above and six below. The canine or bridle teeth, which are found mostly in the horse or male animal and are often absent in the mare, are four in number, two in the upper jaw and two in the lower jaw, one on each side, a couple of inches back from the front teeth. These teeth are from a quarter of an inch to three-quarters of an inch above the gum, they are round and pointed and are of no particular use. They resemble the eye teeth in other animals. The molars or back teeth are twenty-four in number, six on each side in the upper jaw, and six on each side in the lower jaw, their use is to grind and masticate the food.

### WOLF TEETH.

The wolf teeth are two small, round, pointed, temporary teeth which vary in size in different animals, and are situated one on each side in front of the molars or back teeth in the upper jaw. They commence to grow soon after the birth of the animal and if not pulled or knocked out usually decay between the ages of five and eight years and drop out. These teeth affect the eye causing it to look dull and run water, and should be pulled out when first noticed.

The horse has two sets of teeth, the milk teeth are temporary and are the ones that the colt sheds, while the ones that come in and remain without being shed, are called the permanent teeth. The cutting of the teeth in the foal varies some, but at or within nine days after birth the foal has four front teeth two in the centre above and two below, and in the back part of the mouth he is found to have twelve molars, at from seven to nine weeks he gets four more incisors or front teeth, one at each side of the two centre teeth in each jaw; at nine months old he gets the last of his milk or temporary teeth, these being the corner teeth, two in the upper side of the jaw and two in the lower side of the jaw. Now he has his full set of milk or temporary teeth, consisting of twelve molars or grinders and twelve incisors or front teeth, six above and six below, making twenty-four teeth in all. As the colt advances in age he must shed all these teeth. After this age the colt commences getting his permanent teeth; when the age of one year is reached he gets four permanent molars, two in each jaw one on each side behind the three temporary ones. At two years old he gets four more back molars, one on each side of each jaw. When the age of two years and nine months has been reached the two middle teeth of the temporary incisors or front teeth of each jaw fall out and are replaced by two permanent incisors in each jaw, so at the age of three years these four permanent incisors are up and in wear. At this age, the first eight molars, two on each side of each jaw, are shed and replaced by eight permanent molars. At four years old he sheds four more front teeth next to the ones shed at three years old, and are replaced by four more permanent incisors or front teeth. Also at this age it sheds the four remaining temporary molars or grinders, which are replaced by four more permanent molars, and he also gets four more permanent molars at the back of the mouth, thus at the age of four years the colt has a full set of permanent molars, consisting of six on each side of each jaw, making twenty-four in all. At five years old he sheds the four remaining temporary incisors or front teeth, which are replaced by four permanent incisors, these are known as the corner teeth. It is well to become familiar with the time the colt sheds his different teeth, for sometimes the caps or shells of the teeth do not fall off when they should; these should be watched, for they greatly interfere with the animal feeding and should be removed by a pair of pincers. At five years of age the canine or bridle teeth make their appearance, so at the age of five years the colt has all his teeth or what is known as a full mouth.

**TABLE OF THE TEETH.**

| Age. | Incisors or Front Teeth. | | Molars or Grinders. | |
|---|---|---|---|---|
| | Temporary or Milk Teeth. | Permanent. | Temporary or Milk Teeth. | Permanent |
| The foal at or soon after birth..... | 4 | 0 | 12 | 0 |
| " nine weeks............ | 8 | 0 | 12 | 0 |
| " nine months .......... | 12 | 0 | 12 | 0 |
| The colt at one year.. . ......... | 12 | 0 | 12 | 4 |
| " two " ............. | 12 | 0 | 12 | 8 |
| " three " ............. | 8 | 4 | 4 | 16 |
| " four " ...... ..... | 4 | 8 | 0 | 24 |
| " five " ............ | 0 | 12 | 0 | 24 |

At five years old he gets his bridle, or canine teeth, which are four in number; thus at the age of five years a horse has a full mouth of teeth, numbering forty in all.

How to tell the age of a horse by his teeth is fully explained hereafter in this book in connection with examination for soundness of horses.

It is advisable for everyone to become familiar with the anatomy, or structure of the horse, which has been explained in very simple language, for the better the anatomy is understood the easier diseases and treatments are to understand.

# PART II.

# DISEASES AND TREATMENT OF THE HORSE.

## CHAPTER I.

### QUESTIONS TO ASK AND HOW TO EXAMINE A SICK HORSE.

IT IS always well to inquire into the history of the case by asking the following questions before making an examination, as it will help to tell what is wrong with the horse as well as giving an idea as to what part is affected. First—find out how long the animal has been sick. Second—find out how he has been acting. Third—find out the cause of the trouble if you can. If you find that the animal has been standing with his head hanging down, eating very little, breathing heavily, and coughing, you may come to the conclusion it is some trouble of the breathing organs. Again, if you find out the animal is in severe pain, getting up and down and rolling about in the stall, and slightly bloated, you may conclude it is some trouble of the stomach or bowels. Again, if you find out the animal has a straddling gait, attempting to make water often, and allowing the penis to pass out and then draw it in, and sometimes stamping with the hind legs, you may then come to the conclusion it is some trouble of the urinary organs, such as the kidneys or bladder. This will give you an idea, after hearing the history of the case, what set of organs to examine first. In all cases try the pulse; it is generally taken on the small artery which crosses the under side of the jaw about the middle; when the animal is in good health it should beat from 88 to 40 times per minute, or, in other words, this is the number of times the heart beats per minute. The oftener the pulse beats above its regular beat the more serious the case is. It will be necessary to examine closely the organs which you have come to the conclusion are affected.

CHAPTER II.

# DISEASES OF THE RESPIRATORY OR BREATHING ORGANS.

### SIMPLE CATARRH OR COLD IN THE HEAD.

THIS disease means a running or discharge from the nose and sinuses of the head. It first commences with congestion of the mucous membrane, or what is commonly known as the lining membrane, of the nose and head. Then this congestion is followed by an inflammation and dryness of the membrane, and in a few days this is followed by a discharge of a watery nature, which later on in the disease turns to a thicker fluid of a whitish or yellowish color, varying according to the severity of the case.

**Causes.**—The most common, perhaps, is exposure or sudden changes in the weather, such as we have in the fall and spring; or standing in a draft, while warm, after driving. It is generally found in horses that are in poor condition, the system being run down by poor feeding or over work.

**Symptoms.**—The animal is noticed to be dull and not eating very well. The hair is standing out and looks rough; pulse not much affected; throat shows soreness when you press on it; also discharges freely at the nostrils. The breathing is about natural, and usually the animal does not make quite so much urine, or water.

**Treatment.**—As a general thing the treatment is quite simple. Make the horse as comfortable as possible in his stable; see that plenty of pure air can get in, and that his stall is kept nice and clean. Feed him plenty of soft food such as warm bran mashes, boiled oats, or scalded chopped oats, and it is a good plan to boil up some flax seed and put a teacupful of the juice and boiled flax seed in with his food two or three times a day, according to how much it seems to loosen the bowels. In all cases of this kind it is well to keep the bowels open with soft food. In bad cases it is best not to work the animal very much—just enough for good exercise. Give him a teaspoonful of the following mixture, three times a day in his food, and if he will not take it in his food put it on his tongue with a large spoon, which can be done by drawing the tongue out with one hand and putting the spoon well back into the mouth with the other and turning it over; then hold the mouth shut until the medicine gets wet, so that he cannot spit it out.

Saltpetre or Nitrate of Potash........................¼ pound.
Sulphur..............................................¼ "
Ground Gentian Root...................................¼ "

Mix thoroughly and give a teaspoonful three times a day, morning, noon and night.

In some cases where the discharge continues very long it is well to change the treatment and give a teaspoonful of ground sulphite of iron twice a day, night and morning, in his feed until the discharge stops. If his throat is sore rub it well two or three times a day with white liniment. In cases where this discharge is not checked it terminates in what is known as chronic catarrh, or nasal gleet.

## NASAL GLEET OR CHRONIC CATARRH.

In this disease there is a glary discharge from one or both nostrils. It is a chronic inflammation of the sinuses of the head, and the discharge varies much according to where the disease is situated and the length of time it has been going on.

**Causes.**—First, neglected catarrh, especially if the animal has not been treated properly and allowed to run out in the cold without being properly fed, such as being allowed to run out at a straw stack. It may be caused by a severe blow on the bones of the head over the sinuses, and also from a bad tooth. Sometimes a tumor will cause it or some foreign substance, such as food or a piece of stick becoming worked up through the nose into the sinuses; or it might be caused by coughing and the food fly up into the sinuses in this way.

**Symptoms.**—This disease is sometimes mistaken for glanders. The animal at first may be in pretty good spirits, but if the disease is allowed to run on he will soon get thin and run down on account of the constant discharge from the nose. There is a discharge from one or both nostrils of a yellowish color, the lining of the nose will be slightly reddened and in some cases is of a yellowish color. The way to tell if the sinuses are much affected is to tap on the bone over the sinuses with the finger, if it gives a dull sound like as though the sinuses were full, you may conclude that they are diseased, but if it gives a hollow drum-like sound, you may come to the conclusion that the sinuses are not much affected and there is more hope of recovery. As the disease runs on, the animal gets very much weaker, the discharge has a very bad smell; this shows that the bones of the head are becoming affected. If it is a bad tooth that is causing the trouble, the discharge will only come from one nostril; the animal will not eat very well; sometimes he will be noticed, after taking a bite, to throw the food out of his mouth. The breath has a very bad smell and the horse will soon run down in condition. The way to distinguish this disease from glanders is first, that the discharge in glanders is of a greenish color and will sink in water, while the discharge in nasal gleet will float on water; second, by examining the lining inside the nose which, if the animal has glanders, will be found to be covered with small ulcers.

**Treatment.**—This disease is not, as a general thing, easily treated. If the animal is thin and run down in condition, it is well to build him up with good food, regular exercise, pure air and the following mixture:

Ground Sulphate of Iron ................................ ¼ pound.
Ground Sulphate of Copper.............................. ¼ pound.

Mix thoroughly and give a teaspoonful twice a day, night and morning, on his tongue or in his feed. Wash the nostrils twice a day with warm water. If you conclude that the sinuses are much affected or that some food or foreign substance is causing the trouble, the treatment will then be different. The animal will have to be thrown down and tied with a rope, securely, then strip a piece of skin as large as a copper off the bone with a knife, and take a trephine or an inch auger and bore a hole through the bone into the sinuses, which can be easily done for the bone is soft and thin and does not cause much pain to the animal, and can be done with very little risk. When the hole is through the discharge will come out of it, and with it will generally come the seat of the trouble. Keep the hole open as long as you can by passing your finger into it, and also injecting warm water with a few drops of carbolic acid into it with a syringe once a day; use 10 drops of carbolic acid to the pint of water. If the disease is caused by a bad tooth, have the tooth pulled out, and follow up with medicines mentioned above.

## ABSCESSES OF THE BONES OF THE HEAD.

The only treatment is to bore a hole with a trephine or auger through the bones of the head into the abscess, allowing the matter or discharge to escape. Inject into the hole with a syringe a little warm water with a few drops of carbolic acid twice a day until healed; use the carbolic acid—10 drops to the pint of water.

### NASAL POLYPI OR SMALL TUMORS.

These are situated around the nose and chambers of the head, and are sometimes found around the throat, they are usually attached by a neck to the membrane lining these organs.

**Causes.**—They are said to be due to some change in the system, but the cause of them is not clearly understood.

**Symptoms** are rather peculiar. The animal has difficulty in breathing, and sometimes acts as though suffocating and may even fall down, and in some cases will get up and for a while appear better, then the above symptoms come on again.

**Treatment.**—Examine the throat carefully by looking into the mouth; have something to hold the mouth open and pass the hand back in the mouth and feel for these little tumors. Look into the nose, and if they are in reach remove them by tying a thin, strong string tightly around the neck or roots of the tumor, allowing it to drop off of its own accord. They are also removed by twisting them off with the fingers. If you cannot get at the tumors it is then a hopeless case.

### LARYNGITIS OR INFLAMMATION OF THE THROAT.

**Causes.**—They are similar to those of catarrh: Exposure, standing in a draft while warm, or from a number of horses being kept in a poorly ventilated stable; or from injuring the throat in giving a ball of medicine. Sometimes it occures in the form of an epizootic, or where a number of animals become affected at the same time without any apparent cause.

**Symptoms.**—The animal appears dull, the throat swoolen; if he goes to drink it will be noticed the water will run out through the nostrils when he tries to swallow. When you examine the throat and press on it, it causes him pain which is shown by the animal jerking back and if pressed hard will cause him to have a fit of coughing. It will be seen that he swallows frequently and holds his head in a peculiar position, as if trying to favor his throat. The animal does not care to eat much and what he gets should be soft food, as it hurts him to swallow. If the pulse is very much quicker than natural, and the above symptoms present, you have then a very severe case. The mouth is hot and dry and has a sticky feeling; he is noticed to breathe a little heavier than natural. The bowels are usually a little costive and the urine or water a little scanty, as in most all the diseases of the air passages. In two or three days, if the case is going on favorably, it will be noticed he will have a discharge from the nostrils, which is a good sign. This disease usually takes from ten to twelve days to run its course, but after this the animal should not be put to hard work for some time as it is apt to bring on what is known as roaring or bronchitis.

**Treatment.**—As in all other diseases of the air passages give plenty of pure air; have the stable well ventilated; clothe the body according to the season of the year and if the legs are cold have them well hand-rubbed and bandaged, and give the following mixture:

Saltpetre or Nitrate of Potash.........................¼ pound.
Clorate of Potash...... ..... .........................¼ pound.

Mix the two thoroughly together and give a teaspoonful on his tongue three times a day. Rub the throat well with white liniment three times a day. In severe cases it is also well to take three tablespoonfuls of mustard, two tablespoonfuls of flour and enough vinegar to make it like a paste, rub this around the throat every night and leave on till morning. Give the animal plenty of cold water, a little at a time but often, feed lots of hot mashes with a little boiled flax seed which will have a laxative effect on the bowels. Instead of putting on mustard a hot poultice of linseed and bran might be used and left on all night. In very severe cases, and when the animal's life is threatened by choking to death, it is well to perform an operation known as tracheotomy, which is done on the windpipe, about six inches from the larynx or Adam's apple, by first cutting through the skin and then cutting three of the rings in the windpipe and using the regular tracheotomy tube which can be obtained at almost any drug store. When this operation is performed it gives the horse immediate relief for he draws the air through the tube instead of the nostrils. The way to tell when to take the tube out is to place your hand over the tube which makes him breathe through the nose. When he breathes clear enough through his nose to suit you, take the tube out of the windpipe and draw the skin together over the wound by a couple of stitches and then treat as an ordinary wound; of course, this operation is seldom needed except in extreme cases.

### CHRONIC COUGH.

It is called this when no other disease can be seen.

**Causes.**—It may result from laryngitis, or inflammation of the throat. Sometimes a horse is noticed to have this kind of cough for some time before he takes heaves, or broken wind. A cough of this kind is generally worse during changeable weather and is sometimes shown more after eating and drinking or after being brought out of the stable.

**Treatment.**—Give the following powder:

Tartar Emetic........................................$\frac{1}{4}$ pound.
Camphor.....................................................$\frac{1}{4}$ "
Ground Digitalis..........................................$\frac{1}{4}$ "

Mix thoroughly and give one teaspoonful night and morning in feed or on tongue with a spoon. A teaspoonful of oil of tar in his feed three times a day is also recommended.

### ROARING.

This disease is breathing with a loud and unnatural sound upon any violent exertion. More air passes into the nostrils than can pass into the lungs, and this is due to the wasting of the muscles of the larnyx, or Adam's apple; this condition causes the passage through the larynx to be smaller than natural. The air rushing through this small passage into the lungs acts on the vocal cords; this is what causes the peculiar sound.

**Causes.**—This disease sometimes follows laryngitis, distemper and influenza by being put to work too soon after recovering from them. It may also be hereditary, that is where the sire or dam of the horse has been affected with roars. It may come on from tight reining. Horses with very long necks and narrow jaws are apt to become roarers.

**Symptoms.**—As long as you do not excite the animal he is almost free from it, but if he is worked or driven hard he will show it quickly.

**Treatment.**—If this disease is once well established it is incurable, but in cases where the disease is just coming on give

Iodide of Potash...........................................$\frac{1}{4}$ pound.
Nitrate of Potash or Saltpetre.....................$\frac{1}{4}$ "

Mix thoroughly and give a teaspoonful twice a day, night and morning, in his feed, and blister the throat with

Ground Spanish Flies, or Cantharides...............$\frac{1}{2}$ dram.
Red Precipitate, or Biniodid of Mercury.............$\frac{1}{2}$ "
Vaseline, or Lard..................................................1 ounce.

Mix thoroughly together and apply around the throat, rub it in well and grease in three days afterwards. If he is not better in a month, blister again.

### SPASMS OF THE MUSCLES OF THE LARYNX OR ADAM'S APPLE.

This disease is not very often met with but we find it sometimes affecting old horses.

**Causes** of this disease are not known.

**Symptoms.**—The animal may appear in perfect health when all at once he will be seized with a violent fit of coughing, will reel, stagger, and sometimes even fall to the ground ; after a few minutes it will pass off and the animal will seem as well as ever.

**Treatment.**—Give the animal a good dose of physic:

| | |
|---|---|
| Bitter Aloes | 1 ounce. |
| Ginger | 1 teaspoonful. |
| Soda | 1 " |

Dissolve in a pint of lukewarm water and give as a drench, and allow the animal to stand quiet the next day after giving this drench ; follow up after this with a teaspoonful of bromide of potash every morning in his feed.

### BLEEDING FROM NOSE (EPISTAXIS).

This disease is not so commonly met with in horses as it is in man.

**Causes.** It is generally the result of some injury, or in running or fast trotting horses when they are put to violent exertion, rupturing some of the blood vessels in the nose ; it is also more frequently met with in horses in high condition.

**Symptoms.**—How to tell whether the blood just comes from the nose or from the lungs. If the blood comes from both nostrils it is generally from the lungs, while if the bleeding is from the nose it is usually only from one nostril. You can also tell by putting your ear to the windpipe and listening, if it is coming from the lungs you can tell by the gurgling sound heard in the lungs ; if it is in the nose you will not hear this sound in the lungs. When the blood is coming from the lungs the breathing is affected, which is not the case when bleeding from the nose.

**Treatment.**—If not bleeding very much bathe with cold water until it stops, but if bleeding much plug the nostrils with cotton batting saturated with white lotion and remove in twelve hours.

### BLEEDING OF THE LUNGS (HÆMOPTYSIS).

This is generally a symptom of some other disease, but it may come on, in a horse in high condition and not used to work, by putting him to severe exertion when he is not used to it. It is mostly found in trotting and racing horses when they are not properly taken care of. The blood may come from the lining of the air cells or from the lining of the tubes of the lungs.

**Symptoms.**—As a usual thing it is not very hard to find out where the blood is coming from. It comes out of both nostrils, the animal also coughs, breathes quickly, and is generally very weak, and in taking the pulse you will find it beats very quick, but weak. By applying the ear to the windpipe you will hear that peculiar gurgling sound every time the animal breathes.

**Treatment.**—Keep the animal standing very quietly. Apply cold water or ice to the sides and chest. Be very careful how you drench the animal in this disease for they are so easily choked in diseases of the lungs. Give as a drench—

Turpentine .................... 1 ounce, or 4 tablespoonfuls,
Raw Linseed Oil .................. ½ pint,

which acts as a styptic to stop the bleeding; or you might give

Tincture of Chloride of Iron......... 1 dram, or 1 teaspoonful.

Mix in a pint of cold water, shake well, and give as a drench every night and morning. If the legs are cold rub well and bandage them. Allow plenty of fresh air to get to the animal and blanket according to the season of the year. In some cases this disease is treated by giving

Sweet Spirits of Nitre............ 1 ounce, or 4 tablespoonfuls,

in a pint of cold water, three times a day, morning, noon and night, until the animal has relief. This disease is sometimes followed by inflammation of the lungs.

## CONGESTION OF THE LUNGS.

This is where the blood rushes to the lungs from a chill or some other cause, filling up the arteries and veins in the lungs and causing them to become gorged with blood. This disease is always found before inflammation of the lungs, that is, if the congestion is not relieved it terminates in inflammation.

**Causes** of this disease are standing in a draft while warm, getting a cold drink of water while the animal is warm, or by putting the horse to severe exertion, such as running, trotting, or drawing heavy loads, when the system is not in proper shape to stand it. It sometimes follows other diseases such as catarrh or influenza.

**Symptoms** vary much according to the cause. If from fast or hard work the symptoms are well marked. The animal becomes sluggish, trembles at the flank, breathes heavy, nostrils are dilated or enlarged, pulse is quick and weak, the lining around the eyes and nose becomes very much reddened ; by placing your ear to the sides of the chest or to the windpipe there is a peculiar gurgling noise ; the legs and ears will be cold. If the disease follows a case of catarrh or influenza he then will be noticed to refuse his food, tremble all over the body, ears and legs cold, mouth hot, pulse quick and weak, and by placing your ear at the sides you will hear the peculiar sounds. The animal stands, as he is generally found to do in all lung troubles, and if he does go to lie down will get up immediately. If the animal does not soon get relief the disease will run into inflammation of the lungs.

**Treatment.**—The treatment must be quick. This is not a very fatal disease, but a simple one to treat if taken in time. Keep him well supplied with good, fresh air, and always be careful he does not stand in a draft. Keep the body well covered according to the time of the year, and give

Sweet Spirits of Nitre.......... .1 ounce, or 4 tablespoonfuls.
Laudanum......................½ " 2 "

Put in a pint of cold water, shake well and give as a drench, being careful not to choke the animal in giving it. Have the legs well hand rubbed, if cold, and bandaged ; and if the animal is in high condition and full of blood it is well to give, with the above drench,

Fleming's Tincture of Aconite.................10 to 15 drops.

Also put a mustard plaster on the sides over the lungs. Take ¼ lb. of mustard, with 3 or 4 tablespoonfuls of flour and enough vinegar or warm water to make it into the form of a paste. Rub this well over the sides of the chest with the hand, and in some cases it is well to rub the front of the chest as well as the sides. In some cases we find great benefit in having cloths wrung out of warm water held up to the sides. If he is not relieved in a couple of hours give another drench, same as the one mentioned above ; repeat the drench every two hours until the animal has relief. It is also well to give him only a mouthful of cold water at a time, but give it to him often. Feed him soft food, and after he begins to get better, exercise him a little by walking around ; also give the following mixture :

Ground Gentian Root..............................¼ pound.
Nitrate of Potash, or Saltpetre........ ... .........¼ "

Mix thoroughly together and give a teaspoonful three times a day in his feed, and gradually bring him back to his natural feed and work again.

### INFLAMMATION OF THE LUNGS (PNEUMONIA).

This disease is inflammation of the lung substance itself and is quite a common disease. The lungs in their natural state will float on water. In the first stage of this disease the lungs are filled with blood and serum, and they are of a dull reddish color ; at this stage, if the animal dies, the lungs will still float on water. As the disease runs on, if not cured, the lungs change to a dark, grayish appearance ; if the animal dies and the lungs are put in water they will sink.

**Causes**—Are much the same as those of congestion of the lungs and generally come on horses kept in a poorly ventilated stable, which has a tendency to weaken the lungs ; sudden changes in the weather, such as we have in the fall and spring, are liable to bring the disease on. It is also sometimes caused by keeping a horse in a warm stable and then turning him out to pasture to lie on the cold ground ; or, if while he is out, a cold rain storm comes on and he gets wet and is chilled through ; or, if a horse is clipped and exposed to the cold ; standing in a draft while warm ; inhaling smoke ; also by driving against a cold wind too soon after he has had influenza, distemper, or any of those weakening diseases. It also frequently follows congestion of the lungs.

**Symptoms.**—Are much the same as congestion of the lungs, only plainer. There is, generally, very little trouble in telling the disease. It commences first by the animal shivering, after the shivering ceases heat takes place ; the ears and legs will first be hot and then cold ; the mouth is sticky and the breathing affected, but not so much as in a pure case of congestion of the lungs ; the pulse is quick, ranging from fifty to seventy-five beats to the minute, which is stronger than in a case of congestion ; the horse does not care to eat ; stands up all the time, with head hanging down and ears lopped over, and in hot weather perspires freely around the chest ; the eyes have a glossy appearance and, around the inside of the eyelids, are very red. As the disease passes on the horse breathes heavier and sometimes is noticed to sigh, as if in distress ; the bowels become costive, and the manure has a glossy appearance; placing your ear to the side of the chest you will hear a grating sound similar to that produced by taking some hair of your head, just above the ear, and grating it between the thumb and finger. If the horse is loose in a stall he will work around till he gets his head to a door or window, which shows that he wants fresh air. By tapping the finger on the chest over the lungs it will produce a dull sound, which, if the lungs were not affected, should give more of a hollow sound. If the disease is going to terminate fatally the pulse runs up to 100 beats per minute and is so weak yòu can hardly feel it; he will breathe very heavy ; the nostrils make a flapping noise, and his flank draws in and out almost like a heavy horse ; the appetite is entirely gone ; the breath smells very bad ; he still persists in standing, and notices nothing. As death approaches the mouth becomes cold, the pulse cannot be felt ; he may, near the last, lie down, which will cause him to breathe very much heavier ; he again staggers to his feet, breaks out into a cold, clammy sweat all over the body, and finally staggers, falls, and dies. If the case, on the other hand, is more favorable, the animal eats a little and notices things around him, and the above symptoms gradually disappear. It generally takes from 9 to 12 days to run its course, and, as a usual thing, is treated with satisfaction if taken in time.

**Treatment.**—Clothe the body according to the season of the year. If the animal affected is in high condition use sedatives, such as

Fleming's Tincture of Aconite..........8 to 10 drops.
Laudanum............................½ oz. or 2 tablespoonfuls.

Mix in a pint of cold water and give as a drench every three hours until the distressing symptoms have ceased and inflammation seems pretty well checked. If the animal seems weak after this give

Sweet Spirits of Nitre...........1 ounce or 4 tablespoonfuls.
Whisky.................... .....1 wine glass full.

Mix with pint of new milk or gruel and give three times a day, morning, noon and night, until the animal seems stronger. After this, when the animal is getting better and he needs a tonic to build up his system and to keep his kidneys in good action so as to relieve his lungs as much as possible, give

Nitrate of Potash or Saltpetre........¼ pound.
Ground Gentian Root................¼ "

Mix thoroughly and give a teaspoonful three times a day. While the animal is sick feed on soft food, mixing a little boiled flax seed with it to keep the bowels regular. If the animal is very thin in condition it is best not to use much of the aconite and laudanum, but commence the stimulants sooner than if the horse was in high condition. Apply mustard to the sides and chest and also cloths wrung out of hot water, and be sure to allow the horse to have plenty of fresh air by having the stable well ventilated, but do not allow any draft to strike him.

### PLEURISY.

This disease is inflammation of the lining found inside the ribs and over the lungs. This is a serious disease if not taken in time and allowed to run on.

**Causes** are similar to those of inflammation of the lungs, and we often find this disease and inflammation of the lungs coming together. The chief causes are exposure to cold, standing in a draft, washing the body and not properly drying it, injuries to the ribs in any way.

**Symptoms.**—The animal is first noticed to shiver, the pulse quick and strong—much stronger than with congestion or inflammation of the lungs—and is sometimes called a wiry pulse. He seems in great pain, breaths heavily, which is noticed very much at the flanks. At the commencement of the disease he will lie down, but, as a general thing, he will stand up most of the time ; if you make him cough he will suppress it as much as possible, and instead of coughing out loud, as in other lung troubles, it will be more like a long, heavy groan. The idea of this is he tries to keep from moving his chest as much as he can. The ears and legs are cold, but sometimes you will find one ear hot and the other cold ; he has a tucked up appearance, and there will be a hollow line right along the bottom of the false ribs and up towards the point of the hip ; if you press on his sides it causes him great pain, and in turning him around short he will groan with pain. If this disease is not soon checked it will terminate in what is known as hydrothorax, which means a filling up of the chest cavity with a watery fluid.

**Treatment.**—This disease is treated very much similar to that of inflammation of the lungs. Apply cloths wrung out of hot water to the sides if in warm weather, but if in cold weather mustard is best and easiest kept on. Clothe the body well and see that he is allowed plenty of fresh air without being in a draft. At the commencement of the disease, if the horse seems to be in much pain, give

Fleming's Tincture of Aconite .......8 to 10 drops.
Fluid Extract of Belladonna.........½ dram, or 30 drops.
Tincture of Laudanum .............½ ounce, or 2 tablespoonfuls.

Mix in a pint of cold water and give as a drench. Continue the above drench every two hours until the horse is relieved of the pain. If the horse seems weak after this give

Sweet Spirits of Nitre ................1 ounce, or 4 tablespoonfuls.
Whisky ...........................1 wine glass full.

Mix in a pint of gruel and give as a drench three times a day, morning, noon and night, until the animal begins to recover nicely, then use the following medicine to keep the kidneys working freely, as this will have a tendency to keep water from forming in his chest :

Saltpetre, or Nitrate of Potash ........................¼ pound.
Ground Gentian Root...................................¼ "

Mix well and give a teaspoonful three times a day. During sickness it is well to feed the animal with soft foods, with a little boiled flax seed in it to keep the bowels free, give a little cold water to drink in small quantities, and give it often—every hour or two.

## WATER IN THE CHEST (HYDROTHORAX).

This generally follows a case of pleurisy. In some cases you will find several pails of a watery fluid around the lungs in the chest cavity. When there is such a large quantity as this it generally ends fatally.

**Causes.**— Pleurisy.

**Symptoms.**—After the pain and soreness from pleurisy has passed off the fluid then commences to collect around the chest, which causes him to breathe very heavily, the nostrils becoming large, and sometimes make a flapping noise. He breathes quick, and draws in and out at at the flank worse than he does in a bad case of heaves ; the pulse becomes quicker than in pleurisy, and very weak, beating from 75 to 100 beats per minute ; the blood in the jugular vein seems to flow back towards the head instead of flowing down, causing the vein to move every time he breathes. By putting your ear to the chest you cannot hear anything except above the water. If the animal is loose he will be noticed to try to get to the door or window to get the pure air. In some cases the animal will go on like this for several days, not eating very much, and gradually getting worse. Near the last his ears and legs get very cold, and all the other symptoms keep getting worse, and the animal tries to stand on his feet to the very last.

**Treatment.**—If the animal will take food give him good, strong food, such as oats and hay, and it would be well to mix a little of the boiled linseed along with the oats to keep the bowels regular; keep the animal quiet; blister the sides well with a strong mustard plaster; give him of the following:

Saltpetre or Nitrate of Potash..........................¼ pound.
Ground Gentian Root..........................¼ "
Ground Sulphite of Iron..........................¼ "

Mix thoroughly and give him a teaspoonful every five hours. It is also recommended in some works to puncture near the bottom of the chest, between the eighth and ninth rib, with a small trocar and cannula, and let the fluid out of the chest, but this operation is not very successful in the horse, and we would not advise it to be done, although it is sometimes successfully performed in human beings.

If the horse dies and you cut into the chest you will find a great quantity of the fluid collected around the lungs, in some cases as much as three pailfuls.

### BRONCHITIS.

This disease is inflammation of the lining of the bronchial tubes.

**Causes.**—This disease is sometimes caused by driving a horse fast when he is in a weak condition, inhaling smoke, or from a sudden change in the temperature, choking from food passing down the windpipe, and sometimes it is caused from giving a drench which, instead of passing down into the stomach, some of it goes down the windpipe. This disease is oftener seen in the city than in the country.

**Symptoms** are a peculiar dryness of the throat, increased breathing, and if you place your ear to the windpipe, you will hear a wheezing noise. The animal seems quite dull, and does not eat as well as he should. If the disease is allowed to run on the pulse becomes quick and weak, and the legs and ears, after a time, become cold; he seems very much depressed and weak, and if the weather is warm perspires freely around the chest and flanks; on account of not eating very much he will become very gaunt, and it will be noticed that he does not lie down, and if you make him stir around it will start him to cough.

**Treatment.**—If the animal is in good condition and strong, give the following mixture:

Fleming's Tincture of Aconite.....5 to 8 drops.
Laudanum ........................½ ounce or 2 tablespoonfuls.
Sweet Spirits of Nitre..............1 ounce or 4 tablespoonfuls.

Mix in a pint of cold water, shake well and give as a drench. Give a drench every two or three hours until he seems relieved. If the animal is very weak, instead of this, it is best to give

Whisky............ ........................1 wine glass full.
Ale or Beer........................................½ pint.

Give every three hours until he seems relieved, then proceed with the following:

Saltpetre or Nitrate of Potash..........................¼ pound.
Tartar Emetic........................................¼ pound.
Ground Gentian Root............ ......................¼ pound.

Mix thoroughly and give a teaspoonful three times a day. Feed soft feed with a little boiled flax seed to keep the bowels free, give cold water in small quantities, but often, which will have a good effect on the throat; apply mustard or hot cloths wrung out of hot water to the chest, clothe the body well, according to the season of the year, and see that the animal has pure air to breathe without being in a draft. In all lung diseases be careful in drenching so as not to choke the animal.

### BROKEN WIND OR HEAVES.

It is similar to asthma in man.

**Causes.**—It is generally seen in horses that are ravenous feeders and overload the stomach and are inclined to carry a large belly. This affects the stomach, and the same nerve that helps to supply the stomach with nerve power also helps to supply the lungs; this is how we account for it affecting the lungs when the stomach is affected. The kinds of food that are apt to produce it are inferior foods, such as musty hay (clover hay being the worst), or musty oats, or it may be caused from a neglected cold. This disease is rarely found in cavalry horses as they are fed on the very best of food.

**Symptoms.**—This disease is easily detected when it is well established. There is a peculiar way of breathing, a long inspiration followed by a short expiration and a jerking motion at the flank; the nostrils are enlarged and the muscles of the belly come into play greatly in this disease. On a damp, hot, sultry day the symptoms are greatly increased, and may become very alarming, and you might be led to think that the animal was suffering from inflammation of the lungs; but when you examine the pulse you will find it beating about natural. In heaves there is a loud, hacking, painful, internal cough which seems to come from the lungs; this is noticed more after eating or drinking, and is noticed particularly after being brought out of the stable in the morning, but after working a while will not be so bad. If you suspect this disease when you are examining him for soundness give him a good feed, a pail of water and a good gallop. If affected with this disease you can then notice it plainly. In some cases it can be relieved for a short time by giving a large dose of Fleming's tincture of aconite, 10 to 15 drops in a drench, mixed with a pint of raw linseed oil. This is sometimes done by horse traders to relieve the animal while trading, and in some cases they even pour shot into the animal, which relieves him for a time.

**Treatment.**—Where heaves once get well established it is incurable, but it can be helped by careful feeding. By feeding him regularly, and giving him lots of oats to eat and very little hay, so as to keep him gaunt ; water him often—four or five times a day—not more than a pail at a time, and never allow him to get a large feed of hay or a large drink of water at a time. The best treatment of medicine is give first a physic of from 8 to 10 drams of bitter aloes dissolved in a pint of water, with a tablespoonful of ginger and soda given as a drench, and allow him to stand in the stable for a day. This will clean his bowels and stomach out ; after this feed him carefully and give of the following mixture :

| | | |
|---|---|---|
| Ground Gum Camphor | ¼ | pound. |
| Powdered Nux Vomica | ¼ | " |
| Common Soda | ¼ | " |
| Nitrate of Potash or Saltpetre | ¼ | " |

Mix well and give a teaspoonful three times a day in his feed, or on the tongue with a spoon.

## PLEURODYNIA.

This disease is a rheumatic condition of the muscles around the chest. This is not a very common disease.

**Causes.**— Exposure to cold when the animal is recovering from pleurisy or other weakening lung diseases.

**Symptoms.**—There is great pain and difficulty in breathing and shows symptoms somewhat similar to pleurisy. In pressing on the sides he shows even more pain than he does in pleurisy, and when you examine his pluse you will find that they are about regular, whereas in pleurisy they would be beating quick and hard. By putting the ear to the animal's side you cannot hear the grating sound heard in pleurisy.

**Treatment.**—If it is warm weather, or in a warm stable, wring a woolen blanket out of hot water, wrap it around his chest and cover up with a dry blanket to keep the heat in. This blanket would be better heated up by wringing it out of the hot water every hour. While the blanket is being changed rub the sides well with white liniment. Give him

Sweet Spirits of Nitre..............1 ounce, or 4 tablespoonfuls.

In a pint of cold water three times a day, morning, noon and night, until the soreness has passed off pretty well, then follow up with a diuretic to act on the kidneys.

Nitrate of Potash or Saltpetre............¼ pound.
Ground Gentian Root....................¼ "

Mix thoroughly and give a large teaspoonful three times a day in his feed or on his tongue with a spoon.

### SPASMS OF THE DIAPHRAGM.

This disease is sometimes called Thumps on account of the peculiar thumping noise the animal makes in breathing.

**Causes.**—It generally results from an animal being put to very severe exertion, such as in running, trotting, heavy drawing, or any thing of that kind too soon after eating. It is also said to be caused by buckling the girth of a saddle too tight. This disease is more liable to be noticed shortly after the animal has recovered from some weakening disease.

**Symptoms.**—It gives rise to a thumping noise which is plainly heard by listening at the back part of the lungs. In most cases you can hear the noise while standing near the animal. This disease is often mistaken for palpitation of the heart, but by examining with the ear along the side you will find that the noise made is too far back to be affecting the heart, and you would almost think by the peculiar noise made that some person was inside tapping with a hammer. The animal has difficulty in breathing, sweats freely and seems in pain.

**Treatment.**—In an ordinary case give an anti-spasmodic.

Sweet Spirits of Nitre ............1 ounce, or 4 tablespoonfuls.
Tincture of Laudanum ..............1 " " "

Mix in a pint of luke warm water and give as a drench every two hours until the animal is relieved of the thumping noise. If warm weather, apply a woolen blanket wrung out of hot water to the chest, with a dry one outside of it. If cold weather it is better to apply a mustard plaster around the back part of the chest and cover the body well. In some cases, after giving a few doses of the above mixture, if he does not get relief, it would be well to give

Spirits Turpentine..................1 ounce, or 4 tablespoonfuls.
Raw Linseed Oil....................1 pint.

Mix and give as a drench. After the animal is getting better feed well and give regular exercise, and bring the animal gradually back to his regular work.

## RUPTURE OF THE DIAPHRAGM.

This is rupture of the curtain which separates the lungs from the bowels, and if the rupture is large enough to let the bowels pass through in on the lungs and heart death soon relieves the animal.

**Causes.**—Pulling a heavy load up a steep hill, or by high jumping. Sometimes it occurs when the animal is suffering from acute indigestion, when the stomach is full of gas, and in getting up and down sometimes lies down a little too heavy, causing a great strain on the curtain, which causes it to become ruptured.

**Symptoms.**—There is a frothy spume comes from the nostrils, breathes very heavy and quick, breaks out in sweat over the body, the pulse runs up very high—sometimes as high as 100 beats per minute—and gets very weak, the animal appears as though it was suffocating, and if the rupture is to any great extent the symptoms gradually get worse, the legs and ears get cold, when death relieves him.

**Treatment.**—Not much can be done in this case, only give

Laudanum....... ..................1 ounce, or 4 tablespoonfuls.

Mix in a pint of water and give as a drench. This may be given once in a while just to relieve the pain.

## CHAPTER III.

# DISEASES OF THE MOUTH, TEETH, SALIVARY GLANDS AND GULLET.

### IRRITATION OF SHEDDING THE MILK TEETH.

This trouble is generally at its worst when the horse is between three and four years old.

**Symptoms.**—The horse seems weak at work, sweats easily, his hair is standing and looks rough, he does not feel well and he gets gaunt and thin, his bowels get costive and the oats come through almost whole.

**Treatment.**—In a case of this kind always examine the teeth—both front and back—for shells or caps, and if there is any remove them with a pincers or forceps. Give a mild laxative, such as

Raw Linseed Oil .......................................... ½ pint

in a drench. After this feed on soft food, and follow up with the following tonic powders:

Ground Gentian Root.................................. ¼ pound.
Ground Sulphate of Iron .............................. ¼ "

Mix thoroughly and give a tablespoonful twice a day in his feed or on his tongue.

### LAMPAS.

This is very common in young horses. It is not really a disease itself, but simply an irritation of the gums caused by shedding his front teeth.

**Symptoms.**—This swelling is found in the gums behind the front teeth in the upper part of the mouth. The gum looks red, and if you press your finger on it it seems sore.

**Treatment.**—Do not use any harsh treatment for, after the horse gets all his teeth shed and his new ones in, the swelling generally disappears of its own accord. It is well sometimes to take a sharp knife and cut the gum in a few places, which relieves the congestion and soreness, then rub the gums a couple of times a day with alum water—about two teaspoonfuls of alum to a pint of water. Be careful in cutting the gum not to cut back of the third bar or ridge in the roof of the mouth, for there is danger of cutting the large artery in the roof, which, if cut, will bleed freely. If by accident you should cut this artery, the way to stop it is: Take a large piece of cotton batting, place it in the roof of the mouth over the cut and have it held firmly by winding a bandage through the mouth and over the nose, tie the animal up so he cannot eat, and leave the bandage on for twenty-four hours, when it can be safely taken off.

### PARROT MOUTH.

This is not a disease, but simply a deformed mouth, where the upper jaw is longer than the lower jaw, and the teeth in the upper jaw projects out over the teeth in the lower jaw which generally get very long. It is always well to examine a horse's teeth before buying him for fear of his having some trouble like this. These horses should never be turned out to pasture for they cannot graze as their front teeth do not come together. But in some cases they make very good work horses if they are kept in the stable and fed on hay and oats. This is considered an unsoundness in horses.

### WOLF TEETH.

These are two small teeth found in the upper jaw in front of the grinders, one on each side. These teeth have an affect on the horse's eyes, causing them to run water and look dull, and, in some cases, if they are very large, will interfere with the animal's feeding.

**Treatment.**—Sometimes the teeth are knocked out with a punch, but they are best, removed by putting a twitch on the horse's nose and pulling them out with a forceps.

### SHARP EDGES ALONG THE TEETH.

The grinders on the upper jaw are wider than those of the under jaw, and pointed to the outside, while those of the under jaw are narrower than those of the upper jaw and pointed in towards the tongue. You will notice these points about the teeth by opening the horse's mouth and drawing his tongue out to one side. From the continual grinding the outer edges of the upper teeth become sharp, and will sometimes cut the cheek, while those of the under side will become sharp on the inside and cut the tongue. If you suspect the teeth are sharp the best way to examine them is to place a twitch on the horse's nose, have an assistant to hold the twitch, and hold his head up slightly while you take the tongue out with one hand and hold the cheek out with the other, then look back and see if the cheek or tongue is cut, and also if the edges of the teeth are very sharp, if they are, the horse's mouth needs what is called floating or filing the teeth, which can be easily done by leaving the twitch on and running a float or tooth rasp along the outer edge of the upper row of teeth and the inner edge of the lower row of teeth. It is not best to file them too much, just enough to take off the sharp edge of the teeth so they will not cut the tongue and cheeks, for if you file them too much the horse cannot grind his hay so well.

### DECAYED TEETH (CARIES).

You do not find decayed teeth so often in the horse as you do in the human being. Horses rarely, if ever, suffer from tooth ache.

**Causes.**—It generally comes from biting some hard substance and either breaking or cracking the tooth, which then begins to decay.

**Symptoms.**—The horse in eating his feed will be noticed, all of a sudden, to throw his food out of his mouth, fumble his tongue around a little and then commence to eat again. If in drinking sometimes if the water is cold it takes him a long time to drink, having to stop several times in drinking a pailful. In driving he is noticed to hold his head to one side, favoring the side that has the decayed tooth in it. His breath smells bad, and he falls off in condition. If the tooth is in the upper jaw, and the roots affected, there is sometimes a running from the nostril over the tooth.

**Treatment.**—Open the mouth with a speculum or any iron that will answer the purpose, pass the hand back and examine the teeth and find out which tooth it is. Always in examining the mouth it is best to put a twitch on the horse's nose as it assists in holding him quiet. When you are sure which tooth is affected take a large forceps and remove the tooth. After pulling the tooth out keep the tooth opposite the one pulled out filed down so it will not irritate the gum on the opposite jaw. After the tooth has been removed feed on soft food for a few days until the gum gets healed up. If the horse is run down in condition it would be well to give some of the following tonic powders to build him up:

Ground Gentian Root....................................¼ pound.
Ground Sulphate Iron. ...................................¼ "

Mix thoroughly and give a teaspoonful twice a day in his feed or on his tongue

## SPLIT TEETH.

The teeth that become split are generally the molars or grinders on the upper side of the mouth, though, in some cases, it might occur in the lower jaw. If the split tooth occurs in the under jaw the split part is generally found on the inside of the tooth and turned in against the tongue. If it is one of the grinders on the upper jaw the split part is on the outside and turns out and cuts the cheek.

**Causes.**—Generally from getting some hard substance into the mouth and grinding heavily on it, such as a nail or stone.

**Symptoms.**—The animal can scarcely eat, seems very much afraid when you go to handle his mouth, and will sometimes jerk back. If you run your hand along the upper jaw on the outside where the tooth is split and turned out it will be very sore, and the animal will jerk his head away when you press over the tooth. In eating he is noticed to fumble his food around in his mouth, and after having it in a few minutes he will throw it out half chewed, stop a few minutes, and then try to eat some again. Place a twitch on the horse's nose and proceed to examine the mouth by drawing the tongue out with one hand and holding the cheek back with the other and look carefully back along the grinders, and if the split is in the upper side of the jaw you will see it worked out, and, in some cases, stuck into the cheek. If the split tooth is in the under jaw you will find the split part stuck into the tongue.

**Treatment.**—After you have found where the split tooth is, it is easily treated by taking hold of the split piece with the forceps and pulling it out, which is not hard to do in most cases. Then take a float or tooth rasp and run it along that side, and if there is any sharp teeth smooth them off. If the horse is very thin follow up with tonic powders.

## HANGING THE TONGUE OUT OF THE MOUTH.

This is not a disease, but a miserable habit, and if it is once formed you cannot cure it.

**Cause** is generally from the teeth getting sharp and cutting the tongue, or from some injury to the tongue.

**Symptoms.**—At the first start of this habit the horse just holds the end of the tongue between his teeth while he is at his work. After a time it becomes worse, and while the bit is in his mouth the horse hangs his tongue out three or four inches on either side.

**Treatment.**—Examine the mouth as to the state of the teeth, if they are found sharp, float or file them so as to stop them from cutting the tongue. If the cause is a soreness on the tongue dress it with alum water—two teaspoonfuls of alum to one pint of water. This will generally effect a cure if the case is taken in time. There are certain kinds of bits recommended for this habit, but as a general thing they prove a failure.

### CRIB SUCKERS.

This is where a horse takes hold of the manger or anything around him and sucks wind.

**Causes.**—Sometimes a colt will learn this habit from seeing its mother or other horses doing it. It is also caused by soreness of the front teeth at first, and he commences biting at the manger to relieve him, when afterwards it becomes a regular habit.

**Symptoms.**—In examining the front teeth you will find them worn off from biting, and the horse, if you watch him, is continually hanging on to the manger. In some cases he will suck himself full of wind, and sometimes will take severe colic from sucking so much, while in other cases he will simply hang on to the manger with his teeth.

**Treatment.**—When first noticed it is best to put the animal in a box stall and feed him his hay off the floor, and his grain in a pail, which should be removed as soon as the animal is through eating so he has nothing to take hold of with his teeth. Examine the teeth and see if there is anything wrong with them, if they are sharp, causing soreness, file them down, or if it is a milk tooth not properly shed, it is well to remove it. If it is in the spring, and the grass good, he will sometimes get over it by turning him out to pasture. If the animal is old and has been a cribber for some time the best thing to do is to get a muzzle for him, and only leave it off while he is eating.

### FOREIGN SUBSTANCES IN THE MOUTH.

Sometimes we find a piece of stick caught across the roof of the mouth, which will be noticed by the animal not feeding and he will be continually working the tongue around in his mouth, and if this obstruction be not removed the animal will fall off in flesh. In this case examine the mouth well and remove any substance found caught in the mouth with the fingers or with a pincers.

**Barley or Wheat Beards.**—When horses are fed on barley or wheat straw, or chaff that has beards in it the mouth should be examined every week or two, for in a great many cases the beards get so lodged in the mouth that the animal cannot remove them with his tongue.

**Symptoms.**—The horse does not feed well, his breath is bad and his mouth seems sore when you handle it; he becomes gaunt and thin.

**Treatment.**—In all cases where you are feeding this kind of food examine the mouth carefully, and if you find any beards remove them with the finger and wash the sore place with alum water twice a day until it heals up. Use two teaspoonfuls of alum to a pint of water.

## INJURIES TO THE TONGUE.

The tongue is sometimes injured by a person pulling too hard on it when it is drawn out of the mouth, which paralyzes the tongue. There cannot be much done for this, only give very soft foods, which he can almost drink down, and give him a teaspoonful of powdered nux vomica three times a day on his tongue with a spoon. The tongue is often injured by the horse pulling back when tied by the bit. In some cases the tongue is almost cut off. If you think there is no chance of the tongue healing it is best to remove it with a knife and apply Monsel's solution of iron to stop the bleeding, if any. Afterward bathe the tongue with a little alum water three or four times a day for a few days, until the tongue heals up. Use one teaspoonful of alum to one pint of water; also feed the horse on soft food for a few days, while the tongue is healing. If the tongue is not cut enough to remove, treat it same as treatment after removal.

## INFLAMMATION OF THE TONGUE (GLOSSITIS).

This is not a very common disease.

**Causes.**—It is sometimes caused by handling the tongue rough, by pulling too hard when taking it out of the mouth, or by giving irritating medicines which are not diluted enough with water; by eating poisonous grasses and sometimes by a thorn sticking in the tongue.

**Symptoms.**—There is a flow of saliva from the mouth; the animal cannot chew his food well, and there is difficulty in swallowing and breathing, the tongue gets red and is painful when pressed upon, is very much swollen, and in some cases sticks out of the mouth, the horse seems generally feverish, and after a few days there will be seen small boils forming around the tongue which have matter in them, the lining covering the tongue becomes dry and cracked in several places. If the animal does not get relief he cannot eat nor drink, and will soon die of starvation.

**Treatment.**—If it is a thorn or any foreign substance, remove it and give a dose of laxative medicine, such as one pint of raw linseed oil. Bathe and gargle the tongue with the following:

Tincture of Laudanum ............... 1 ounce, or 4 tablespoonfuls.
Pulverized Alum..................... 1 teaspoonful.
Water............................... 1 pint.

Gargle or bathe the tongue three or four times a day, and blister him in the space under the jaws with a mustard plaster. If the tongue is swollen very much it is well to lance it with a knife and allow the watery matter to escape, also open the little boils that have matter in them and let it escape. Feed the horse soft food with lots of boiled flax seed in it, as it has a soothing effect on the tongue. Follow up with the following powder:

Nitrate of Potash or Saltpetre......... ¼ pound.
Sulphur.............................. ¼ "
Pulverized Alum...................... ⅛ "

Mix thoroughly and give a teaspoonful on his tongue three times a day. These are to gargle and heal the tongue, as well as help the animal to thrive.

### APHTHÆ OR THRUSH.

At certain times this is a very common disease.

**Causes.**—Certain kinds of food will cause one kind of aphthæ, but the kind we usually see is the infectious kind. This is where the disease is communicated from one horse to another through the air, from stable to stable.

**Symptoms.**—The horse is usually first noticed to be dull and not feeding well, has a slight cough, runs down in condition, and sweats easily when he is working. Upon opening his mouth to examine it you will find a lot of little pimples, like small blisters, all over the tongue and the insides of the lips and cheeks, these pimples or blisters are found all the way through the lining of the gullet, stomach and intestines, and in some cases the animal gets quite feverish and unable to do any work.

**Treatment.**—Give a half pint of raw linseed oil in a drench to start on. This will loosen up the bowels and help to carry off the disease. A tablespoonful of ginger is a good thing to give with the oil, then give the following:

Common Soda............................................¼ pound.
Sulphur..........................................................¼ "
Nitrate of Potash or Saltpetre............................¼ "

Mix thoroughly and give a teaspoonful three times a day on the tongue with a spoon, also gargle the mouth out with alum water—two teaspoonfuls to a pint of water—with a sponge three times a day until the pimples disappear. While he has the sore mouth feed on soft food, and put a lot of boiled flax seed in the feed which will keep the bowels loose.

### INJURIES TO THE LIPS AND CHEEKS.

The lips and cheeks sometimes become bruised and cut in various ways. If the skin is broken to any extent it is best to stitch it up with a needle used for sewing up wounds, but it may be done with a darning needle. In sewing use carriage trimmer's twine, or a piece of white wrapping twine. Put a stitch about every half inch, and in stitching it is best to tie a separate knot for each stitch, then bathe with warm water two or three times a day and apply, after bathing, the white lotion. In sewing the wound it is best to place a twitch on the animal's nose to keep him quiet. If the skin is not broken bathe and apply the white otion same as mentioned above.

### PARALYSIS OF THE LIPS AND CHEEKS.

This is not a very common disease, although it is met with occasionally.

**Causes.**—It is an injury to the nerves which supply the lips and cheeks with motion. For instance, it is done sometimes by using a heavy poke on a horse in the pasture, also in the stable by having him tied with a heavy halter, or any other kind of an injury that would affect the nerve will produce this ; cold weather will sometimes bring it on.

**Symptoms.**—It is first noticed that the animal is not able to use his lips in eating or drinking, or in any other way, and they hang flabby and loose, and in most of cases they look as if swollen, but it is only the looseness of them that gives them that appearance. When the horse tries to drink he has to put his head deep into the pail so that the water covers up his lips and nose, for this is the only way he can drink.

**Treatment.**—Keep the horse's strength up as best you can by feeding soft food which can be easily chewed. In this disease it is best to keep the animal in the stable and give him the following medicine :

| | |
|---|---|
| Powdered Nux Vomica | ¼ pound. |
| Ground Gentian Root | ¼ " |

Mix thoroughly and give a teaspoonful three times a day, hand rub his cheeks three times a day and apply the white liniment after each rubbing. This disease generally takes from two to six weeks to recover and in all cases remove the cause of the trouble.

---

# DISEASES OF THE SALIVARY GLANDS.

## SLAVERING OR FROTHING AT THE MOUTH.

**Causes.**—It is generally caused from something in the feed that the animal is eating, or a heavy dose of aconite will cause it, or in fact anything that will stimulate the secretion of the salivary glands.

**Symptoms.**—A continual dripping of the saliva from the lips.

**Treatment.**—Change his feed and wash his mouth out with alum water two or three times a day—two teaspoonfuls to a pint of water. If this does not help it give him a good physic:

| | |
|---|---|
| Bitter Aloes | 8 drams. |
| Common Soda | 1 teaspoonful. |
| Ginger | 1 " |

Mix in a pint of water and give as a drench ; let the horse stand in the stable the next day. This will generally stop the slavering.

## THICKENING OF THE PAROTID GLAND.

This is usually called thick glands.

**Causes.**—Tight reining, or sometimes it remains thick after distemper, or from inflammation of the gland.

**Symptoms.**—A heavy thickening behind the jaw bone and below the ear.

**Treatment.**—Blistering is the best way to get rid of this. Use the following mixture as a sweat blister:

| | |
|---|---|
| Pulverized Catharides, or Spanish Fly | 1 dram. |
| Vaseline, or lard | 1 ounce. |

Mix thoroughly together and there will be enough to blister the glands on both sides of the throat. In putting this on always rub it in well, then tie his head up so he cannot rub his neck, leave it for three days, then grease it with some lard and keep greasing every third day till the blister is off, and if by this time the swelling has not gone down it would be advisable to repeat the blister.

### INFLAMMATION OF THE PAROTID GLAND.

This is the salivary gland situated below the ear and between the back part of the jaw bone and the neck.

**Causes.**—Generally from a bruise of some kind.

**Symptoms.**—There is a large, painful swelling just below the ear, on the affected gland of either side. It is so painful the horse can hardly eat or drink, and he stands with his head poked out.

**Treatment.**—Give it lots of bathing with vinegar, hot water and saltpetre; after bathing, rub dry, and rub well with white liniment, then apply a poultice of boiled turnips and bran, or linseed meal and bran, about half and half. Change the poultice three times a day, and bathe and rub with liniment each time the poultice is changed. This will check it and drive it away. If it does not check the inflammation the gland will fester, form matter and come to a head. It is well to let it come pretty well to a head before attempting to open it. The way to tell when it is ready to lance or open is, you find a soft spot where the hair generally falls out, and when you press your finger on the spot and take it off the matter presses the skin back to its place quickly. You can easily tell when it is fit to let. Take your knife or lance and give the skin over the soft spot a little nick, which may be done without any danger of bleeding. After this is done press the matter all out and keep on bathing and poulticing till the swelling has entirely gone down. After it is healed up, and if the gland remains a little thick, blister with the following :

Cantharides, or Spanish Fly.............. ...............1 dram.
Vaseline or lard.........................................1 ounce.

Mix thoroughly together and apply one half of the mixture, rub it well and tie the horse's head up so he cannot rub it, leave till the third day, then grease with some lard, and keep on greasing for a few days until the blister gets healed up, then take some warm water and soap and wash the grease off and, after drying, use the other half of the blister same as first half. During the time the animal is sick with this, feed and water him from a high manger On account of his throat being so sore feed him mostly soft feed which would be easily swallowed. Give him the following powder for a tonic and diuretic to act on the kidneys:

Nitrate of Potash, or Saltpetre ............ ...........¼ pound.
Ground Gentian Root...................................¼ "

Mix together and give a teaspoonful twice a day in his feed or on his tongue.

### PARALYSIS OF THE GULLET OR PHARYNX.

This is a very serious disease, for the animal cannot swallow neither food or water, but fortunately it is not often met with in the horse.

**Causes.**—It is generally cause by some injury to the throat.

**Symptoms.**—The animal will take food into his mouth, chew it and prepare it for swallowing, and then spit it out, not making any effort to swallow; he will try to drink, but cannot make any effort to swallow it. If you examine the throat you cannot see anything wrong; no swelling, and it does not seem the least bit sore; the horse seems eager to eat and drink, but cannot; he falls off greatly in condition, gets very weak and will soon die from astrvation.

**Treatment.**—If he is a very valuable animal and worth going to the expense, treat him with a stomach pump by taking the juice got by boiling hay and making gruels made out of chopped oats, new milk and eggs and pumping it down into the stomach; in this way he is kept alive until the muscles of the gullet have regained their strength of swallowing. In giving the gruels put teaspoonful of nux vomica in it three times a day. This is a nerve stimulant, and will help the muscles to regain their strength; also hand rub the throat well around the gullet and apply white liniment five or six times a day until the animal regains the power of swallowing.

### CHOKING WITH OATS.

This is generally found in old horses that are very greedy feeders and not used to getting oats.

**Causes.**—Are generally where a horse has been out at pasture and brought in and given a feed of oats and he goes at it so greedy he fills his mouth and tries to swallow it without chewing it properly.

**Symptoms.**—The horse is noticed not to be eating his oats, and if you examine there is very little of the oats gone out of the box and the horse is slobbering at the mouth and coughing, and if you watch him for a few minutes he will be noticed to gag and and draw the muscles of the neck stiff and bend the neck down as if he was trying to force it up out of his throat, then he will take a violent fit of coughing for a few minutes, and in some cases he will throw out a frothy substance probably mixed with a few oats. He will keep on doing this until he gets relief

**Treatment.**—In some cases by drenching him with raw linseed oil the oil will work around the oats and make it slippery and he will be able to cough it out or swallow it down. It is also well to hand rub him along the neck, which will sometimes help to start the oats down to the stomach. He generally gets entirely over it in a few hours. After a few hours, if he has not got relief, it is then best to put a twitch on his nose, put a gag in his mouth, and pass a probang down his throat (the probang is a long tube used for relieving horses and cattle in choking); relieve him without using the probang if you can. It is advisable, after bringing a horse from pasture, when giving him his first feed of oats, to dampen it with water, as he is not so apt to choke when the oats are damp. After a horse has once choked he is more liable to choke again, and to prevent this, scatter his oats well, and have a few large, round stones put in the feed box so he cannot get a large mouthful at a time. It is very seldom a horse will choke on any kind of fruit or vegetables, such as apples, potatoes and carrots, but if he does, use the above treatment.

## DILITATION OR ENLARGEMENT OF THE ŒSOPHAGUS.

The Oesophagus is the name of the tube which carries the food from the gullet to the stomach.

**Causes.**—From choking which causes the tube to become enlarged, forming a pouch or sack where the food often becomes lodged and causes symptoms of choking.

**Symptoms.**—This enlargement can be seen if in the neck region.

**Treatment.**—It can sometimes be relieved by rubbing on the enlargement with the hand, which causes the food to pass down into the stomach, thus relieving him. It also can be relieved by drenching with raw linseed oil.

## SWELLING AROUND THE HEAD AND THROAT.

This swelling is noticed around the horse's head and throat just after he has been turned out to pasture for a few days, and looks very alarming when first noticed.

**Causes.**—It is caused by an increased flow of blood to the head when the animal has his head to the ground and not being used to it.

**Symptoms.**—It is loose swelling around the jaws and throat of the horse. When you examine it, it is not at all painful, and the animal seems perfectly healthy in every other way. If the animal is brought in for a night, and is fed where he will hold up his head, the swelling will all disappear and will be all right in the morning.

**Treatment.**—In some cases he does not need any treatment at all, but if the animal's blood seems very bad it would be well to give him a few doses of the following:

Nitrate of Potash or Saltpetre........................¼ pound.
Sulphur..............................................¼ "

Mix thoroughly and give a tablespoonful in a little grain night and morning.

---

## CHAPTER IV.

# DISEASES OF STOMACH AND BOWELS.

The stomach of the horse is very small in proportion to that of other animals, and digestion takes place very quickly.

### ACUTE INDIGESTION.

This is one of the most common diseases of the horse.

**Causes.**—If the horse is not used to being fed very heavy, and he gets a large feed of rich food, such as bran, oats or hay, it is apt to set up the disease; or eating wheat may bring on a bad case of it; it is sometimes caused by a horse being put to work too soon after a large meal. A change in the food will cause it, or even a change of work, such as a horse being used to the farm and then put on the road driving. When a horse is turned in on a field of clover, after a rain or frost, it is apt to bring on a bad case of indigestion.

**Symptoms.**—We will take a case as we often see it on a farm. The farmer intends to take a journey of fifteen or twenty miles, and the night before he gives the horse an extra good feed, and, if he is used to feeding on cut feed he will probably change the food to hay. The next morning he is up early and gives the horse another good feed before starting on his journey. The horse starts off lively, but after a few miles he becomes dull, sweats freely around the belly and chest, and is noticed to pass manure in small quantities, and in some cases he will even scour; if he is stopped he shows signs of cramps or pains in his belly, and attempts to lie down, he looks around at his side, and when you examine him he is slightly bloated; as he is driven on, he becomes duller and more bloated, but finally reaches his journey's end, and you can hardly unhitch him from the rig, he is in such pain. As soon as he is taken into the stable he shows violent symptoms by laying down and trying to roll on his back, then he will jump to his feet, look at his flank, and again lie down, and sometimes he is noticed to paw first with one foot then the other, and keeps on repeating these symptoms until he gets relief ; he perspires freely all over the body from the pain ; tapping with your finger on his side it will produce a drum-like sound, showing that there is gas there. If you sound the stomach and bowels with the ear you will not hear much noise, any more than the tinkling of gas. He is noticed to pass gas per anus, which is a good sign. The heat of the body, the legs and ears are about natural. This is a good point to note, for in inflammation of the stomach or bowels, the legs and ears are always cold and clammy, his pulse beats from 50 to 75 beats per minute, and beating moderately strong, whereas if there was inflammation it would be beating very strong and wiry ; he is also noticed to breathe heavy and quick, which is caused from the stomach being so distended with gas that it presses heavy on the lungs. By placing your ear to the bottom part of his neck you will hear him belching wind. This disease generally takes from twelve to twenty-four hours to run its course. In some cases, where the disease is not so bad as the one mentioned, the symptoms will not be so distressing, he will be noticed to leave his feed and commence pawing, then lie down quiet, probably for a few minutes, then get upon his feet again, take a few bites of feed, paw, and lie down again. In some cases he will seem easy for a few minutes, when the pain comes on again. If you examine his bowels you will hear them working pretty well, but will hear gas passing through them. The pulse will not be changed much, probably 5 or 10 beats faster than they should be, and the heat of the body, and of the ears and legs will be about natural. In a case of this kind, if the animal does not get relief he will probably show these symptoms for days, or even a week, before the symptoms will get worse. It is not well to allow the animal to suffer too long if he does not get relief himself.

**Treatment.**—This disease is found more in heavy horses than in light, and also is more fatal in the heavy horse. In some cases, where the animal does not get relief in this disease, it will terminate in rupture of the stomach, or inflammation. If you are driving or working the animal, the first thing to be done is to stop working him and get him to the stable as soon as possible and treat him as soon as you can. Give him the following:

| | |
|---|---|
| Spirits of Turpentine................ | 1 ounce or 4 tablespoonfuls. |
| Tincture of Laudanum.............. | 1 " 4 " |
| Raw Linseed Oil................... | 1 pint. |

Mix, shake well and give as a drench. Give an injection of half a pail of luke-warm water and a little soap, with a teaspoonful of turpentine in it. Have his belly well hand rubbed and apply a mustard plaster. Half pound of mustard, four tablespoonfuls of flour and enough vinegar to make it into a paste, apply this well over the stomach—also clothe the body according to the season of the year, and do not allow the animal anything to eat till he gets relief, for it will only make him worse. In bad cases it is best to have some person stay with the animal to keep him from hurting himself in rolling about; also keep the animal quiet as you can, and never, in any case, run him or keep him walking around the yard, for it is best to keep him quiet as you can. If he does not get relief from the above dose in an hour and a half, give him the following:

| | |
|---|---|
| Bitter Aloes......................... | 8 drams. |
| Sweet Spirits of Nitre.............. | 2 ounces or 8 tablespoonfuls. |
| Ginger.............................. | 1 tablespoonful. |
| Common Soda....................... | 1 " |

Dissolve in a pint of luke warm water, shake well and give as a drench. If he does not get relief in two hours after this drench, follow up every two hours by drenching with the following.

| | |
|---|---|
| Sweet Spirits of Nitre................ | 1 ounce or 4 tablespoonfuls. |
| Ginger.............................. | 1 tablespoonful. |
| Common Soda........................ | 1 " |

Dissolve in a pint of warm water and give as a drench every two hours. Also give an injection every couple of hours, and have the belly and legs well hand rubbed.

In severe cases take a half pail of salt, heat well and put in a grain bag, tie half way down, and place it over the small of his back, then cover him up with a blanket to keep the heat in; keep this changed by more hot salt every hour for heat is a good thing to keep down the pain and keep inflammation from setting in.

In a case where the horse is only slightly affected, take him from work and give him the following:

Bitter Aloes..........................8 drams.
Sweet Spirits of Nitre................1 ounce or 4 tablespoonfuls.
Common Soda.... ....................1 tablespoonful.
Ginger...............................1 "

Dissolve in a pint of warm water and give as a drench, and follow up with the following drench:

Sweet Spirits of Nitre................1 ounce or 4 tablespoonfuls.
Common Soda.........................1 tablespoonful.
Ginger...............................1 "

Dissolve in a pint of warm water and give as a drench every four hours until the animal is relieved. Blanket him well and allow him to stand in the stable for a few days, till the physic is worked off. In all cases, when he is recovering, feed on soft food, such as bran mashes, and give him all the luke warm water he will drink. In all cases of stomach trouble, warm the drinking water, as it has a good effect on the stomach. In severe cases, where the animal is going to die, and the treatment does not do him any good, he gradually keeps getting worse, begins to bloat very bad and breathe very heavy, and his ears begin to droop over. If you examine the pulse, it is up to 90 or 100 beats per minute, and you can hardly feel it. The animal does not lie down so much, but walks around his box, trembling very much all over the body; his legs, ears and nose gradually become cold as death approaches, also the inside of his mouth is cold and clammy, and he will be noticed to strain, as if wanting to pass something, but nothing comes; he finally will stagger, fall and die. All through this disease, the animal will be noticed to make water in small quantities, but often. You must not be misled by this symptom, for it is caused from the swollen stomach and bowels pressing on the bladder. The above disease is one which should be well studied, for it is met with so often in the horse.

### RUPTURE OF THE STOMACH.

This disease is generally caused during the time the animal is suffering from acute indigestion.

**Causes.**—A horse suffering from acute indigestion, will, from the pain, sometimes throw himself down on his side, and the stomach being so distended or swollen with gas will cause it to become ruptured.

**Symptoms.**—The animal will be noticed all at once to become very bad—much worse than before—the pulse will get very quick and weak, he will perspire all over the body, and in a very short time will show symptoms of approaching death by his bowels and stomach becoming very much swollen, his legs, ears and nose will begin to get cold, which indicates approaching death, and the animal generally dies very quickly.

**Treatment.**—There is no treatment that will give him relief, but it is best to give the animal an ounce of tincture of laudanum to relieve his pain till he dies.

### CHRONIC INDIGESTION.

This disease is sometimes called dyspepsia. This is quite a common disease among horses, especially about the time they are shedding their teeth.

**Causes.**—It may result from a disordered state of the bowels or liver, or from a young horse while shedding his teeth, not chewing his food properly, from being very high fed, from feeding dusty clover hay, and from irregular feeding and watering.

**Symptoms.**—The symptoms of this case is not alarming, the horse gradually falls off in condition and gets weak, sweats very easily while at work, his coat is dry and dusty, and he does not seem to have any ambition, his manure is of a dark clay color, he is sometimes attacked with little fits of colic or pains in the bowels, and he is noticed to be licking the walls and manger and seems to have a craving appetite.

**Treatment.**—Give him a change in feed and a dose of physic consisting of the following:

| | |
|---|---|
| Bitter Aloes | 8 drams. |
| Common Soda | 1 tablespoonful. |
| Ginger | 1 " |

Mix in a pint of luke warm water and give as a drench, allow the animal to stand in his stall in the stable until the physic passes off; feed on soft food and follow up with the following mixture as a tonic for him:

| | |
|---|---|
| Ground Gentian Root | ¼ pound. |
| Common Soda | ¼ " |
| Ginger | ¼ " |
| Sulphate of Iron | ¼ " |

Mix thoroughly and give a teaspoonful three times a day in his feed or on his tongue with a spoon. This will build him up and start him to gain.

### BOTS.

These are found clinging to the inside of the stomach of the horse, and we have rarely ever examined a horse's stomach but what we found some in it. In some cases they are found in large numbers, and in these cases they injure the animal

**Causes.**—Bots are caused by the bot-fly stinging and laying its eggs under the horse's head, neck and legs. These flies, while they are laying their eggs on the horse, seem to annoy him very much, and he will be seen to jerk his head as if they had stung him. These little eggs are taken into the horse's mouth by him biting and rubbing his legs with his mouth in the fall of the year. They pass from the mouth down into the stomach, where they lodge for the winter by hanging on to the lining membrane of the stomach and develop and grow until spring comes, when they will let go their hold, pass through the bowels and out with the manure, where they soon become developed into a regular bot-fly, and fly off to bother the horse during the summer. This is the way they reproduce.

**Symptoms.**—The horse will be noticed not to be doing well, and has a ravenous appetite, but the feed does not seem to do him much good. He will also be noticed to hold his head up and be moving and twisting his upper lip.

**Treatment.**—Give him the following mixture :

Raw Linseed Oil......................1 pint.
Spirits of Turpentine.................1 ounce or 4 tablespoonfuls.

Shake well and give as a drench. Give this drench once a week until the animal seems relieved, allowing him to stand a day after each drench; also give him a teaspoonful of ground sulphate of iron in his feed twice a day.

---

# DISEASES OF THE BOWELS.

## SPASMODIC COLIC.

This is a spasmodic contraction of the muscular fibres of the coats of the bowels, or in other words, cramps of the bowels. The small bowels are the ones usually affected, but the large ones might, too, be affected.

**Causes.**—The principal cause is a change of food, such as giving a feed of roots when the animal is not used to it, and especially when frozen, or a cold drink of water when the animal is hot. Sometimes it comes on after giving the horse a physic. A sudden change in the weather will sometimes bring it on. Some horses become, as it were, subjected to this disease, more especially if the stomach is not digesting the food properly. Although this disease is very painful while it lasts, it is not very al, and usually passes off quickly. If not attended to, the chief danger is, that it will set up inflammation, which then becomes very serious.

**Symptoms.**—The attack comes on very quickly, in some cases the horse will begin to paw, cringe, look at his side, and throw himself to the ground as if in great pain, will roll around and try to balance himself on his back. If the weather is hot, he will sweat very much. He may lie quiet for a few minutes, get up, and may appear all right, may eat some, then the pain will come on again, and act in the same manner. He will generally pass manure in small quantities, and also make his water, which is a good sign. On putting your ear to his side, the bowels will be working just about natural, except when severe pain comes on. The pulse, when he is at ease, is about natural, but when in pain, it beats very much faster, probably 60 or 65 beats per minute, while in inflammation the pulse gradually goes up and stays up all the time. If you press on his bowels it seems to relieve him in colic, while in inflammation it gives him more pain. As a general thing it does not last very long, probably two or three hours, although, in some cases, we have seen it last as long as ten or twelve hours, but the symptoms in that case would not be so severe.

**Treatment.**—A student was once asked at a certain veterinary college, what he would do if he were sent for in a case of this kind. He said "he would go as fast as he could for fear the case would be all over before he got there."

The favorite remedy for colic is:

| | |
|---|---|
| Sweet Spirits of Nitre................ | 1 ounce or 4 tablespoonfuls. |
| Tincture of Laudanum................ | 1 " 4 " |
| Ginger................................ | 1 tablespoonful. |
| Common Soda ......................... | 1 " |

Mix in a pint of luke warm water and give as a drench. Also another good remedy is:

| | |
|---|---|
| Raw Linseed Oil...................... | 1 pint. |
| Spirits of Turpentine................. | 1 ounce or 4 tablespoonfuls. |

Mix and give as a drench. Another good remedy is:

| | |
|---|---|
| Fleming's Tincture of Aconite .................. | 10 drops. |
| Whisky......... ............................. | 1 wineglassful. |

Mix in a pint of luke warm water or beer and give as a drench. In cases of this kind it is best to have some person stay with the animal for fear he would get cast on his back or hurt himself. In severe cases have the belly well hand rubbed, and have cloths wrung out of hot water applied to the belly, or you may find good results from a mustard plaster applied over the bowels. Any of the above drenches, except the oil and turpentine, may be given every hour until the animal gets relief. After the animal has got over the pain, to get rid of the irritation in the bowels, or whatever has caused it, give a dose of physic consisting of:

| | |
|---|---|
| Bitter Aloes.... | 8 drams. |
| Ginger ... | 1 tablespoonful. |
| Common Soda.... | 1 " |

Mix in a pint of luke warm water and give as a drench.

In each of the drenches mentioned it is well to put in a tablespoonful of ginger and common soda.

### WIND OR FLATULENT COLIC.

This and acute indigestion are a great deal the same. The stomach is generally affected in this disease as well as in indigestion.

**Causes.**—This disease is caused from a derangement of the digestive organs, and the food in the bowels forms gas which fills the bowels and causes extreme pain while it lasts; but in the majority of cases it is caused from a change of some kind in the food, and is also found chiefly in aged horses.

**Symptoms.**—The symptoms of this disease resemble acute indigestion somewhat, only that the gas is formed in the large bowels instead of the stomach. The animal lays down and rolls, the belly becomes bloated with gas, and if you notice there will be no belching of wind from the stomach as there is in acute indigestion.

**Treatment.**—Give injections freely of one-half pail of luke warm water, a little soap and a tablespoonful of turpentine, and for a drench give:

| | |
|---|---|
| Spirits of Turpentine.... | 1½ ounces or 6 tablespoonfuls. |
| Tincture of Laudanum.... | 1 " 4 " |
| Raw linseed Oil.... | 1 pint. |
| Common Soda.... | 1 tablespoonful. |
| Ginger.... | 1 " |

Shake well together and give as a drench, and follow up with the following drench:

| | |
|---|---|
| Sweet Spirits of Nitre.... | 1 ounce or 4 tablespoonfuls. |
| Ginger.... | 1 tablespoonful. |
| Common Soda.... | 1 " |

Mix in a pint of luke warm water and give every hour until the horse gets relief. It is best in this case to have someone to stop with the animal to keep him from hurting himself or from getting east. Hand-rub the belly well, and in severe cases apply a mustard plaster to the bowels, and also have one-half pail of hot salt in a bag placed over the animal's kidneys; this will have a tendency to move the gas in the bowels and helps to keep down inflammation. After the animal has been relieved, if it has been a bad case, it is best to follow up with a dose of physic, consisting of:

| | |
|---|---|
| Bitter Aloes | 8 drams. |
| Common Soda | 1 tablespoonful. |
| Ginger | 1 " |

Dissolve in a pint of luke warm water and give as a drench, and allow him to stand in the stable the next day. Feed on soft, light feed. This will generally prevent him from having another attack of colic.

### INFLAMMATION OF THE BOWELS (ENTERITIS.)

This is a very common disease in the horse and is, perhaps, one of the most fatal.

**Causes.**—The disease sometimes follows a severe case of colic, where the animal does not soon get relief. It may be caused from eating food which has clay or sand in it, which causes an irritation of the bowels. Eating pea straw will cause the disease sometimes; drinking stagnant water, exposure to cold after a long, exhausting drive, the animal getting a chill which rushes the blood in upon the bowels and sets up congestion, which is followed by inflammation.

**Symptoms.**—The horse is attacked very suddenly, begins to tremble, paws with one foot and then with the other, and turns the head around to the side, cringes and lies down, and does not get a minute's ease as he does in colic, but will get up, walk around, look at his side, and if his pulse is taken at this stage of the disease, it will be found about 45 beats per minute, full and bounding. His legs and ears will be hotter than natural. He passes manure in small quantities, which looks slimy. The pain keeps on increasing, the symptoms get worse, and he does not get a minute's peace; his pulse is up to about 75 beats, and it is still full and bounding and does not vary as it does in colic, but keeps getting higher as the disease advances. He sweats freely, and the lining in his eyes becomes very much reddened and angry looking; his legs and ears change from hot to cold, and the pain keeps on increasing. At this stage his ears begin to lop over and he gets a very haggard look on his face, as if in extreme agony. After a few hours he is a pitying sight to see. If you sound his bowels at this stage there is not the slightest movement to be heard, only a jerking and trembling all through his insides. He begins to breathe heavy, and his ears and legs have a cold, clammy feeling. He keeps on in very great pain, lying down, getting up, and walking around his box, and, if seen to make water, it will be red and bloody looking, and if there is any passage from the bowels, it will be mostly slime. If he does not get relief in the course of eight or ten hours, mortification then sets in, and the animal becomes quiet and easy, but he still keeps sweating and breathing heavy, and in some cases will try to eat and once in a while he will be noticed to walk around his box. In this stage he does not lie down. The surface of his body, his ears, his nose, his lips and legs get colder and have a death-like feeling. If you take his pulse now, it will be up to 100 beats per minute, and so weak you can hardly feel it, showing that his heart is just fluttering, and that was all; the haggard look on his face becomes more marked; he will be noticed to strain a few times, as if trying to pass something, but nothing comes. He will keep on his feet as long as he can, but will finally stagger, fall and die. This disease generally runs a course of from 10 to 15 hours, but in some cases we have known them to live as long as two or three days, where there was not much of the bowels affected.

**Treatment.**—This disease, if taken as soon as the animal is noticed sick, may be sometimes cured, but the treatment must be quick and careful, for, if the disease once gets a couple of hours the start, it is then a hopeless case. Give the following:

| | |
|---|---|
| Tincture of Laudanum.............. | 2 ounces or 8 tablespoonfuls. |
| Fleming's Tincture of Aconite........ | 10 to 15 drops. |
| Common Soda........................ | 1 tablespoonful. |
| Ginger.............................. | 1 " |

Mix in a pint of luke warm water, and give as a drench. This drench is to relieve the pain and try and check the inflammation. This drench may be repeated every hour until the animal gets relief. Apply lots of heat to the body in the form of large woolen blankets, wrung out of hot water and held up to the belly, and half pail of hot salt in a grain bag to the back. In every case, after you are through using the hot blankets, apply a mustard plaster, consisting of:

| | |
|---|---|
| Mustard............................ | ½ pound. |
| Vinegar............................ | Enough to make it like paste. |

Rub this well over the belly. Do not give any physic or injection in a case of this kind, for they only irritate the bowels and make the case worse. It is always best to keep the bowels quiet in this disease. Where the animal is in high condition it is well to take a half pail of blood, if in the first stage of the disease, but if you do this, do not give so much aconite. If the animal gets relief, it is best to feed him on soft feed with flax seed in it, which has a soothing effect on the bowels. The horse's bowels will be found, after death, to be black and thickened and full of watery fluid.

## CONSTIPATION OF THE BOWELS.

This is where the bowels become sluggish and loaded with food and manure.

**Causes.**—This disease is often a symptom of another disease, as in liver trouble, or where the stomach is not doing its work properly. It also may come after distemper or influenza, where the bowels become weak and cannot do their work; also paralysis of the bowels, where the bowels are paralyzed; sometimes from eating over ripe and inferior food, such as pea straw or barley straw. In some cases it is caused from a large tumor growing on the inside and pressing on the bowels, not allowing them to act.

**Symptoms.**—There will be very little manure pass, and what comes will be in little hard balls. The animal will look unnaturally full and show signs of pain, but not much. He is sometimes noticed to lie down and roll, and look around at his sides. His pulse is not much changed, and when listening at his side there will be very little movement in the bowels. He does not eat much and looks dull and dumpy, and his water is of a thick, yellow color. If you examine his rectum or back bowel by oiling your hand and passing it in through the anus, which can be easily done without any danger, you will find it full of hard, dry manure.

**Treatment.**—To start with give him a good dose of physic, consisting of:

Bitter Aloes..........................8 to 10 drams.
Sweet Spirits of Nitre.................1 ounce or 4 tablespoonfuls.
Powdered Nux Vomica...............1 teaspoonful.

Mix in a pint of luke warm water and give as a drench, then follow up with the following drenches:

Sweet Spirits of Nitre................1 ounce or 4 tablespoonfuls.
Powdered Nux Vomica...............1 teaspoonful.
Common Soda.................. .....1 tablespoonful.
Ginger..............................1 "

Mix in a pint of luke warm water and give as a drench every five or six hours until relieved of the pain. Twenty-four hours after you give him the drench with the aloes in it. If the bowels have not begun to move, follow up with a drench of one pint of raw linseed oil. Clean the manure from the rectum or back bowel with your hand twice a day, and give an injection of one half pail of luke warm water and a little soap. After the pain is relieved, and the bowels working, it is well to follow up with the following powders :

| | |
|---|---|
| Ground Gentian Root........................................ | ¼ pound. |
| Ginger................................................ | ¼ " |
| Common Soda........................................ | ¼ " |
| Powdered Nux Vomica.................................. | ¼ " |

Mix thoroughly and give a teaspoonful three times a day in his feed. The powders will strengthen and tone the bowels and start the animal to thrive. Feed the animal on soft feed with plenty of flax seed in it, which will have a good effect on the weak bowels. In cases of constipation, where the physic does not seem to be acting right, a little exercise will often start it to work.

## DIARRHŒA.

This is the very opposite to constipation, and is a disease where the animal passes a large amount of fluid manure, which is due to the congested state of the bowels, and is seen most in horses of weak confirmation, as narrow chested and gaunt looking horses.

**Causes.**—Where the animal gets a few feeds of rich food after being used to poor food for a length of time. Sometimes from a feed of roots, such as turnips and carrots, especially if they are frozen. Also drinking stagnant water, which acts as a blood poison ; and we sometimes have very bad cases caused by an animal feeding on a sandy pasture, where the grass is short, and in grazing the short grass takes up sand with it which causes an irritation of the bowels. An over dose of physic medicine will cause this, and when it is thus caused the diarrhœa is called superpurgation. Diarrhœa is a disease very easily treated, as a general thing, that is if the animal is in a healthy, strong condition. It is sometimes seen in nervous horses when they are put in races and get excited. It is also noticed in excitable road horses. In these cases it is due to excitement.

**Symptoms.**—This disease is very easily told. The animal passes a lot of watery looking manure. If you examine the pulse it will not be much affected at first, but if the disease is allowed to run on it will become quick and weak. The animal does not eat, and becomes very gaunt and weak looking, and if allowed to run on the legs will become colder than natural; after a time there will be slight pains, caused by the irritation and spasms in the bowels. When the animal begins to get in pain—if he is not soon relieved—the disease begins to get more serious, for the congested state of the bowels would soon run on to inflammation of the bowels.

**Treatment.**—In many cases all you have to do is to change the food, clothe the body according to the season of the year and give a little medicine. Endeavor to find out the cause of the trouble, and if it is caused by some irritation in the bowels, such as irritating food or sand, then give the following:

| | |
|---|---|
| Raw Linseed Oil.................... | ½ pint. |
| Tincture of Laudanum.............. | 1 ounce, or 4 tablespoonfuls. |
| Ginger.............................. | 1 tablespoonful. |
| Common Soda...................... | 1 " |

Mix and give as a drench. In this case give the drench to assist nature in throwing off the cause of the disease. Where the disease is not caused from an irritation of that kind give the following:

| | |
|---|---|
| Tincture of Catechu................ | 1 ounce, or 4 tablespoonfuls. |
| Ginger.............................. | 1 tablespoonful. |
| Common Soda...................... | 1 " |

Mix in a pint of coffee about the same strength as what is used at the table and give as a drench, repeat the drench every four or five hours until the animal has relief. The animal will be very dry; give him the water luke warm, in small quantities, but often, and in the water mix a handful of flour. Feed him on dry feed and keep him quiet. In severe cases it is well to apply a mustard plaster over the bowels and put a half pail of hot salt in a bag and place it over the kidneys.

### DIARRHŒA IN YOUNG ANIMALS.

**Causes.**—The causes of this disease in young animals are generally exposure to cold, or where the mother's milk is either too rich or too poor, or where the young animal is allowed to lie out on the damp ground in the spring or fall of the year. This chills the bowels and sets up diarrhœa, or in some cases where the mother is working hard and becomes heated and the foal drinks a large amount of the hot milk.

**Symptoms.**—He will pass a large amount of fluid manure, which will stick around his legs and tail; and will become weak and sickly and very gaunt, and his coat will be staring.

**Treatment.**—If in a strong foal, give :

Castor Oil ..........................1 ounce, or 4 tablespoonfuls.
Tincture of Laudanum..............10 to 15 drops.

If the mother's milk is weak try and improve it by giving lots of nourishing food. If it is caused from drinking the milk from the mother when hot, always milk a little out before the foal gets to the mare. If this does not relieve him in five hours, follow up with

Tincture of Laudanum....10 to 20 drops.
Brandy or Whisky.......½ to 1 ounce, or to 2 to 4 tablespoonfuls.

Mix with some of the mother's milk and give as a drench three times a day. If in severe cases keep the foal warm and apply a light mustard plaster to his belly.

## BLOODY FLUX (DYSENTERY).

This disease affects the lining of the large bowels, in which large ulcers are formed, which bleed, causing the manure to be streaked with blood.

**Causes.**—Often from a severe attack of diarrhœa, from pasturing on wet, marshy lands, or from eating hay grown on such lands, or using impure water.

**Symptoms.** The passages from the bowels are streaked with blood and have a bad smell ; sometimes slime comes away with the manure. He does not feed much, and in some cases his appetite is entirely gone. The pulse will be at about 50 beats per minute and weak, and there will be colicky or cramping pains in the bowels.

**Treatment.**—This disease is in some cases not treated with success, and in bad cases is considered very serious. Give the following :

Raw Linseed Oil......................½ pint.
Castor Oil.................. .........½ "
Tincture of Laudanum................1 ounce or 4 tablespoonfuls.

Mix, shake well, and give as a drench ; then follow up with the following drenches :

Tincture of Laudanum.... ...........1 ounce or 4 tablespoonfuls.
Tincture of Catechu ..... ..... .....1 " 4 "

Mix in a pint of luke warm water and give as a drench three times a day—morning, noon and night until he gets relief. Clothe the body well, according to the season of the year ; feed on soft feed with lots of boiled linseed in it, and take the chill off his drinking water for a few days. After the first symptoms have passed off some, follow up with the following powder :

Ground Gentian Root .................. ..... .........¼ pound.
Sulphate of Iron.......................................¼ "

Mix well together, and give a tablespoonful twice a day.

### TWIST IN THE BOWEL (VOLVULUS).

This is not a very common disease, although sometimes it is met with.

**Causes.**—It may be due to colic, when the animal is in very bad pain, and the bowels cramped, and the animal rolling about. It is generally noticed in young animals.

**Symptoms.**—It is very hard to be sure it is this disease, for the animal shows similar symptoms to inflammation of the bowels ; there is no passage on account of the twist in the bowels ; the animal is noticed to sit upon his haunches, and he sweats, and seems in great distress ; the pulse runs up and gets weaker and weaker ; he is generally slightly bloated, on account of no passage. In two or three hours the bowels, where the twist is, become inflamed and the animal stays in severe pain until he dies.

**Treatment.**—In a case where you are sure it is this disease, there is no treatment, except to relieve the pain by giving one ounce or four tablespoonfuls of laudanum every hour. If the animal is opened after he dies, there will be found a half hitch on the bowel and for about a foot or so on each side of the twist it will be black and inflamed.

### INTUSSUSCEPTION.

This is where you have one part of the bowel to slip inside of the other. This disease is generally found in foals living on milk.

**Causes.**—It is hard to say just what causes it, but it is supposed to be due in some cases to cramps in the bowels.

**Symptoms.**—They are similar to those of colic ; the animal will have pains for a few days, will not eat ; the bowels do not work very well, and after a few days in some cases the part of the bowel that was slipped inside the other will begin to sluff and pass off in the manure, the bowel will heal and the animal will gradually recover. In cases where you suspect this, give to a good sized foal :

Raw Linseed Oil...................... ½ teacupful.
Tincture of Laudanum..................1 dram, or 1 teaspoonful.

Mix in some of the mother's milk and give as a drench. The dose must be given in proportion to the size of the foal. After this follow up with the following:

| | |
|---|---|
| Tincture of Laudanum.................. | 1 dram, or 1 teaspoonful. |
| Sweet Spirits of Nitre............ ..... | 1 " 1 " |
| Common Soda.................. ...... | 1 teaspoonful. |
| Ginger ................................ | 1 " |

Mix in a little of the mother's milk and give as a drench every four hours until the animal gets relief.

### BALLS FOUND IN THE BOWELS (CALCULI).

They are chiefly made up of lime, and vary from the size of a marble to twenty pounds. They generally commence by the animal drinking or eating a piece of brass or iron, or anything which will have a tendency to collect the lime.

**Causes.**—Generally from feeding the sweepings of a mill floor, or such like. It may take the ball a long time to collect before it gets large enough to stop the passage, the balls are generally found in the large bowels.

**Symptoms.**—The first symptoms of this are, for a while, the animal is subjected to wind colic, which afterwards, sooner or later, as the ball gets larger, terminates in a complete stoppage of the bowels, which sets up inflammation and generally terminates in death from twenty-four to forty-eight hours after the complete stoppage in the bowels.

**Treatment.**—In a pure case of this kind nothing can be done, only give doses of laudanum, 1 ounce or 4 tablespoonfuls every couple of hours to relieve the pain while the animal lives. In the first symptoms, give a good dose of physic, and in all cases where you suspect this disease it is best to examine the rectum or back bowel by oiling your hand and passing it into the rectum, and if you can feel the ball, remove it with your hand.

### RUPTURE OF THE RECTUM OR BACK BOWEL.

This is a very serious injury, but in some cases they will recover, especially if the rupture ison the upper side of the rectum.

**Causes.**—Are generally from some foreign substance, such as the shaft of a buggy in a runaway, or any other such cause, entering in at the anus into the back bowel and rupturing it. It has also been caused from what is known as mal-address. This is when the stallion, serving a mare, enters into the wrong passage.

**Symptoms** are generally stoppage in the bowels, and also bleeding from the anus after the stallion has served the mare, or if a shaft in a runaway, this will be plainly seen. You can tell the extent of the injury by oiling your hand and passing it into the rectum and examining it.

**Treatment.**—In a case where it was done from any substance, such as a stick or shaft, it is best to examine and see if there is any sliver or piece of the stick left in the wound, and if there is remove it and clean the manure out of the rectum by passing your oiled hand up three or four times a day. After you have removed the manure give an injection of a little warm water and soap, which will have a soothing effect on the wound, and also help to keep the bowels regular. Feed on soft food, and only allow him to have a small amount of it, so as to keep the bowels empty as possible, without starving the animal, and give him the following drench.

| | |
|---|---|
| Raw Linseed Oil ...................... | ½ pint. |
| Tincture of Laudanum ................ | 1 ounce or 4 tablespoonfuls. |

Mix and give as a drench. After this, if the animal seems in pain, it is well to follow up with 1 ounce or 4 tablespoonful drenches of laudanum every four hours in a little luke warm water. If the animal is very fleshy, it is well to give five to ten drops of Fleming's tincture of aconite to keep down inflammation.

## TUMORS OR ABSCESSES IN THE RECTUM OR BACK BOWEL.

This disease is not very common.

**Causes.**—When the bowels are costive it has a tendency to cause this; or from rudely inserting the hand or an injection pipe.

**Symptoms.** The symptoms are noticed most at the time of making manure. He will be noticed to be in severe pain when passing anything, and in some cases he will lie down from the pain. If the tumor or abscess is large you will notice the animal straining, but pass nothing.

**Treatment.**—If it is an abscess, and you think there is matter in it by feeling it, take a small knife or lance and pierce it, allowing the matter to escape. If it is a tumor, and can be got at, then remove it with a knife by cutting it out. In some cases the ecraseur (which is an instrument with a chain that squeezes it off) comes in very useful in removing tumors here. The after treatment is giving raw linseed oil, and feeding boiled flax seed in the feed to keep the bowels loose.

## PROTRUSION OF THE RECTUM OR BACK BOWEL.

This is often met with and is a miserable looking sight, especially if it is left out long and it becomes swollen.

**Causes.**—From a horse jumping a fence and getting caught and lying on his belly half over the fence; or in cases where the animal gets very much bloated, as in wind colic or acute indigestion, and the bowels get pressed out. It has been seen in cases of diarrhœa, and one great cause of this is constipation of the bowels, where the animal strains to pass manure and, in so doing, turns the rectum out. It is turned out sometimes when the mare is foaling, and sometimes in castrating an old stallion, where he forces very much.

**Symptoms.**—You will see from three or four inches, and, in some cases, two feet of the bowel.

**Treatment.**—Wash the bowel well with luke warm water, and then place a twitch on the horse's nose and have one of his front legs held up so he cannot kick; have the tail held out of the road, then take sweet oil and oil the bowel all over; commence returning the bowel at the anus, as you have to turn the bowel inside out; shove the parts well back in with the hand and arm, and have someone to hold his tail down tight for an hour or so until he quits forcing, after the bowel is replaced. It is also well to have his hind end raised, by means of straw or boards under his hind feet. Give the following dose of medicine to work on his bowels and relieve the pain:

Raw Linseed Oil......................1 pint.
Tincture of Laudanum .... ..........1 ounce or 4 tablespoonfuls.

Shake well together and give as a drench. Feed the horse on soft feed with lots of boiled linseed in it to keep the bowels loose. If the case is caused by constipation of the bowels, give:

Bitter Aloes....................................8 drams.
Ginger............ ................. ...........1 tablespoonful.
Common Soda ......... .. ........ ...........1 "

Mix with a pint of luke warm water and give as a drench.

## PERITONITIS.

This is inflammation of the serous membrane found lining the inside of the belly and over the outside of the bowels.

**Causes.**—It is generally caused from exposure to cold after some weakening disease. This disease is also noticed to come after the colt has been castrated and he has been left out in a cold rain or walked through a river when warm after castration, or left standing in a cold east wind. It may also be caused from the belly being bruised.

**Symptoms.**—The animal will be noticed to be in slight pain, will lie down, stretch himself out and moan, sweat freely if the weather is warm, then get up and move around and seems very weak, breathes very heavy, almost as heavy as in a case of inflammation of the lungs. The pulse runs up to 70 or 80 beats per minute. If he is not soon relieved his legs and ears become cold ; his ears lop over, and he seems very weak—hardly able to get up when he is down. By listening with your ear to his side you will find his bowels do not work very much, and if you press over the bowels it causes him pain. The animal refuses food, and, in some cases, the manure will have a very glossy appearance.

**Treatment.**—Give the following :

| | |
|---|---|
| Raw Linseed Oil.................... | 1 pint. |
| Tincture of Laudanum............... | 1 ounce, or 4 tablespoonfuls. |
| Fleming's Tincture of Aconite........ | 5 to 10 drops. |

Mix and give as a drench. Apply lots of heat to the belly in the form of woolen blankets wrung out of hot water, and follow up afterwards with a mustard plaster over the bowels, and a half pail of hot salt in a bag over his back. Feed on soft food with lots of flaxseed in it to keep his bowels regular. After the first drench, if the pain is not relieved, follow up with the following drench :

| | |
|---|---|
| Tincture of Laudanum ............... | 1 ounce, or 4 tablespoonfuls. |
| Fleming's Tincture of Aconite........ | 5 to 10 drops. |

Mix in a pint of luke-warm water and give every two hours until the pain is relieved. Sometimes this disease is followed by dropsy or a collection of water in the belly.

### DROPSY OF THE BELLY.

This is a collection of a watery fluid in the belly around the bowels, and is generally the result of inflammation of the membrane mentioned above.

**Symptoms.**—The animal is very weak, the pulse quick and weak, the muscles of the body soft and flabby, the belly looks swollen, and i. you press on it you can tell it is caused by some fluid inside. The bowels are constipated, but in some cases of this disease the animal will eat pretty well.

**Treatment.**—Give remedies that will tend to absorb the fluid, encourage his appetite by giving him lots of good food to eat. Give the following :

Iodide of Potassium.................................¼ pound.
Ground Gentian Root..................................½ "
Nitrate of Potash or Saltpetre..........................¼ "

Mix thoroughly together and give a tablespoonful three times a day. Give the animal a little exercise every day. It is recommended in bad cases to tap the lower part of the belly with a trocar and cannula, but this operation does not prove very successful in the horse.

### HORSE EATING TOO MUCH WHEAT.

This is a very dangerous thing, especially if the horse is not used to getting wheat and eats a quantity of it. The wheat swells and forms gas in his stomach and after a time gets like dough, which sets up acute indigestion and often terminates in a case of acute founder.

**Treatment.**—As soon as you have found out the horse has eaten wheat, do not let him have any water or feed to eat for twenty-four hours and keep him quiet, after that give him plenty of luke warm water and soft feed. If acute indigestion is set up, give the treatment given for acute indigestion ; if it turns to founder, give treatment given for acute founder.

### LONG ROUND WORMS (LUMBRICI).

**Symptoms.**—The animal will not do well, will fall off in condition, and every once in a while will pass some long, round worms in his manure. Sometimes when they are in large quantities they will set up spells of colic, and we have known cases where they form a ball in the small bowels and stop the passage, killing the animal.

**Treatment.**—Get rid of the worms by giving :

Raw Linseed Oil......................½ pint.
Spirits Turpentine ..................1 ounce or 4 tablespoonfuls.

Mix and give as a drench once a week. As well as this give a teaspoonful of sulphate of iron in his feed twice a day. This is the best remedy known for the worms.

### PIN WORMS.

These are short, fine worms about an inch or two long and only affect the rectum or back bowel.

**Symptoms.**—The horse generally feeds well, but will fall off in condition. His coat will be dry and dusty ; he will rub his tail and there will be a white, slimy stuff around the anus.

**Treatment.** --If the animal is in good condition, give a physic of

Bitter Aloes.................................. 8 drams..
Common Soda. .................................. 1 teaspoonful.
Ginger.............................................1 "

Mix in a pint of luke warm water, and give as a drench. After this clean out the rectum with your hand and inject the following:

Quassia Chips........................................½ pound.
Rain Water............................................1 gallon.

Mix and boil down to one half gallon, then strain off the chips and inject with a syringe. After the injection keep it in the rectum one half hour by holding the tail down. Clean the rectum out and give an injection once a week. This is a cheap and sure cure for them. Medicine given by the mouth will not do much good, for it will never reach the worms.

---

CHAPTER V.

# DISEASES OF THE LIVER AND SPLEEN.

## CONGESTION AND INFLAMMATION OF THE LIVER.

These two diseases are so much alike, having the same causes, symptoms and treatment, that we will treat them both together.

**Causes.**—It is usually caused from feeding very high and getting little exercise, or an abscess in the liver will cause these diseases.

**Symptoms.**—The animal shows pain, looks around at his sides, lies down, but does not roll as he does in bowel troubles. He will then get up and stand awhile, breathes heavy and quick, the pulse is quick and the bowels, as a general thing, are costive and the manure is black and slimy looking. The lining of the mouth and eyes are of a yellow color, similar to jaundice. Other symptoms are that he will be lame in the off front leg; and his urine is green, resembling the bile of the liver.

**Treatment.**—If the animal is in good condition, give

Bitter Aloes........................ 8 drams.
Sweet Spirits of Nitre . .........1 ounce, or 4 tablespoonfuls.
Tincture of Laudanum ..............1 " " "

Mix in a pint of luke warm water and give as a drench. Apply a mustard plaster well rubbed in on the under part of the the belly and clothe the body well, according to the season of the year. After this follow up with the following medicine:

Iodide of Potassium, ........................................ ¼ pound.
Nitrate of Potash or Saltpetre........................... ½ "

Mix thoroughly together and give a teaspoonful three times a day in his feed or on his tongue with a spoon. In cases where the cause is from high feeding and little exercise, feed light, soft food and give regular exercise.

### YELLOWS (JAUNDICE.)

This is, perhaps, the most common disease of the liver.

**Causes.** From inflammation of the liver, from gall stones stopping up the tube which leads from the liver to the bowels, from weakening diseases, such as influenza or distemper, from an abscess forming in the liver or from any other disorder of the liver, where the bile is not taken from the blood.

**Symptoms.**—The bowels become constipated and the manure is of a dark clay color, the animal is dull and does not feed well, the lining of the mouth and around the eyes is of a yellow color, from which it gets the name, jaundice.

**Treatment.**—If it is in the spring of the year, by turning him out on the grass it will often effect a cure itself, if not, and the animal is in fair condition, give

Bitter Aloes ........................................ 4 drams.
Calomel ........................................ ½ dram.

Mix in a pint of luke warm water and give as a drench, or it is sometimes mixed in the form of a ball (as to how to make a ball refer to the receipts in the back of this book). After this follow up with the following powder:

Iodide of Potassium........................................ ¼ pound.
Nitrate of Potash or Saltpetre........................... ½ "

Mix together and give a teaspoonful three times a day. Feed the animal on soft, nourishing food, such as boiled oats, scalded chop stuff and bran, with plenty of boiled flaxseed, and allow him gentle exercise every day, this will generally effect a cure.

### BILE STONES (BILIARY CALCULI.)

This disease is not so common in horses as it is in man, but they may exist in great numbers, and if they do they stop the flow of the bile out of the liver and cause the bile to be absorbed back again into the blood, then it sets up jaundice. Persons living high and taking little exercise are liable to these stones. This same rule holds good in horses.

**Treatment.**—Give either a pint of raw linseed oil, or 8 drams of bitter aloes dissolved in a pint of luke warm water to physic the bowels, and give the following :

Diluted Hydrochloric Acid.............. ½ dram, or ½ teaspoonful.

Mix in a pint of water and give as a drench two or three times a day. The action of this acid is to dissolve the stones and get rid of them. Feed the horse light and give regular exercise

## ENLARGEMENT OF THE LIVER (HYPERTROPHY)

This is usually seen in old horses, and is caused by faulty feeding. This disease is also seen in man, which is generally caused from taking intoxicating liquors.

**Symptoms.**—The animal falls off in condition, sometimes has diarrhœa, while again he is costive. This continues changing from one to the other. There will also be yellowness of the lining of the mouth and eyes, and the animal will die a lingering death.

**Treatment.**—There is no cure, but it may be helped sometimes by regular feeding and regular exercise.

## INFLAMMATION OF THE SPLEEN.

This disease is chiefly found in the southern parts of the United States, where it is very warm.

**Causes** are unknown.

**Symptoms.**—Similar to colic, the animal is dull and languid and has a tendency to hang the head and lop the ears. He will lie down and roll, get up, stand easy for a while, will not eat much, and the pulse runs up and is quick and weak. This is a hard disease to form a positive opinion before death.

**Treatment.**—If you suspect it is inflammation of the spleen, give the following :

Raw Linseed Oil..... ..............1 pint.
Tincture of Laudanum..............1 ounce, or 4 tablespoonfuls.

Shake well, and give as a drench. Apply a mustard plaster over the left side of the belly, opposite the stomach and spleen. Clothe the body well to keep him warm, and give the following drenches :

Tincture of Laudanum................1 ounce or 4 tablespoonfuls.
Sweet Spirits of Nitre................1 " 4 "

Mix in a pint of water and give every two hours until the animal gets relief.

CHAPTER VI.

# DISEASES OF THE URINARY ORGANS.

## INFLAMMATION OF THE KIDNEYS (NEPHRITIS).

This disease is divided into two kinds—acute inflammation and chronic inflammation.

### ACUTE INFLAMMATION OF THE KIDNEYS.

**Causes.**—Exposure to cold or standing out in cold rain storms, such as we have in the fall and spring; lying on the ground when it is cold and damp; by giving large quantities of medicine which acts on the kidneys; from carrying a heavy weight on the back; or in running horses, from violent exertion in racing.

**Symptoms.**—The animal seems feverish, the pulse is full and bounding and runs from 60 to 80 beats per minute, the mouth is hot and dry, he sweats freely and breathes heavy, he looks around to the sides, and, in some cases, puts his nose right upon the side opposite the kidneys; the animal will sometimes cringe and lie down easy, stretch out, and will be heard moaning, as if in great distress; sometimes he will lie for half an hour at a time, but will lie quiet and will not try to roll on his back as he does in bowel diseases; also, by pressing over the loins it causes him more pain. If you listen at the bowels you do not hear much movement or rumbling in them, and there is very little passage from the bowels. He will try and make water often, but passes very little at a time, and it is generally of a red color and tinged with blood. If the animal does not get relief after two or three days, all the symptoms gradually grow worse, and when he tries to make water he passes nothing but blood; in this case he generally dies in a day or so.

**Treatment.**—The treatment must be quick if you want to save the life of the animal. Give the following drench:

| | |
|---|---|
| Raw Linseed Oil .................... | 1 pint. |
| Tincture of Laudanum ............... | 1 ounce, or 4 tablespoonfuls. |
| Fleming's Tincture of Aconite ........ | 10 to 12 drops. |

Mix and give as a drench. Apply woolen blankets, wrung out of hot water, over the small of the back and cover this over with dry blankets so as to keep the heat in and cause the animal to sweat. Always apply mustard plaster over the back after the blankets are taken off. It is also recommended to apply a newly flayed sheep skin over the loins and leave on for twenty-four hours.

Keep him quiet as you can and, after the first drench, if he does not get relief, follow up with the following drench :

| | |
|---|---|
| Tincture of Laudanum ....... . ... | 1 ounce, or 4 tablespoonfuls. |
| Extract of Belladonna.............. | ½ dram, or 30 drops. |
| Fleming's Tincture of Aconite ....... | 10 drops. |

Mix in a pint of luke warm water and give as a drench every two hours until the animal gets relief. Now, always remember in this disease never give any medicine to act on the kidneys, such as saltpetre or sweet spirits of nitre, for the kidneys should be left as quiet as possible. After the animal has relief give a teaspoonful of common soda in his feed three times a day. Feed on soft food with lots of boiled flaxseed mixed with it. If the bowels are very costive it is well to give injections, two or three times a day, of warm water and a little soap, which will help to regulate the bowels.

## CHRONIC INFLAMMATION OF THE KIDNEYS.

**Causes.**—This is caused by using too much medicine that has a stimulating effect on the kidneys, or from eating food that has a tendency to act on the kidneys, such as pea or oat straw.

**Symptoms.**—The horse is uneasy, lies down and gets up, stands with his hind legs spread apart, as far back as he can get them, and they are generally swollen ; his pulse is not much affected ; he passes urine but in small quantities, sometimes quite natural and sometimes streaked with blood ; he generally walks stiff, and by pressing on his back he will show signs of soreness.

**Treatment.**—Give a pint of raw linseed oil as a drench, and if there is much pain it is well to put in one ounce or four tablespoonfuls of Tincture of Laudanum and five drops of Fleming's Tincture of Aconite, feed on soft food with lots of boiled flaxseed in it, and give him a teaspoonful of common soda three times a day in his feed. Place a half-pailful of hot salt in a bag over his kidneys, keep this changed every hour until he gets relief. After he gets relief give him regular exercise and follow up with the following powders :

| | |
|---|---|
| Nitrate of Potash or Saltpetre.......................... | ¼ pound. |
| Ground Gentian Root.......................... | ¼ " |
| Sulphate of Iron .......................... | ¼ " |

Mix and give a teaspoonful three times a day in feed.

## DIABETES.

This is where the food is converted into sugar and passes off through the kidneys.

**Causes.**—Are from a weakened state of the system following some weakening disease, and it may also be caused from eating musty feed.

**Symptoms.**—There is great thirst, the animal drinks large quantities of water. In one case an animal was known to drink thirty-eight gallons of water in five hours. The horse is dull, feeds poorly and passes an abundant amount of water or urine which is of a clear color; his coat looks dusty, and he becomes hide-bound, and he will gradually pine away till he dies.

**Treatment.**—Change his feed, and if in summer time, let him have a run to grass; if in any other time of the year give him plenty of cooked feed, such as boiled oats or scalded chop stuff; give him pure water to drink in small quantities, but often, and give the following:

Tincture of Iodine......................½ dram, or ½ teaspoonful.

Mix in a pint of water, give once a day for four or five days, until he seems better. After that follow up with the following powders:

Sulphate of Iron.......................................¼ pound.
Ground Gentian Root...................................¼ "

Mix and give a teaspoonful three times a day in his feed or on his tongue with a spoon.

### ISCHURIA.

This is where the horse does not pass urine or water.

**Causes.**—From the kidneys not acting properly.

**Symptoms.**—The horse will not pass any urine.

**Treatment.**—Give one ounce, or four tablespoonful doses of sweet spirits of nitre three times a day until he makes water, then follow up with the following powder:

Ground Gentian Root...................................¼ pound.
Nitrate of Potash or Saltpetre..........................¼ "

Mix and give a large teaspoonful three times a day in feed until the animal is all right. We also have the disease in another form, the kidneys secrete the urine or water all right, but it is held in the bladder:

**Causes.**—Spasms or contraction of the neck of the bladder, or calculi or bladder stones will cause it by working up into the neck of the passage, or sometimes from a horse holding his water on account of having no bedding under him, being afraid of splashing his legs.

**Symptoms.**—The animal attempts to urinate or make water often, but nothing comes; he groans with pain, stamps his hind feet, and will sometimes lie down and get up. In the horse you will notice him passing his penis in and out. In this case, if you are not just sure of it, oil your hand and pass it up into the back bowel or rectum, and by passing down towards the bladder you will find it greatly distended with water.

**Treatment.**—If it is a case where the animal has no bedding under him, always shake straw under him, and give the following drench:

| | |
|---|---|
| Tincture of Laudanum................ | 1 ounce, or 4 tablespoonfuls. |
| Sweet Spirits of Nitre................ | 1 " 4 " |

Mix in a pint of warm water and give as a drench. If after half an hour the horse is not relieved and makes water, you will have to take it away, which is very easily done in the mare; oil the hand and pass it in along the floor of the vulva about four inches, when you will feel a small hole on the under side. In some cases, by passing the finger about an inch into this hole, it will cause her to strain and she will make water. If this does not have the effect, then pass the catheter, which is a limber tube made for the purpose, down into the bladder. In the horse, take hold of the penis, enter the catheter in the hole in the penis and gently pass it up into the bladder and allow the urine to drain out through the catheter. Put a little sweet oil on the catheter before using it. Generally after taking the water away once it will be all right.

### INFLAMMATION OF THE BLADDER (CYSTITIS).

**Causes.**—The bladder is sometimes injured in difficult cases in foaling; Exposure to the weather, being out in cold rains, or lying on the damp ground.

**Symptoms.**—The animal walks with a straddling gait, and makes water often in small quantities, which is sometimes streaked with blood.

**Treatment.**—Give the following:

| | |
|---|---|
| Tincture of Laudanum............... | 1 ounce, or 4 tablespoonfuls. |
| Raw Linseed Oil.............................................. | 1 pint. |

Mix and give as a drench. Apply heat over the small of the back in the form of a mustard plaster, also apply mustard around the back part of the belly. Keep the animal quiet and feed boiled linseed to act on the bowels. After the first dose, if the animal has not relief in two hours, give the following drench:

Tincture of Laudanum...............1 ounce, or 4 tablespoonfuls.
Fleming's Tincture of Aconite ............ .....5 drops.

Mix in a pint of luke warm water and give as a drench. Give this drench every two or three hours until the animal seems relieved, and then give the following powders :

Ground Gentian Root................ .................¼ pound.
Common Soda ......................................¼ "

Mix and give a tablespoonful twice a day in his feed.

### STONES IN THE BLADDER (CALCULI).

These stones are found in the kidneys or in the tubes that lead down to the bladder, or in the bladder, or may be found in the tube that leads out of the bladder, but they are mostly found in the bladder.

**Causes.**—Are generally from the kind of food and water the animal uses, turnips being one of the worst, or drinking water that contains a great deal of lime.

**Symptoms.**—The animal is uneasy and has colicky pains. He will be in more pain just after passing water. In some cases where there are a number of these stones, the animal after making water will pass blood, and for a few times after this he may make water all right, and then will pass blood again.

**Treatment.**—Give lots of soft feed with boiled linseed in it to loosen the bowels and give the following medicine :

Diluted Hydrochloric Acid ............................½ dram.

Mix in a pint of water and give as a drench three times a day. The action of this acid is to dissolve the stones. Continue the use of this until the animal is better.

### INVERSION OF THE BLADDER.

This is only met with in mares, and generally at the time of foaling, when the mare is straining violently.

**Symptoms.**—In straining, the bladder becomes forced back and turns out through the tube, and hangs out of the vulva inside out.

**Treatment**—If it is noticed at the time it is done, before it becomes swollen and enlarged, it can generally be turned back to its place by pressing it in with the hands and fingers. After you have returned it, if the animal is in pain, give one ounce or four tablespoonfuls of tincture of laudanum every two hours until the animal stops straining and seems relieved.

### DRIBBLING OF THE URINE OR WATER (ENURESIS).

**Causes.**—Sometimes from an irritation of the bladder or from the neck of the bladder becoming paralyzed.

**Symptoms.**—The animal is noticed to have urine dribbling away from it all the time.

**Treatment.**—Give a teaspoonful of powdered nux vomica twice a day. If the animal's water seems thick it is well to give a teaspoonful of nitrate of potash or saltpetre once a day in the feed.

### PARALYSIS OF THE BLADDER.

**Causes.**—From a weakened state of the system or being exposed to the cold, and is noticed most in animals that are very poorly kept.

**Symptoms.**—In a pure case of this the urine or water is not passed on account of the bladder not being able to contract, and thus becomes very full, causing much pain. By oiling the hand and passing it in the back bowel or rectum and pressing it down you will find the bladder to be very full of water.

**Treatment.**—Draw the water off with a catheter (which is a limber tube made for the purpose) every night and morning and give the following:

Powdered Nux Vomica..................................¼ pound.
Ground Gentian Root..................................¼ "

Mix thoroughly and give a teaspoonful three times a day in his feed until he has regained the strength to pass urine.

---

CHAPTER VII.

# DISEASES OF THE GENITAL ORGANS OF THE HORSE.

### INFLAMMATION OF THE TESTICLES.

**Causes.**—Are sometimes from a kick or blow of any kind, or from their swinging around and striking the legs in the trotting stallion. In some cases they are bruised in lying down.

**Symptoms.**—The symptoms are very plain, the horse seems in pain, and testicles become swollen and very sore to touch; he walks with a stiff, straggling gait and is generally noticed to stand.

**Treatment.**—Give physic drench, consisting of

Bitter Aloes..........................8 drams.
Fleming's Tincture of Aconite..........10 to 12 drops.
Tincture of Laudanum.................1 ounce, or 4 tablespoonfuls.

Mix in a pint of luke warm water and give as a drench.

Bathe the testicles with warm water and then apply a hot poultice to them, consisting of half linseed meal and bran. This poultice can be held to its place by means of cords over his back. Change the poultice every two or three hours and keep bathing well with warm water. Also clothe the body well and, if he wishes to eat, give him plenty of soft feed with boiled flax seed in it. After the first drench, if he does not get relief, give him the following :

| | |
|---|---|
| Tincture of Laudanum............... | 1 ounce, or 4 tablespoonfuls. |
| Sweet Spirits of Nitre................. | 1 " 4 " |

Mix in a pint of luke warm water and give as a drench three times a day.

### DROPSY OF THE SCROTUM, OR BAG (HYDROCELE).

This is where there is a large amount of fluid collects in the scrotum and makes it look large and flabby.

**Causes.**—It often follows a case of inflammation of the testicles.

**Treatment.**—Give iodide of potassium—a teaspoonful twice a day in his feed. In some cases it is recommended to draw the fluid off by tapping the scrotum with a small trocar and cannula, but it is best to try and absorb it, anyway, by using medicines mentioned.

### INJURIES TO THE PENIS.

Either in the stallion or in the gelding, by being kicked or struck with a whip or stick while the penis is out of the sheath, or it may be caused in the stallion by putting him to too many mares, or handling the penis rough, or sometimes from becoming frost bitten in very cold weather.

**Symptoms.**—The penis will be swollen and the animal will not be able to draw it back into the sheath, it hangs out and, if you press on it, it seems sore.

**Treatment.**—Bathe well with warm water until the swelling seems to go down, also take a small pen-knife and tap the penis where it is swollen a few times and allow the water and blood to run out ; this will often reduce the swelling. After this, oil the penis and try and pass it back into the sheath. After the penis is put back, plug the opening of the sheath well with cotton batting, which generally holds it to its place for a few hours at a time, thus giving the penis the natural heat of the body, which is very helpful to it. Bathe, oil, and replace the penis two or three times a day until it regains its strength, and give the following :

Ground Sulphate of Iron .............................. ¼ pound.
Powdered Nux Vomica .............................. ¼ "

Mix and give a teaspoonful three times a day in his feed until the animal can draw the penis back into the sheath himself.

### SWELLING OF THE SHEATH.

**Causes.**—This is generally caused from what is known as a dirty sheath, also from bad blood and disordered kidneys.

**Symptoms.**—Swelling around the sheath and hind legs, his water or urine is thick and yellow.

**Treatment.**—Take warm water and soap and wash out the inside of the sheath and grease it with lard. Give the horse a physic ball, or drench, mentioned in the receipts at the back of this book, to clean him out, and give the following powders:

Nitrate of Potash or Saltpetre .............................. ¼ pound.
Sulphur .............................. ¼ "

Mix thoroughly and give a teaspoonful three times a day in his feed. Give the horse a little exercise every day and the swelling will soon disappear.

### GROWTHS ON THE END OF THE PENIS.

These growths are of various kinds and prove very troublesome and painful when the animal is urinating or making water.

**Causes.**—It is hard to tell the cause, but sometimes from a slight injury not being treated the sore place will throw out a growth.

**Treatment.**—If the growth is not very large wash it off with warm water and soap, then touch the parts with a stick of caustic potash, which will burn it; after the scab falls off, wash, and then burn again, repeat this until you have the growth entirely removed. Dress it every day until it is entirely healed with the white lotion.

### WARTS AROUND THE SHEATH.

We frequently see warty growths around the sheath which can generally be got rid of by tying a small strong cord around the wart very tightly, which will stop the blood circulating; by leaving the string tied tightly the wart will soon drop off. This is the best way to remove them, if it can be done. Another very good method of removing them is to cut them off with a knife and burn the spot with a stick of caustic potash. Warts, as a general thing do not bleed much.

CHAPTER VIII.

# DISEASES OF THE GENITAL ORGANS OF THE MARE.

## DISEASES OF THE OVARIES.

Enlargement of the ovaries is the most common disease we have to deal with in connection with the ovaries.

**Causes.**—The cause is not known.

**Symptoms.**—The animal is noticed to be very irritable, falls off in condition, and is continually in season, and if put to a horse does not get with foal, and this is one cause of a mare being bar ren.

**Treatment.**—If the mare is in good condition give her a physic drench consisting of the following :

| | |
|---|---|
| Bitter Aloes | 8 to 10 drams. |
| Ginger | 1 tablespoonful. |
| Common Soda | 1 " |

Mix in a pint of luke warm water and give as a drench, allow the animal to stand in the stable a couple of days after the drench, and follow up with the following powder :

| | |
|---|---|
| Iodide of Potassium | ½ pound. |
| Nitrate of Potash or Saltpetre | ½ " |

Mix thoroughly and give a teaspoonful three times a day.

## WHITES (LEUCORRHŒA.)

This disease is met with mostly in old mares that are poor in condition and that have ceased to breed. There is a white glary discharge from the vulva behind, which looks like curdled milk. This discharge has a bad smell, the animal falls off in condition becoming thin and weak.

**Treatment.**—Oil your hand and pass it in behind with a cloth or sponge saturated with hot water and soap, wash out the passage thoroughly clean, or this can be done with an injection pump and several pails of luke warm water and wash it out that way. After bathing, wash the womb with the following lotion :

| | |
|---|---|
| Sulphate of Zinc | 1 teaspoonful. |
| Sugar of Lead | 1 " |
| Powdered Alum | 1 " |

Mix in a pint of luke warm water, and with a sponge rub the inside of the womb with this lotion every second day till the discharge stops, and give the following powders :

Sulphate of Copper.... ......... ......................¼ pound.
Sulphate of Iron........................................¼ "

Mix and give a teaspoonful twice a day in her feed ; feed her on rich food and give regular exercise, and, as a general thing, she will soon be all right. It is a very dangerous thing to put the mare to a horse while she has this disease, as the horse is apt to catch it. If she is, and the horse catches, it is called clap or gonorrhœa.

### CLAP (GONORRHŒA.)

**Causes.**—From a horse being put to a mare that is diseased in some way.

**Symptoms.**—The horse's penis becomes sore and swollen, and there is a slight discharge of a mattery appearance. If the horse is put to the mare while in this way, he will give the disease to the mare he is put to.

**Treatment.**—See that the horse is put to no mares until he recovers from this disease. Wash the penis off with luke warm water and a little castile soap and dry with a soft, cotton cloth, then apply the following :

Sulphate of Zinc................... . ...2 drams or 1 teaspoonful.
Sugar of Lead.........................2 " 1 "

Dissolve in a pint and a half of luke warm water and shake well ; saturate the penis well with the lotion by the use of a sponge. Bathe and apply every day for a week or so until the animal is all right. Also give the following powder :

Nitrate of Potash or Saltpetre............... .........¼ pound.
Sulphate of Iron........................... ........¼ "
Ground Gentian Root... ................. ............¼ "

Mix and give a teaspoonful three times a day in his feed.

### BARRENNESS IN THE MARE.

This is when the mare will not breed.

**Causes.**—From enlarged or diseased ovaries ; from a contraction or closure of the neck of the womb, or from the neck of the womb being twisted off to one side.

**Symptoms.**—The mare may be repeatedly put to the horse without becoming in foal.

**Treatment.**—In a case of this, examine the neck of the womb by passing your hand into the passage to the neck of the womb, and, if you find it contracted, or closed, open it by working your fingers around in it until it dilates or opens. If it is a little hard to dilate, saturate a sponge with extract of belladonna and carry the sponge into the neck of the womb with your hand and squeeze out the medicine around the neck of the womb. Leave the mare quiet for an hour after this, then pass in your hand and you will find that the medicine has relaxed the fibres in the neck of the womb and you can easily open it with your fingers. After the neck of the womb is dilated, or opened, put the mare to the horse, and you will generally find she gets in foal. If it is a case where the neck of the womb is turned off to one side, try and straighten it with your hand and have the mare immediately put to the horse. If it is a case where the ovaries are diseased, there can be nothing done. Sometimes after you have put the mare to one horse several times, and she does not get in foal, by changing the horse will often catch her. Mares will start to breed as young as two years old and will breed as old as twenty years. Some have been known to breed older than this.

---

## CHAPTER IX.

# FOALING (PARTURITION) AND THE DISEASES FOLLOWING IT.

The mare, after being put to the horse and gets in foal, usually carries her foal eleven months, but some vary a few days less, while others may go as long as twelve months. The covering around the foal is called the cleaning, placenta, or after-birth, and is attached to the inside of the womb to the little processes called villi, connecting the after-birth, and the foal is the navel string or umbilical cord. Between the foal and the after-birth is a fluid (the use of which is to protect the foal from being injured while its mother is moving around). This fluid is called the amoniotic fluid.

### HOW TO TELL WHEN A MARE IS WITH FOAL.

The mare becomes quieter in disposition, and thrives better; the belly gradually becomes distended, and at the end of the sixth or seventh month, after the mare has taken a drink of cold water, the foal will move around. On account of the foal lying to the left side, the moving of it can be noticed plainer on that side of the mare than it can be on the right side. Also, another way to tell is by oiling the hand and passing it into the passage and find out if the neck of the womb is sealed and tight. In some cases you can feel the foal in the womb, at the same time you are examining the neck. This is the surest method of telling, especially if the mare is only in foal a short time and it is very small. During the time of carrying the foal the mare does not come in season every three weeks, although mares have been known to take the horse and still be with foal.

### THE WAY TO USE A MARE WHEN WITH FOAL.

Keep her in her natural condition as nearly as possible. Feed fairly well, although it is not well to have her too fat. Keep her out running around every day if it is fine, so as to have good exercise. It does not hurt a mare to work her as long as the work is light and steady, but never pull her too heavy or back her up suddenly, for mares often lose their foal by doing this.

### SIGNS OF IMMEDIATE FOALING.

There is a falling away at each side of the tail very noticeable, and, as a general thing, wax or milk will run from the teats for a day or so before foaling. A few hours before foaling the mare seems to be very uneasy ; labor pains come on, and with the pains she is noticed to strain. Very soon the water bag appears, and as it comes the pains become worse; she strains and lies down. If the foal is coming as it should, the head and front feet will make their appearance, after this the mare should be delivered of it in a very few minutes. If the mare has much difficulty, it is well to pull upon the legs while she is straining. Generally the cleaning comes away with the foal. The foal has known to be smothered when the cleaning comes away without being broken, so it is better for someone to be around during the time the mare is foaling, and if anything should occur like this, break the cleaning, or after-birth, and save the foal from smothering.

### NAVEL STRING (UMBILICAL CORD).

If this cord does not break immediately after foaling, take a piece of cord and tie it very tightly one inch from the belly, then cut the navel string off an inch below where it is tied and leave the string on until it drops off ; this is to keep it from bleeding.

### THE NATURAL WAY FOR THE FOAL TO COME.

The foal should come with front end first, with the front feet and head coming together.

No. 1.

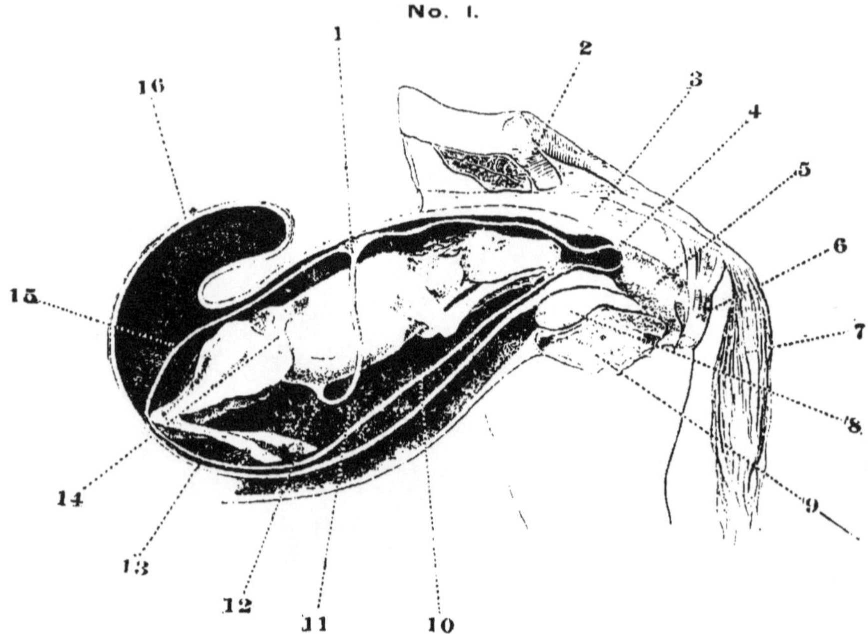


PLATE IV.—POSITION OF FOAL IN WOMB.

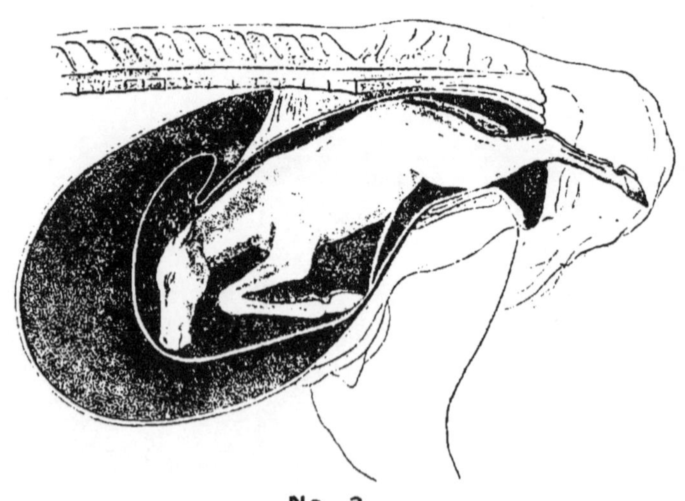

No. 2.

# EXPLANATION OF PLATE IV.

## POSITION OF FOAL IN WOMB.

### No. 1.

This cut shows the natural position of the foal lying in the womb, showing the natural way it should come out—front legs and head first—when the mare is foaling.

1. Navel string.
2. Rump bones.
3. Back bowel, or rectum.
4. Neck of womb.
5. Vagina, or passage out from womb.
6. Vulva.
7. Tail.
8. Bladder.
9. Mare's milk bag, or Mammary glands.
10. Cleaning, placenta or afterbirth.
11. Fluid, or water around the foal. This is what comes out when the water-bag breaks.
12. Navel.
13. Womb.
14. Foal, or fœtus.
15. Cleaning, placenta or afterbirth.
16. Womb.

## POSITION OF FOAL IN WOMB.

### No. 2.

This cut shows the position of what is known as a breech presentation; or, in other words, where the mare is foaling and the foal is coming backwards. This is important to note, for if a mare is foaling and the foal coming backwards, always straighten the hind legs out first, and the foal will come away all right. When the foal is coming backwards, it is not wise to try to turn it, but bring it away as shown in this cut.

SPECIAL NOTICE.—Every place where Tablespoonful is mentioned in this book should read: SMALL, or DESERT TABLESPOONFUL, which is equal to two Teaspoonfuls.

DIFFICULTIES MET WITH IN A MARE FOALING.

Sometimes the labor pains come on and the mare wants to foal, but the neck of the womb remains closed. In this case, oil your hand, enter it into the passage to the womb and, in some cases, you can open the neck of the womb very easily by pressing it open with your fingers. If it is hard to open, saturate a sponge with extract of belladonna and rub it around the neck of the womb, leave it for a little while and you will soon find it easily opened. In a case where the front legs and head appear, and the foal does not seem to come any further with her straining, use gentle force by pulling on the legs and head, and if this does not bring the foal you may come to the conclusion there is something wrong. In this case, oil your arm, or take warm water and soap, so as to make your hand and arm slippery, shove the foal back into the womb and give it a thorough examination and find out what is wrong. If it is a case where there is dropsy of the belly (that is where the belly is swollen up with water), take a sharp knife and pass it back in (guarded by the hand so as not to injure the mare) to the foal's belly and then cut a small hole, large enough for the water to run out, start the legs and head again into the passage and she will soon foal if she is given a little assistance.

Sometimes the foal's head is enlarged with water on the brain. In this case the foal will come out so far that only the legs and the point of the nose will be seen. After using a little force, and it will not come, shove the foal back and feel the state of the head. If it is enlarged take a sharp knife and cut a hole in the softest part of the enlargement and allow the water to escape from the brain. Now, by giving the mare a little assistance, you will find it to come all right.

The foal may come in various positions. We sometimes have a case where the head and one front leg comes out and the mare cannot foal. In this case shove the foal back and bring the leg that is not up, forward, and you will find she will foal all right. The front legs sometimes come without the head, the head being turned back. In this case shove the foal back, take a small piece of rope, four or five feet long, make a noose and slip it over the foal's head onto his neck, have an assistant to pull on the rope while you use your arm and hand in helping to straighten the neck and head, when this is straight, have the assistant to pull on the rope while you pull on the legs, and you will find the foal to come all right. Sometimes the head and neck will come, and the front legs are turned back. In this case shove the foal well back and catch the front legs and bring them forward, then pull on the legs and the foal will come all right.

Sometimes its four legs will come together and the head turned back. In this case shove the front legs back into the womb as far as you can with your hand and arm, and then bring the foal out backwards by pulling on the hind legs. Never, under any consideration, attempt to bring the foal head first in this case.

The foal sometimes comes backwards, and if the hind legs come out first, the foal generally comes out all right. When the foal is coming backwards, and the legs do not come out as they should, and you feel nothing but the tail, rump and hips of the foal, the hind legs being turned in under, it is a rather difficult job, but it can be done all right by shoving the hind end of the foal upwards and towards the front of the womb, then slipping the hand down and getting hold of the foot of the hind leg and lifting upward and backward until you can bring the leg out into the passage, then reach down the hand and get hold of the other leg and bring it out as you did the first one, then the foal will come away backwards all right.

As well as different positions, we meet with what are known as deformities, or freaks of nature. In cases of this kind, on account of never meeting hardly two cases of the same kind, you will have to make a thorough examination and use your judgment as to the best way of getting out, and act on the plans given in the different positions.

In case of having difficulty with twin foals always examine the case thoroughly, and which ever foal is the nearest to the back deliver him first, and as well as this, before pulling on the legs, always be sure you are not pulling on a leg of each foal. After delivering the first foal the last one comes very easily. In any case where you have to shove the foal back into the womb it is well to raise the mare's hind end up by making her stand on manure or boards, as this has a tendency to help the foal to go forwards. Avoid using hooks and knives, as you are apt to injure the womb. It is always best to use your hands and pieces of fine rope.

You may always make up your mind that if a mare is not delivered of her foal in half an hour, or thereabouts, after the water bag has come out that there is something wrong; for a mare, if everything is all right, foals in a very few minutes. It is best then, to make an examination, and, if anything is wrong, fix it right away, for it is a great deal easier to do it then than after the mare has been sick a couple of hours. If anything that is wrong is attended to at once you often save the mare's life as well as the life of the foal.

---

## TROUBLES AND DISEASES WHICH FOLLOW AFTER FOALING.

### CLEANING PLACENTA OR AFTER-BIRTH.

In some cases we find that the mare does not clean properly, and part of the cleaning remains attached to the inside of the womb, and it will be left hanging out behind.

**Treatment.**—Try the following medicine. Give her:

Raw Linseed Oil..................... ½ pint.
Sweet Spirits of Nitre................ 1 ounce or 4 tablespoonfuls.
Fluid Extract of Belladonna.......... 25 to 30 drops.

Shake well together and give as a drench and place a half-pail of hot salt in a grain bag over the kidneys, also throw a blanket over this to keep in the heat, and keep the mare quiet for a few hours and she will generally clean herself and save the bother of removing it with the hand. If this does not cause it to come away, leave it for twenty-four hours after foaling, and have the mare held by the head and also one of the front legs held up so she cannot kick, have another assistant to hold the tail out of the way, roll up your sleeves and oil your right arm and hand, take hold of the cleaning with your left hand while you pass the right one into where the cleaning is attached to the womb, commence at the top, gradually forcing the cleaning off the inside of the womb. When once you get it started, work it all off by gently pulling, and the mare, as a general thing, will be all right.

### AFTER PAINS.

These are the labor pains continued after the foal is out of the womb.

**Symptoms.**—The mare will be in pain, lie down and get up and strain some.

**Treatment.**—Keep the mare quiet, and watch her that she does not force her foal bed out, and give the following to relieve her pains:

Tincture of Laudanum..............1 ounce, or 4 tablespoonfuls.
Sweet Spirits of Nitre.............1 " 4 "
Ginger...............................1 tablespoonful.

Mix in luke warm water and give as a drench every hour and a half or two hours until the mare is relieved. Cover the body well and place a half-pail of hot salt in a bag over the kidneys and give her a hot bran mash.

### FOAL BED TURNED OUT.

This trouble is mentioned in connection with putting out of the calf bed in the cow, where it is specially dealt with. The causes and treatment are the same in both cases. This trouble is rarely met with in the mare, but is very common with the cow.

### RUPTURE OF THE WOMB.

This is generally a very serious trouble.

**Causes.**—From a mare while the labor pains are on, and she is in great distress, throwing herself heavily on the ground or floor. This is generally the way this trouble is caused.

**Symptoms.**—Sometimes, even before the foal is born, you will find the mare's bowels hanging out, and, in some cases, dragging on the ground, and the mare will be in severe pain. While in other cases you will find the rupture while you are examining around the foal in the womb, while again, if you suspect rupture after the mare has foaled, you can find it by examining the womb, it may be large or small. In cases where the bowels are hanging out and she is in severe pain, it is best to destroy the animal to relieve her of her pain as soon as you can, for there can be nothing done to save her life In other cases of rupture, after the foal has been taken away and you put in your hand to see that everything is all right, you will find the womb to be ruptured. It may be ruptured in any part, or of any size. This is very serious, but cases have been known where they got better when the womb was torn on the upper side.

**Treatment.**—Try and keep down the inflammation by applying heat to the body in the way of blankets and hot salt in a bag over the back, and it is well in some cases to apply a coat of mustard over the belly and give the following medicine :

Tincture of Laudanum..............1 ounce, or 4 tablespoonfuls.
Sweet Spirits of Nitre..............1 " 4 "
Fleming's Tincture of Aconite.... ...10 drops.

Mix in a pint of luke warm water and give as a drench every six hours until the mare is relieved. It is well to change the salt every hour, so as to keep up the heat. If she wishes to eat, feed her soft feed with plenty of boiled flax seed in it to keep the bowels loose. In regard to the treatment of the rupture, it is best to leave it alone and let nature itself heal it up, because, in opening up the passage and bathing the womb and putting in medicine, it allows the air to pass through the tear onto the bowels, and also the medicine and the water will leak through onto the bowels, and with these things getting in around the bowels it does a great deal more harm than good. If the bowels get costive, give raw linseed oil in drenches of a pint, and also give injections of warm water and a little soap twice a day until the bowels commence to work freely.

### RUPTURE OF THE PASSAGE.

**Causes.**—It generally occurs in foaling and is done by the legs of the foal in coming out through the passage if not coming straight, or where the foal is too large for the passage.

**Symptoms.**—In some cases, where the mare tries to foal herself, you will find the legs out—one out the natural passage, while the other is out through the anus. In other cases, where the mare has foaled and is apparently all right, you will first notice the manure coming out through the vulva instead of its natural passage. Oil your hand and arm, pass them into the vulva and examine the passage thoroughly, and if it is ruptured you will find a tear somewhere on the walls of the passage, and it will generally be found on the upper side of the passage, between it and the rectum.

**Treatment.**—If it is in a case where the foal is coming out, and is caught with one leg in each passage, oil your hand and shove the legs back into the womb and then bring them both out of the right passage. After this keep the mare very quiet and feed on soft feed with plenty of flax seed in it, and if it is in the spring of the year let her have grass. If the bowels seem to be costive, give her small doses of linseed oil, half a pint at a time, so as not to make the bowels too loose; also, if the mare seems feverish and swollen behind, give one ounce or four tablespoonfuls of tincture of laudanum along with each dose of oil Don't try to stitch up the rupture, but simply inject the passage with a half-pailful of warm water and soap with twenty drops of carbolic acid in it. When this is injected in the rectum, it flows down through the rupture and cleanses and helps it to heal; repeat the injections twice a day and feed on soft food. In cases of this kind it is best not to depend on medicine, but allow nature to do most of the work.

### INFLAMMATION OF THE WOMB (METRITIS).

This is inflammation of the substance of the womb.

**Causes.**—From difficult cases of foaling, where there is a great deal of force used. Exposure after foaling, lying on the damp ground, or it might be caused from a mare getting a chill while she is warm after foaling.

**Symptoms.**—It usually occurs three or four days after foaling. There is uneasiness, pain, and sometimes straining, and in straining sometimes a dark red fluid will pass from her. She will also be noticed to lie down and moan and will look around at her flanks, her pulse is quick and wiry, her ears and legs will be hot, and then change to cold, cold sweats will appear around her sides and flanks ; the vulva will be swollen, and she will pass urine or water in small quantities.

**Treatment.**—Give

Fleming's Tincture of Aconite ........10 to 12 drops.
Tincture of Laudanum ...............1 ounce, or 4 tablespoonfuls.

Mix in a pint of luke warm water and give as a drench ; then apply blankets, wrung out of hot water, over the kidneys; after this is taken off, apply a mustard plaster. If the bowels are costive, give injections of warm water and soap, and give a physic, consisting of one pint of raw linseed oil. Wash the womb out once a day with warm water with a few drops of carbolic acid in it, about ten drops to a pint of water. After the first drench, if she does not seem relieved, follow up with the following drench :

Sweet Spirits of Nitre.................1 ounce, or 4 tablespoonfuls.
Tincture of Laudanum.................1 " 4 "
Fleming's Tincture of Aconite.........10 drops.

Mix in a pint of luke warm water and give as a drench every three hours until the animal seems relieved. Give her plenty of warm water to drink, feed on soft food, keep her warm, and see that the foal keeps her well sucked out. This disease sometimes terminates in leucorrhœa, or whites.

### INFLAMMATION OF THE VAGINA AND VULVA, OR PASSAGE OUT FROM THE WOMB.

**Causes.**—It generally comes after difficult cases of foaling.

**Symptoms.** –The vulva is generally very much swollen, and the inside of the lining is of a red color.

**Treatment.**—Give the animal a pint of raw linseed oil to loosen up the bowels, and keep them loosened up well by feeding soft feed with lots of boiled flax seed. If in the spring of the year, feed her mostly on fresh grass. Bathe the outside of the vulva with warm water, and tap the lips of the vulva with a sharp knife in a few places where they are swollen; this will let the water run out and bring down the swelling. Each time after bathing, three times a day, apply the white lotion.

### INJURIES TO THE HIPS DURING FOALING.

**Symptoms.**—The mare will be stiff, and there will be a falling away of the muscles on the affected side.

**Treatment.**—Bathe the muscles of the hip where it is fallen away, with warm water and salt every second day; rub the parts dry with a cloth and apply the acid liniment. Continue this treatment every second day until the stiffness disappears, and the muscles gain their natural size. During the treatment let her run out to pasture as she needs gentle exercise. In this case, it generally takes some time for complete recovery.

### INFLAMMATION OF THE MILK BAG (GARGET).

This is inflammation of the glands which secrete the milk; it is not so common in mares as it is in cows.

**Causes.**—It is caused from getting cold, by lying on the cold, damp ground in the fall or spring, or from the foal not sucking properly, or from an injury of any kind.

**Symptoms.**—The bag will be hard, hot and tender, and swollen some, she will be noticed not to be feeding, seems feverish and her bag is very sore when handled, and the milk is thin, watery looking, and mixed with curdy looking milk which will come out in jerks when you are milking her.

**Treatment.**—Give as a drench:

Tincture of Laudanum..............1 ounce or 4 tablespoonfuls.
Nitrate of Potash, or Saltpetre......1 teaspoonful.
Fleming's Tincture of Aconite ......5 drops.

Mix in a pint of luke warm water and give as a drench, but if the bowels seem costive, mix in a pint of linseed oil instead of water. Bathe the bag well with warm water and vinegar, if the weather is warm, and apply the white liniment; weaken the liniment down half strength with water; after applying the liniment oil the bag all over with goose oil. Do this three or four times a day, and if the animal is in much pain do it oftener. See that the bag is kept well milked out, either by the foal or by the hand. In cases where it is not taken in time it ends up with festering of the bag. In this case the bag will continue to swell, the milk dries up and the bag will be very sore; matter will form in the bag and it will come to a head. Treat by bathing with warm water and vinegar and apply liniment and goose grease afterwards. This will have a tendency to bring it to a head so it will be fit to lance. The way to tell when it is fit to lance is to feel around until you find a soft spot in the swelling, then open it with a lance or sharp knife at the soft place so as to allow the matter to escape. After this keep bathing with warm water and vinegar four or five times a day and use, instead of the liniment, the white lotion. In severe cases of this kind it is best to take the foal away for a while as the milk is not good, and is apt to sicken the foal.

### SWELLING ALONG THE BELLY BEFORE FOALING (DROPSY).

This swelling commences at the bag three or four weeks before foaling and continues gradually swelling forwards until the swelling extends out between the front legs, but the mare does not seem at all sick. eats and seems as well as ever.

**Treatment.**—If the mare is used to being in the stable, turn her out and let her have lots of exercise every day and give the following powders

| | |
|---|---|
| Nitrate of Potas. or Saltpetre | ¼ pound. |
| Sulphur | ¼ " |
| Ground Gentian Root | ¼ " |

Mix thoroughly together and give a teaspoonful twice a day in her feed until she foals.

This is not a serious disease, but it is well to look after it before it gets too bad. It generally gets all right as soon as the mare foals.

### MARES LOOSING THEIR FOALS (ABORTION).

This is most often seen in mares before the sixth month of being with foal, but may occur after that.

**Causes.**—From a slip; from drawing too heavy, or backing a heavy load; or the smell or sight of blood will cause it.

**Symptoms.**—The mare will be very restless, get up and down, walk around until the labor pains come on, which soon causes the water bag to come out and break, then the front legs and head of the foal will appear, and she will soon be delivered of the foal.

**Treatment.**—In case the foal should not be coming straight, straighten it and take it away, then give the following:

Raw Linseed Oil....................1 pint.
Spirits of Turpentine...............1 ounce or 4 tablespoonfuls.

Mix and give as a drench. In case inflammation of the womb should follow abortion, refer to the treatment of inflammation of the womb.

---

CHAPTER X.

# DISEASES AND TROUBLES OF YOUNG FOALS.

### CONSTIPATION IN YOUNG FOALS.

It is sometimes a troublesome thing in a very young foal to get his bowels to move.

**Causes.**—Foals of a mare fed on very dry feed and foaling early are often troubled in this way, or foals not getting their mother's first milk.

**Symptoms.**—The foal will be noticed to be fuller than usual; he does not seem to care to suck; he will strain as if wanting to pass something but nothing comes. In some cases he will lie down and look around at the side as if in pain.

**Treatment.**—Get a small syringe and give small injections of luke warm water with a little raw linseed oil in it; give these injections three or four times a day. In cases where the manure balls are hard and large, it is necessary to oil the finger, pass it up into the rectum and remove them. Be a little cautious not to irritate the bowels too much. Do this every time before giving an injection. Give him on his tongue a tablespoonful of raw linseed oil and a teaspoonful of whisky mixed well together; give this three times a day until the bowels move and the colt seems better. You will have to use your own judgment as to the amount of oil and whisky given, according to the size and strength of the colt. The above is for a colt a day or two old. Keep him warm and comfortable, and keep the mare milked out so that he will not get too much milk.

### DIARRHŒA IN YOUNG FOALS.

**Causes.**—From allowing the foal to be out and lie on the damp ground too soon after foaling; exciting the foal, caused from the mare being turned in with other horses and running around a great deal; from the mother's milk being too rich or too poor, or sucking the mother while she is hot, after working.

**Symptoms.** The colt is noticed to pass a thin, watery discharge from the anus which sticks around his tail and legs. He does not suck much and soon becomes gaunt and dull, and in some cases he will be griped.

**Treatment.**—Remove the cause if you can find it. Keep the mare and foal very quiet, and if it is from the milk being poor, feed the mare a little heavier, or if you think it too rich, feed the mare a little lighter. Give the foal:

| | |
|---|---|
| Whisky ............. .......................... | 1 teaspoonful. |
| Tincture of Laudanum.......................... | 20 drops. |
| Flour.............................................. | 1 teaspoonful. |

Mix with a little of the mother's milk and give every four or five hours until the foal is relieved. In all cases of this kind keep the body warm, as it has a tendency to relieve the congested state of the bowels. This is a dose for a small foal two or three days old, so you must use your judgment in giving it to an older colt.

### LEAKING OF THE NAVEL AND RHEUMATISM IN YOUNG FOALS.

**Causes.**—It is supposed to be due to a germ getting in at the navel string, while others say it is caused from a cold, a chill or an injury.

**Symptoms.**—The foal will be first noticed to be lame in one hind leg, and some of the joints will be swollen and sore to handle; when he goes to make water it will come out through the navel string. The next day the swelling in that leg may be moved into one of the other legs. After the rheumatism has moved around from leg to leg for a few days it will be noticed to settle in one of the joints, then it will fester and break, and the matter that runs away will smell very bad and seems to come from the bone. After it ceases to run where it festered and broke, and is healing up, it will be noticed to settle in another joint, break and run, and will keep on breaking and running in different places until the colt is so weak that he dies. In other cases the joints swell and do not break, but there is a continual running from the navel, and on account of this running so much he gets weak and soon dies.

**Treatment.**—When the leaking is first noticed, if treated properly it can be cured. Apply Monsell's solution of iron to the navel with a feather four or five times a day, this will stop the leaking; and rub the swollen joints four or five times a day with white liniment, and give the foal half a teaspoonful of salicylic acid on the tongue three times a day. If this is taken in time, before the joints begin to run, this will invariably effect a cure. After the joints break, bathe with warm water three or four times a day, and apply white lotion. Sometimes we have a pure case of leaking at the navel without rheumatism ; in this case, treat by applying Monsell's solution of iron to the navel with a feather four or five times a day until it stops leaking ; also keep the colt very quiet and warm.

### WEAKNESS IN THE LEGS AND JOINTS OF A YOUNG FOAL.

Bandage the legs to support and strengthen them, and also give them a good hand rubbing. If he is down and cannot get up, raise him up four or five times a day to suck, and give him every chance to strengthen his legs.

### CROOKEDNESS OF THE LEGS OF YOUNG FOALS.

In some cases they will be over on the knees, while in others they will be crooked in the hind legs.

**Treatment.**—Leave the foal alone; do not try to do anything to the legs. Feed him well, and as he gets strong his legs will straighten.

### DISTEMPER IN YOUNG FOALS.

This is frequently met with in foals, and is fully dealt with in "Distemper or Strangles."

## CHAPTER XI.

# CASTRATION, DISEASES AND TROUBLES FOLLOWING IT.

The best age to perform this operation is at one year old—during the months of April and May, these being the safest months, after the colt has been turned to pasture for eight or ten days and is shedding his coat.

### PRECAUTIONS BEFORE OPERATING.

Never operate on a colt unless he is in good condition and healthy. See that he has no distemper or any such disease; also have good weather—not too hot or too cold—and avoid east winds and damp weather. Have your hands and instruments thoroughly clean, by washing them in warm water, a little soap and a few drops of carbolic acid. If the horse is older than a year old, it is best to feed very light the day before the operation so his bowels will be empty, as it makes it easier on the horse when he is thrown down and tied. Before throwing the colt down, satisfy yourself by examining and asking if the colt has had any rupture; if he is ruptured he will have to be operated on differently. It is also a good plan to examine if both testicles are down before throwing the animal. In aged horses, if the sheath is dirty it is best to wash it out and oil it well with lard a few days before operating; it often prevents swelling after castrating.

### THROWING OR CASTING THE ANIMAL FOR THE OPERATION.

There are various ways of doing this—some use the belt tackling, others use Farmer Miles' tackling—but the simplest and most used in this country is the rope tackling, as shown in cut below. This tackling consists of about forty feet of rope, ¾ inch

rope will do if good; make a loop by doubling it in the centre, tie a knot and slip the loop over the head, pass the ends of the rope back, one on each side, and run the rope through the ring in the hind hobble, which is made out of heavy, thick straps, and buckles on the hind leg below the fetlock, and has a ring in front of it to run the rope through. This hobble is better than putting the rope around under the fetlock, as the rope burns the leg. Pass the rope forward after it is put through the ring in the hobble up to the neck and through the loop at the neck. Have a man hold the head while one on each side pulls on the rope, same as seen in the cut; this will throw the animal. Now tie him securely so that he will not hurt himself, then proceed with the operation.

The Farmer Miles' tackling is not used much for castrating common colts, but is used to cut ridgling or rig colts. This is where the testicles are up in the belly and cannot be found in the bag. This is the only throwing tackling that can be used in this kind of an operation. The belt tackling is a good one, but it is rather expensive for a man who does only a few colts.

### OPERATING WITH THE CLAMPS.

The clamps are generally made out of cedar, and are about from 4 to 5 inches long, hollowed out in the centre and loaded with green salve, the receipt of which is at the back of the book. Tie one end of the clamp with strong carriage trimmers' twine, leave the other end loose. Now, take hold of the testicle with the left hand and cut into the scrotum, or bag, with a sharp knife, close to the median raphe. Always have your knife sharp, and make a good large cut so as to allow the testicle to slip out, then take hold of the testicle with the left hand and the clamp with the right and slip it over the cord, draw the cord just moderately tight and spread it out in the clamp then tighten the clamp at the end that is not tied with a castrating pincers, also tie this end good and tight with the twine mentioned. Now, take off the pincers and operate on the other testicle the same as the first one, then cut off the testicles and part of the cord below the clamps, leaving about half an inch of the cord below the clamps ; this will help to prevent the clamps from slipping off. The clamps are considered the best way to operate.

### OPERATING WITH THE ECRASEUR.

This is a new method of operating, and has met with very good results in most cases. There are no clamps used in this method. Let the testicles out the same as if you were going to operate with the clamps. Then slip the testicle through the chain at the end of the ecraseur, and screw it up until it squeezes it off. This closes up the end of the artery in the cord and prevents bleeding. It is always best to operate on one at a time, although some operators take off both at the same time, but there is more danger of bleeding. Pour sweet oil, with a few drops of carbolic acid, into the cuts and let the animal up. Use two drops of carbolic acid to one ounce or four tablespoonfuls of sweet oil.

## OPERATING BY SEARING.

This method is used in some places, and is performed by letting the testicles out and putting on an iron clamp in the same place as a wooden one. Cut the cord off close to the clamp, and have a hot iron, dip it in oil, and burn the end of the cord and artery ; this sears it and prevents it from bleeding. Now remove the iron clamp and operate on the other testicle in the same way. By oiling the hot iron it keeps it from sticking to the cord and pulling off the scab.

## OPERATING WITH LIGATURES.

This way is not much used for horses, but is a very good method for bulls and boars. In this method the testicles are let out in the same manner as in the other ways, and the cords are tied with a strong piece of twine to prevent bleeding, and cut off just below where they are tied. But of all these methods of castrating we think the safest and best is the clamps. If the colt is ruptured it requires a more careful operation, which will be fully described in connection with scrotal hernia.

## HOW TO TREAT THE COLT AFTER CASTRATION.

If the weather is chilly, damp or east winds, keep the colt in at nights, in a box stall, clean and well bedded, turn him out every day that is fine and let him have nothing but grass to eat. If the operation was done with the clamps remove them the second day, and in cutting them off cut the string on the front end, spread the clamp well apart in the front so as to allow it to drop off the cord ; in doing this be very careful not to pull down the cord out of its place. If the colt swells very much bathe the cuts with warm water until you soften them, then take butter on your fingers and insert it into the cuts and open them up, allowing the discharge to run out. A little swelling on the point of the sheath, as long as it is not too large and the animal feeds well and has a whitish discharge from the cuts, is not considered serious and need not alarm you. These are natural results of castration.

## BLEEDING FROM THE CORDS (HEMORRHAGE.)

The blood may come from the veins of the scrotum or from the arteries of the cords. If it is coming from the veins you can tell it by its dribbling away from the cuts and its dark color. If it is from the artery of the cord it comes out in spurts, and is of a bright red color.

**Treatment.**—If it is from the veins of the scrotum cast the animal and plug the cut with cotton batting saturated with Monsell's solution of iron and leave the plug in twenty-four hours when it will be safe to remove without danger of bleeding. The batting can be removed by placing a twitch on the animal's nose and take it out without throwing him down.

Bleeding from the artery of the cord is generally after the operation has been performed with the ecraseur and it has failed to close the artery of the cord, and when the animal gets up he will be bleeding. In other cases it occurs from the horse pulling a clamp off by jumping a fence or catching it with his teeth.

**Treatment.**—Throw the animal and secure him, get hold of the cord and place a clamp on the end of it good and tight, this will stop the bleeding, then allow the animal to get up.

The question has often been asked if a one-year old colt would bleed to death by bleeding from the artery of the cord? The answer is: Yes, cases have been known where animals have bled to death by bleeding from this artery. The animal generally bleeds some after castration, but so long as it does not bleed very freely it need not alarm you, and by keeping the animal quietit will generally stop bleeding of its own accord.

### RUPTURE (HERNIA).

This is where the animal has a small rupture that was not noticed before the operation was performed; or, in some cases the animal will rupture himself at the time of operation.

**Symptoms.**—The bowels will be noticed to be hanging out of the cut; there may be only a little—two or three inches, or a foot, and it has been known to be so much that the animal would tramp it under his feet. If the bowel is out very far, and becomes strangulated and inflamed, the animal will be in great pain and act as if he had inflammation of the bowels. The bowel will be of a bluish-red color, and after a time, if left out, will become mortified. The way this generally occurs is: There will be no sign of the rupture during operation, but in a few hours the owner will go back to see the colt and find him in the above-mentioned state.

**Treatment.**—If the bowels are out, and inflamed and bruised with the hind feet, and he seems in great pain, there can be nothing done but to destroy the animal. In cases where there is only a small amount of the bowels out, throw the animal and secure him, oi your hand with sweet oil and force the bowels back through the hole into their natural cavity, sew up the cut that was made to let the testicle out and this will keep the bowels from coming down; keep the animal very quiet and feed on soft food to keep his bowels loose. Swelling will take place, and between the swelling and the stitches it will keep the bowels in their place and the hole will heal up, and there will be no more trouble with the rupture. In a case of this kind it is best to keep the clamp on a day or two longer than you would if the animal was all right.

### THE FATTY LINING OF THE BOWELS COMING DOWN AFTER CASTRATION.

In some cases where there is a very slight rupture which is not noticed during castration, and after the animal is let up and walks around, in a few hours the fatty lining of the bowels comes out—it may be out from six inches to two feet.

**Symptoms.**—When you go to see the colt there will be something hanging from the cut, generally of red color, and when you feel it, it is of a fatty feeling and the animal does not seem to take any notice of it at all; he feeds away and seems in perfect health.

**Treatment.**—Take a scissors and cut it off; there is no danger of bleeding or any trouble, and watch that there is no more comes out.

### ABSCESSES FORMED IN SCROTUM AFTER CASTRATION.

This generally comes in a case where the animal is all healed up and seems to be doing very nicely.

**Causes.**—Some dried matter; a dead piece of the cord or a sliver being left in the scrotum after it is healed. This causes an irritation, starts it to fester and forms an abscess.

**Symptoms.**—There will be heat, pain and swelling around the scrotum or bag, and the animal will walk very stiff in his hind legs.

**Treatment.**—Poultice and bathe well, rub with weak white liniment. This will bring the abscess to a head, then lance it and allow the matter to run out, or in some cases the poultice will bring it to a head and it will break of its own accord.

## SWELLING AFTER CASTRATION.

As we mentioned before, a small amount of swelling is not a bad sign, but when he begins to swell up very bad, especially around the scrotum and cuts as well as in the sheath, the animal seems very stiff and does not want to move nor care to eat; and if the discharge from the cuts is of a red waterish color, the case becomes more serious and needs immediate attention.

**Causes.**—From lying on cold, damp ground or standing out in cold east winds or a cold rain and he gets cold in the cuts, which sets up inflammation and swelling; or allowing the cuts to heal up too soon and dam back the natural discharge; or it may be from the cuts being poisoned from dirt on the hands of the operator or on some of the instruments, or standing in a dirty stable, or it may be from the colt's blood being very bad.

**Treatment.**—Keep the bowels loose by giving small doses of raw linseed oil, bathe the cuts well with warm water and tap the point of the sheath in a few places with a sharp penknife, allowing the watery stuff to keep dropping out; this will help to relieve the swelling. After you give him a good bathing and get the swelling down some, put salty butter on your finger and open the cuts well, this will allow any discharge that is formed above to run out. After the bathing and the opening of the cuts, then apply a poultice of hot linseed and bran, hold this poultice to the cuts by means of strings over the back. It will not be necessary to tap the sheath and open up the cut every time you dress the swelling—once a day will be all that is necessary; but bathe and put on hot poultices three or four times a day, this will allay the inflammation, and if there is poison in the cut, the poultice and hot water will draw it out. Feed the animal on soft food and give gentle exercise, which has a tendency to take down the swelling. As soon as the colt begins to eat, and there is a good healthy white discharge from the cuts, you may then consider him as going to come all right.

## SCIRRHUS CORD.

This is a growth on the end of the cord.

**Causes.**—From the cord being pulled down in taking off the clamp, or from the colt when it is itching from healing, biting it and pulling it down. The cord is pulled down through the cut and if not noticed at the time and put back into the bag, the cut heals tight around it and holds it down, and the cord being exposed to the air becomes irritated and diseased, and a growth is formed on the end of it.

**Treatment.**—If this is noticed right at the time it is very easily checked by bathing it with warm water, which softens the cut, then take your finger with some butter on it, break the cord loose from the skin, shove it back into the bag and it will soon get all right. If it has not been noticed in time, and gets very large, it will soon have to be operated on. Cast the animal and secure him, break the skin from the cord, where it is healed to it, with your finger and thumb, as much as you can; the parts you cannot break with your finger and thumb cut with a knife until you get the cord all loose, place a clamp on the cord so you will be able to cut all the diseased part off below the clamp, leave the clamp on two or three days and remove it, same as after castration, open one end and spread it. The cord may also be taken off with the ecraseur instead of putting on a clamp, and in some cases it works better than the clamp. After the operation, if the animal swells much, bathe with luke warm water three or four times aday, and after each bathing apply the white lotion. If the point of the sheath is swollen much it is well to tap it in a few places with a sharp penknife to let the watery stuff keep dropping out. Feed lots of soft food with boiled flaxseed in it, and give the animal gentle exercise every day. If the case has been allowed to run on until the cord becomes diseased up through the ring in the rim of the belly it is then a hopeless case.

## PERITONITIS FOLLOWING CASTRATION.

This is inflammation of the lining of the scrotum and the lining of the abdominal or belly cavity.

This disease is more fully described in connection with the diseases of the bowels.

**Causes.**—From too severe medicine being used in the clamps; or from a bungling operation; from the animal standing out in cold east winds or rains; or lying on the damp ground. The inflammation first commences in the lining of the bag or scrotum, and extends up through the ring in the rim of the belly and spreads all over the serous membrane lining—the belly cavity. This generally comes on about the third or fourth day after castration.

**Symptoms.**—The animal is very dull, will stand around without eating, and seems as if he was cold. The cuts are not swollen, but there is a bloody, watery fluid keeps dropping away. As the symptoms gradually get worse the animal seems in pain, will lay down, and keep getting up and down, breathes very heavy, as if he had lung trouble. If in warm weather, he sweats freely, his pulse is weak and fast—from 60 to 70 beats per minute ; if he passes anything from the bowels it will be covered with slime, and his water will have a reddish appearance. This is a very weakening disease, the animal gradually gets worse for a couple days, then he dies.

**Treatment.**—For a yearling colt give the following :

Tincture of Laudanum................½ ounce, or 2 tablespoonfuls.
Fleming's Tincture of Aconite........5 drops.
Raw Linseed Oil.....................1 pint.

Mix thoroughly and give as a drench. If this does not give relief in three hours follow up with the following :

Tincture of Laudanum..............½ ounce, or 2 tablespoonfuls.
Fleming's Tincture of Aconite.........5 drops.

Mix in half pint of luke warm water and give as a drench every three hours until he seems better. If he seems weak after recovery give a wine glassful of whisky in a pint of oatmeal gruel three times a day for a few days until he gains his strength. Keep him good and warm by blanketing him, and apply a mustard plaster over the bowels until he seems relieved, poultice the cuts with a hot poultice of linseed meal and bran, which will start a healthy discharge to run from the cuts. Warm his drinking water and feed on soft food. When once this disease gets well started it generally proves fatal.

### LOCKJAW (TETANUS) AFTER CASTRATION.

This disease usually comes on from the ninth to the twenty-first day after castration, and generally follows a case that you think is doing extra well.

**Causes.**—The real cause is not known, but it is liable to follow any kind of an operation, or even a very slight injury. It is frequently noticed in colts that are exposed to the cold, walking or standing in a river for any length of time after being castrated, or allowed to run in a wet, marshy pasture. For treatment and further particulars of this disease turn to "Lockjaw or Tetanus," which is dealt with more fully in the diseases of the nervous system.

### BLINDNESS (AMAUROSIS) OCCURRING AFTER CASTRATION.

This disease is sometimes noticed to come on an animal after he has been castrated, especially if he bleeds freely at the time. This disease is mentioned more fully in connection with the diseases of the eye.

### CASTRATING ORIGINALS OR RIGS.

This is where the testicles do not come down into the scrotum or bag, but remains up in the abdominal or belly cavity. The cause of this is not fully understood. In a case of castrating a rig it is advisable to have an expert who pays special attention to this branch of castration to do it. If the testicles are not down at a yearling it is best to let the animal run over for a year or so, for the testicles often come down themselves. In some cases one testicle will be down and the other up, while again neither of the testicles will be down.

### THE DIFFERENT MEDICINES USED IN LOADING THE CLAMPS.

Some use biniodid of mercury or red precipitate—2 grains to an ounce of lard or vaseline, but the green salve, the receipt of which is at the back of the book, is what we recommend and use.

---

## CHAPTER XII.

# RUPTURES (HERNIAS) AND THE MODES OF TREATING THEM.

### NAVEL OR UMBILICAL RUPTURE.

This is where the navel opening does not become properly closed at the time of birth, and the bowels come down through the opening in the rim of the belly and forms a pouch or sack in the skin, the size of which varies from the size of a hen's egg to larger than a goose egg. This is a miserable blemish and is best to be treated in the spring of the year, when the colt is a year or two old.

**Treatment.**—Prepare the animal by starving it twenty-four hours, then throw and secure him, shove the bowel well back and draw the skin well up and put a heavy stiff clamp on it, secure both ends of the clamp with stout cord and run a few darning needles through the skin below the clamp, this will keep the clamp from slipping off, break off the point of the needles so they will not catch in anything; let the animal up and leave the clamp on until it falls off of its own accord, which is generally from nine to twelve days, by this time the hole will be healed up and the rupture will not be seen any more. After the clamp falls off there will be a raw spot which will need to be watched in warm weather so that maggots do not get into it. In case they do wash it off well with warm water and soap and apply the creolin lotion a few times.

**Warning.**—Be careful not to catch the bowel in the clamp with the skin.

### RUPTURE (VENTRAL HERNIA.)

This is a rupture anywhere in the rim of the belly, and may vary from the size of a hen's egg to that of a man's head.

It is generally due to an injury from a kick of an animal, or an injury of any kind which strikes and bursts the rim of the belly, or it may be caused from an animal pulling very heavy.

**Treatment.**—The way to be sure if it is hernia is that you can shove it up through the hole in the rim of the belly, and as soon as you let go it will come out again; you also can feel the hole in the rim of the belly. These ruptures have often been tried to be treated with the clamp and cutting in and sewing them up, but the best plan is to leave them alone and get as much work out of the animal as you can.

### RUPTURE IN THE BAG (SCROTAL HERNIA).

This is where the bowel and the fatty covering of the bowel comes down along with the testicle.

**Causes.**—Some colts are ruptured at birth and they never get all right. It is also caused by the colt running, jumping or any other such violent exercise, or it may result at the time of castration in severe struggling.

**Symptoms.**—The scrotum has a swollen and enlarged appearance, and you can press the bowel and covering up through the hole into the belly, and when you let it go it will come down again.

**Treatment.**—If it is in a colt that is not castrated, you can get rid of this very easily while you are castrating him. In a case where you are going to castrate and fix the rupture at the same time, have the animal well prepared by starving him a day or so before the operation, then throw and secure him, shove the bowel and fatty lining back into the belly, and in taking up the testicle with your left hand, also allow the hand to rest on the hole where the bowel comes out; make a small cut large enough to allow the testicle to slip out, and slip the clamp on over the cord; also draw up the white covering or tunics you cut through in letting the testicle out, and fasten this tight in the clamp as well as the cord; this will prevent the rupture from coming down. A day or so after the operation it will swell some and fill up the hole where the rupture comes down and the rupture will entirely disappear. After four or five days it will be all right to remove the clamp, and there will be no danger of the rupture coming down. In the stallion it cannot be treated except by castrating in the same method as is mentioned above.

---

CHAPTER XIII.

# DISEASES OF THE EAR.

### DEAFNESS.

If it is of long standing nothing can be done for it, and it is hard to detect it in some cases.

**Causes.**—It generally comes on horses that are used where there is a great deal of noise, such as artillery horses, or it may be caused by a diseased state of the drum of the ear or nerve.

**Symptoms.**—The animal seems stubborn and cannot be taught to obey the word.

**Treatment.**—There connot be very much done to the horse but place a twitch on his nose and pour a little sweet oil in his ear every day; this sometimes helps them.

### INJURIES OR CUTS AROUND THE EAR.

If the skin or cartilage is torn, put a twitch on the horse's nose and take a needle used for sewing skin cuts and draw the wound together with stitches of carriage trimmers' twine, bathe it well with warm water twice a day and apply the white lotion until it is healed.

### DISEASES OF THE CARTILAGE OF THE EAR.

**Causes.**—This disease is generally caused from an injury of some kind.

**Symptoms.**—It will keep festering and breaking every month or so at the place the cartilage is diseased.

**Treatment.**—Place a twitch on the animal's nose and with a sharp knife split the skin and tissues open to the diseased part of the cartilage, scrape the diseased part out and burn around where the diseased cartilage is with caustic potash, which will generally set up a healthy action, and it will heal up all right.

### FROST BITES OF THE EAR.

This is generally caused by keeping the animal in a cold place and allowing the ear to become frozen. This is not so common in horses as it is in young cattle.

**Treatment.** Bathe with cold water, and apply the white lotion after bathing, three or four times a day. If it is taken in time it will save the ear from dropping off. If the ear becomes dead, and drops off, treat the same until it heals up. If you notice the ear just at the time it is frozen, apply snow to it to draw out the frost.

---

## CHAPTER XIV.

# DISEASES OF THE EYE.

Before studying the diseases of the eye it is advisable to study the anatomy of the eye, found in Part I. of this book.

### SIMPLE OPHTHALMIA.

This is inflammation of the outside covering of the eyes and the lining of the lids.

**Causes.**—This is often the result of an injury of some kind, such as being struck with a whip in the eye; or extreme heat or cold will cause it; being kept in a dark stable, or where there is foul air around the stable, or from chaff or any other substance getting into the eye.

**Symptoms.**—The eye is very dull and partly closed, and sometimes the eyelids will be swollen, and water running from the corner of the eye. The eye is sore to handle and it hurts when the animal is brought into the light, and on account of this he will keep it closed. After a day or so there will be a scum gradually come over the eye and it presents a very irritated appearance.

**Treatment.**—Examine the eye carefully and try and find out the cause of the trouble. If it is a piece of chaff, or any substance in the eye, remove it; if it is the fault of the stable, get it fixed. Bathe the eye well with new milk, just from the cow, twice a day and, after bathing it, each time apply the eye wash, mentioned in the receipts at the back of this book, in and around the eye. Keep on with this treatment until the animal is relieved. In cases where the eyelids are much swollen it is well to bleed ; this is done by taking a sharp pen knife and raising the vein below the eye by pressing your fingers on it, then cut it and allow it to bleed until it stops of its own accord, which takes about half an hour.

### MOON BLINDNESS (PERIODIC OPHTHALMIA).

This is inflammation of the inner structures of the eye.

**Causes.**—Extreme heat and cold, poorly ventilated stables or dark stables ; all these have a tendency to bring this disease on. It also comes on hereditary, that is, where the dam or sire has had this disease. It is very apt to follow up in his or her colts.

**Symptoms.**—They are generally well marked, and together with the history of the case, you will have little difficulty in telling if it is this trouble that is affecting the eye. The animal may be put in the stable apparently all right at night, and in the morning you notice the eyes to be running water, seems very weak, and the eye partially closed. It generally affects one eye at a time, although it may affect both eyes at once. It is also noticed to change from one eye to the other. The disease gradually gets worse for two or three days, the eye gets weaker and assumes a yellowish or reddish appearance. The animal can see but very little out of the eye. In two or three days more the eye will return back to almost its natural state, but may look a little bluer than usual. It will appear all right again, but after the course of from three to six months it will show again with the same symptoms. This time it may be in the opposite eye, or in the same one. It will keep coming on in spells like this from time to time, each time leaving the eye a little more blurred, until in time it will completely blind the animal. In some cases the animal will go blind in a year from repeated attacks, while in others it takes as long as three years. The first time the disease attacks the eye you may think it a simple case of inflammation of the eye, but after it comes back a few times you may make up your mind it is a case of periodic ophthalmia.

**Treatment.**—The treatment generally terminates very unsatisfactorily in the end, as the animal goes blind, and when once you have made up your mind it is this disease it is best to part with the animal, but you can help to keep back the disease, when every time the eye is noticed to be sore, to bathe them well with new milk twice a day and apply the eye wash in and around the eye. This will help to check the inflammation and keep the sight in the eye longer, but it finally terminates in cataract of the eyes, that is, blindness. In some cases where the eye looks very irritated it is well to bleed from the vein below the eye, but it does not generally do so much good as it does in simple ophthalmia.

### CATARACT OF THE EYE.

Cataract is a pearly white deposit across the sight of the eye, and it may be what is known as a complete cataract, that is where it covers all the sight, or it may be a partial cataract, that is where the animal can see a little.

**Causes.**—It is caused by repeated attacks of inflammation of the eye, the same as we have in ophthalmia; or foals have been born with cataracts on their eyes. This disease may effect one or both eyes. There is a white deposit over the sight of the eye.

**Symptoms.**—If it is only partial cataract the animal can see a little, but if it is a complete cataract the animal cannot see at all.

**Treatment.**—Very little, if anything, can be done, unless by an operation, and such an operation is not attended with much success in the horse because you cannot regulate his sight with glasses as is done in people. As for medicine, when the cataract is well formed there can be nothing given to help him. The only thing we can advise you is when you are buying a horse be sure he is not affected with this disease.

### STAR GAZER (AMAUROSIS).

This is paralysis of the optic nerve which gives sight to the eye.

**Causes.**—Standing in a stable that is dark, or by striking the head against something. In severe cases of bleeding, horses have been known to go blind, but as the blood returns he gains his sight.

**Symptoms.**—The eye has a large, glassy appearance, and the pupil of the eye is very much enlarged, and if he is taken out of a dark stable into the light the pupil of the eye does not close up the same as if it was all right, but remains large all the time. The animal generally carries his head high and steps high.

**Treatment.**—If the case is of long standing nothing can be done for it; if it comes on from injury by striking the head against something, or from bleeding, it can be treated all right.

Give the animal a teaspoonful of nux vomica in his feed three times a day, this acts as a stimulant to the nerve ; wash the eye with eye wash twice a day for a few days. In buying horses be particular and watch the eyes closely for unsoundness, for some of the best buyers have been nipped in this disease.

### CANCEROUS GROWTH IN THE EYE.

It first commences with an inflammation in the eye, after this there will be a growth noticed to be coming out of the ball of the eye, which gradually keeps on growing until it hangs out over on the cheek, it looks red and angry, and the least little injury will cause it to bleed freely, and gives the animal an unsightly appearance. In some cases the bones around the eye become diseased, and then it smells very bad. This disease is more often met with in the cow than in the horse.

**Causes.**—The causes of this are like all other cancerous growths—from cancer germs getting into the blood and settling in the eye.

**Treatment.**—You may effect a cure in the first stages of the disease by removing the eye, growth and all. Cast the animal and secure him, and have his head held solid, cut around the eye between the eye and the eyelid, and lift the eye up by sticking a hook into it, then cut the structures off at the back part of the eye and have a hot iron to sear it, to stop the bleeding, then allow the animal to get up, and dress it the same as you would an ordinary wound by bathing it with warm water and applying white lotion three times a day. After applying the lotion, if the cut looks angry, apply compound tincture of benzoin, or what is called Frier's balsam, with a feather to keep it from growing again.

### FILARIA OCULI.

This is a small worm, about the size of a pin, found in the humors of the eye, moving around. It is liable to be found in any part of the body; they have been found around the testicles and bowels.

**Causes.**—From a microbe, and is mostly seen in horses that pasture on low, wet land ; it is taken into the system through water and feed, and from the bowels it gets into the blood and is carried into the eye or various parts of the body through the blood.

**Symptoms.**—This worm may grow from the size of one half inch to two inches long, and it sets up considerable irritation in the eye. The eye has a hazy appearance and seems very weak, and by watching it you can see the worm sometimes in the front, then it will move around to the back. If the worm is allowed to remain in the eye it will keep the eye irritated.

**Treatment.**—The only thing that can be done to save the eye is an operation. Cast the animal and secure him, have his head held solid so that he cannot move, take a small lance or sharp pen-knife and make a small incision or cut across the sight of the eye, which will let the humors of the eye run out, and with it the worm generally comes. Keep the animal quiet after the operation and put him in a clean, dark box stall for a few days. Bathe the eye with new milk twice a day and apply the eye wash after bathing until the eye is healed. It generally takes a week or so for the eye to heal and the fluid to form in the eye. After the course of two or three weeks all that will be noticed is a slight scar, which will gradually absorb away in time and the eye will get all right.

## GLAUCOMA.

This is a hardening of the back humor of the eye.

**Causes.**—From a severe injury to the eye, or where the eye is pierced by a stick or anything, and it injures the back part of the eye; or sometimes from natural decay in old animals.

**Symptoms.**—The horse gradually gets blind, and has high, peculiar action in front, the same as seen in all blind animals. If you look into the eye through the pupil you will see that the eye has become hardened and is of a grayish color.

**Treatment.**—In a case where it comes on from an injury, bathe with new milk three or four times a day, and apply, after bathing, the eye wash. It is more frequently noticed in gray horses than in others, and if it comes on in old age nothing can be done to cure it.

## INFLAMMATION OF THE HAW OF THE EYE (MEMBRANA NICTITANS).

This is the diseased or enlarged state of the cartilage in the corner of the eye.

**Causes.**—Generally from chaff or something getting into the eye and irritating it. This causes it to enlarge and stick out of the corner of the eye and gives the animal a great deal of trouble, and also looks very bad.

**Treatment.**—When it is first noticed remove the irritation, if it can be found, and bathe with new milk or luke warm water two or three times a day, and after bathing apply the eye wash. The cartilage will generally return to its natural size in a few days. If this fails to remove the enlargement, place a twitch on the animal's nose to keep him quiet, take a small pincers and catch hold of the cartilage and draw it out of the eye some, then take a scissors and cut it off. This is a very simple thing to do, for it will not bleed and does not hurt the animal very much, for it is only a piece of cartilage. After you cut the cartilage off treat the eye the same way as mentioned above, and in a few days you will hardly be able to tell there was anything wrong with the eye.

### INJURIES TO THE EYELIDS.

This is where the eyelid becomes torn or bruised.

**Causes.**—It may be caused in various ways.

**Treatment**—If the eyelids are torn, draw the edges of it together by putting a few stitches in it with a needle that is used for sewing up wounds, and use the carriage trimmers' twine. It is not always necessary to throw the animal for this, but just put a twitch on his nose, but if the animal is very wild it is best to throw him down; bathe the wound with new milk or warm water three or four times a day; after bathing apply the eye wash, in a few days the stitches will come out and the wound may open up some and look worse, but keep on treating it and it will heal up in a short time. In some cases, where the eyelid is torn and the eye injured, the eye will fall out on the cheek, this may be caused in a runaway. Wash the eye off and set it back to its place and sew up and dress the wound as mentioned above.

### INJURIES TO THE DUCTS OF THE EYE

#### WHICH CARRY THE TEARS DOWN TO THE NOSE.

When this tube becomes injured the tears will flow out over the cheek. It is recommended to take a small syringe and inject this tube or passage with warm water. The best way to inject this is from below by putting the point of the syringe in the small hole found in the bottom part of the nose and force the water up through this until it runs out at the corner of the eye, this will clear the passage out, and sometimes effect a complete cure, but if the case is of long standing it is best to leave it alone, for the horse might be able to work for years.

### NEAR SIGTHEDNESS (MYOPIA).

This trouble is generally found in horses with very full eyes.

**Cause.**—There is no particular cause for this; the animal was foaled with this disease.

**Symptoms.**—The horse has a very full eye, with enlarged pupils, and will shy very much, as he cannot see objects along the road at any distance.

**Treatment.**—There is none, but always be careful in buying a horse that he is not a shyer for it is a miserable thing.

---

## CHAPTER XV.

# CONTAGIOUS DISEASES AND FEVERS OF THE HORSE.

---

### I.—CONTAGIOUS DISEASES.

### SMALLPOX IN THE HORSE (VARIOLA EQUINA).

This disease is not very often met with in America, but is quite common in Europe.

**Causes.**—They are same as smallpox in people, it is contagious, and spreads from one horse to the other by germs. The way the disease is carried from one horse to the other is from the scales, which contain the germs, getting on the groom's clothes, the saddle or harness, and in this way it is communicated to another horse, where it gets into the blood and sets up the disease. This disease has to run a course, and it generally takes from nine days to two weeks before the animal begins to recover. This disease may be communicated from the horse to the man, cows or other animals, so it is well to be careful when around a case of this kind and not allow the scales or scabs, which fall off the horse, to get onto you.

**Symptoms.**—There is more or less fever, the pulse quick and weak, the animal is thirsty and does not care to eat much, red patches appear upon the skin, with a small hole in the center of each blotch, from which oozes a watery serum. These patches vary in size and are apt to be found on any part of the body, the skin becomes very sore and tender, the mouth is sore and saliva keeps running away from it, the throat also seems sore and it appears difficult for him to swallow. In the course of nine to fifteen days the red spots dry up and the scales fall off and the animal gradually recovers. It is these scales which fall off that, when they are carried to other animals, spread the disease.

**Treatment.**—The treatment is simple: Keep the animal from other horses, feed on soft food with lots of boiled flaxseed in it, if in the spring of the year, grass is best. Give

Sulphur ........................ ... ................½ pound.
Nitrate of Potash or Saltpetre.........................¼ "

Mix thoroughly together and give a teaspoonful three times a day on his tongue, which is all the medicine he will need internally. Attend to the animal's general comfort, keep him warm and do not expose him to the cold, for this would be liable to kill the animal if it drove the rash in from the skin. Wash him all over where the scabs are once a day with the following mixture.

Creolin..............................1 ounce, or 4 tablespoonfuls.
Rain Water ........................ ......1 quart.

Shake well together and apply with a sponge or cloth and this will kill all the germs of the disease as they come out on the scabs; rub him once a day and this will keep the disease from spreading. The only danger in this disease is letting the animal get cold and driving the rash in from the skin, which poisons the blood. After the animal gets better it is best to gather all the straw and manure out of the stall he has been in and burn it, then close the stable up and burn sulphur in a dish, which will kill all the germs left in the stable.

### GLANDERS.

This is a very contagious or catching disease in the horse, and one of the most serious and loathsome diseases the horse is liable to. It occurrs in two forms, chronic glanders and acute glanders. This disease has existed for thousands of years, and has been treated by every kind of medicine known, and nothing has ever effected a cure.

### CHRONIC GLANDERS.

This disease was common in this country when it was being cleared up, and is now sometimes found in back townships.

**Causes.**—It is due to germs poisoning the blood, and it is thought that cases of catarrh and nasal gleet have terminated in a case of glanders. Sometimes it has broke out in a severe form while horses are on shipboard during the time of a storm, when the hatches are shut down, but in a case of this kind there must have been one of the animals affected, and when closed up tight the disease began to show itself. If it gets into large, crowded cities where there are large numbers of horses kept in one stable it is very troublesome.

**Symptoms.**—It is most often seen in old and debilitated animals. In the chronic form of the disease it may exist for a long time without showing severe symptoms, and might be mistaken for nasal gleet; but if you take the temperature of the animal, you will find it up to from 103 to 105 degrees, while in nasal gleet the temperature will be normal—about 98 or 99 degrees. There is a discharge from the nose; at first it is rather watery, but afterwards it is of a greenish-yellow color and very sticky; it sticks around the nose, and has no smell—differing from nasal gleet. The discharge in this case will sink in water, while in other diseases it will float. The eyes will have a discharge from them. After the disease has gone on for some time, the inside of the nose becomes full of patches of ulcers which have very little tendency to heal. The lymphatic glands around the head and neck become swollen and hard. The animal falls off greatly in condition; becomes thin and hide-bound, and generally pines away in a lingering death. Man and dogs will take this disease from horses, but other animals are not subjected to it. When it attacks man, it is a terrible disease; so you see the necessity of handling the disease with very great caution.

**Treatment.**—As soon as there is any suspicion of this disease, put him some place where other animals will not come in contact with him; have only one pail and feed box for him, and allow no other animal to eat or drink out of it; also handle him with care yourself, and be sure there are no cuts on your hands, or do not allow your clothes to touch him, as you might carry the disease to other animals. It is best to send for the veterinary government inspector, and if he pronounces it a case of glanders, have the animal immediately destroyed and burned, and have the stable disinfected.

### ACUTE GLANDERS.

This disease is mostly noticed in Euorpean countries, and not often met with in this country.

**Causes.**—This is caused from germs getting in and poisoning the blood, the same as in chronic glanders, only that it runs its course very much more quickly than the chronic form.

**Symptoms.**—Four or five days after the germs have entered the system the animal will be noticed to tremble; his temperature will run up as high as 105 degrees or more, and he will run freely at the nose, the discharge being similar to that of chronic glanders, only it is sometimes streaked with blood. The lungs become affected and inflamed, which frequently causes death. All the other symptoms are the same as in the chronic form.

**Treatment** and precautions are the same as in chronic glanders.

---

## II.—FEVERS OF THE HORSE.

They are four in number, as follows: Influenza, purpura hemorrhagica (or what is commonly called button farcy), strangles or distemper, epizootic cellulitis or pink eye.

### INFLUENZA.

This disease received its name when, at one time, it was thought it was influenced by the stars. It is very common among the horses in America, and is a disease that is liable to affect any organ of the body.

**Causes.**—It is caused from germs floating in the air; this is how it spreads so quickly from one horse to another. It is more common in the spring and fall, when the horses are shedding their coats, than at any other time of the year. In 1874 and 1878 this disease broke out in the form of a regular epizootic, and spread all over the country and caused a great deal of trouble. Horses that are kept in badly-ventilated stables, especially those that are underground, are more likely to catch this disease.

**Symptoms.**—The first symptoms are a dull, languid appearance; sweats freely on the slightest exertion; the coat starry and dirty looking; the mouth hot and dry, and there is a slight cough. After these well-marked symptoms set in, by pressing on his throat it seems very sore and causes him to cough. The pulse will be quick and weak, and in some cases will go as high as 80 beats per minute. The horse will hang his head, and seems to be suffering from severe headache and nervous depression; and he gets so weak you would imagine you could shove him over. At this stage of the disease, he will breathe heavy and have a peculiar rattling noise in the throat, which you can hear by putting you ear to the side of his throat. His eyes look heavy and red; his bowels become costive; and if you take his temperature it will be up as high as 105 degrees. There will be a discharge from the nostrils, and if it is of a yellowish white color, it is a favorable sign. At this stage of the disease it is apt to settle on the bowels, and cause inflammation and death; or, it may settle on the liver, when the symptoms will vary—sometimes it will be diarrhœa and then costiveness; or, it may settle on the lungs and set up inflammation of the lungs; so you see the necessity of careful nursing, not allowing the animal to get cold during the disease. The animal generally persists in standing all through this disease. The legs and sheath become swollen, which is considered rather a good sign as long as it not swollen to any great extent.

**Treatment.**—Clothe and attend to the general comfort of the animal according to the season of the year. If the legs are cold, hand-rub and bandage them, and allow the animal to have plenty of fresh air in the stable as long as he does not stand in a draft. Support the system and assist nature to throw off the disease, for influenza will run its course in spite of medicine. Keep the bowels loose by feeding on soft food with plenty of boiled flax seed in it, but never, under any circumstances, give the animal a physic drench. Give him the following medicine:

Chlorate of Potash.................................¼ pound.
Nitrate of Potash or Saltpetre..........................½ "
Powdered Digitalis...............................¼ "

Mix thoroughly together and give a teaspoonful every six hours on his tongue with a spoon. In feeding, give the animal small quantities, so he will be able to eat it all without leaving any in his manger; but give it to him often. Allow the animal cold water to drink in small quantities but often. Rub his throat well with the white liniment four or five times a day, and if his lungs become affected, apply a light mustard plaster over the sides of the chest every day; continue this treatment until the animal seems relieved. In cases where the animal is very weak, give a wine glassful of whisky in a pint of oatmeal gruel three or four times a day as a drench, being careful not to choke the animal; this is a great thing to keep up the strength. When the animal is getting better, the eye gets clear; he will take notice of things about him; the pulse gets more natural—slower and stronger—and the appetite gets better. After the disease has passed off, and the animal seems very weak and thin, give the following mixture:

—11—

Sulphate of Iron ........................................ ¼ pound.
Ground Gentian Root ........................................ ¼ "

Mix thoroughly and give a large teaspoonful three times a day in nis feed. Feed well and give gentle exercise every day. This disease, if treated this way, and allowed to run its course without getting cold, will terminate favorably. But above all things never bleed, or give aconite or a physic in this disease, for it only tends to make the animal weaker and the disease worse.

### BUTTON FARCY (PURPURA HEMORRHAGICA).

This disease is not contagious or does not spread from one animal to another. It is defined to be some putrid condition or charbonous affection of the blood. The disease mostly affects the capillary blood vessels of the skin, but in some cases it will affect the lining of the lungs and air passages.

**Causes.**—It generally comes after some other disease, such as influenza, distemper or any other weakening ailment, also in horses that are overworked, thin and run down in condition. When the system is in this state by turning the animal out cold nights or allowing him to be out in a cold rain, and he gets a chill, you need not be surprised to see this disease come on. It is noticed mostly in the spring and fall. The blood gets very thin, almost like water, when the animal is suffering from this disease.

**Symptoms.**—The disease generally comes on very suddenly, the horse may be apparently all right at night, and in the morning he will be literally covered with patches of swelling all over the skin. In some cases it affects the animal more in some parts than in others, for instance, we have seen cases where the eyelids were so swollen the animal could not see out of them, his lips will also be swollen so much he cannot eat, in other cases his sheath and legs will be badly swollen, or little spots may be seen all over the skin, especially on the sides of the neck and thighs and over the back. The peculiarities about these swellings are that they come on so quickly, and sometimes disappear from one part of the body and come on in another. By giving the animal exercise it will take them down, but afterwards the lumps will come back worse than ever. Examine the lining of the nose, and if it is affected watch out for lung disease. The general symptoms are the animal will seem dull and breathe a little heavier than natural, pulse not much changed, but weak, and he will generally try and eat some. Cases have been known where the swelling got so bad that patches of the skin would drop off and leave raw sores.

**Treatment.**—It runs its course in three to thirty days. Place the animal in a comfortable box stall, attend to his general comforts and keep him warm. Give him the following drench :

| | |
|---|---|
| Raw linseed Oil................... | ½ pint. |
| Spirits of Turpentine............... | 1 ounce or 4 tablespoonfuls. |

Mix together and give as a drench, repeat this drench every three days until the animal is relieved. Give him a teaspoonful of chlorate of potash three times a day on his tongue with a spoon until he is better. Feed on good, light, soft food with plenty of boiled flaxseed in it to keep the bowels loose. Do not apply anything externally unless the swelling breaks out ; if it does, bathe three times a day, and after bathing apply the white lotion. Never, in any case, attempt to open any of these lumps, for it only does harm. Bleeding is good in the early stages of the disease if the animal is strong enough to stand it. Be very careful in nursing him that he does not get a relapse, for it is liable to cause death. There are cases known where the mouth and nose have swollen so badly that the animal died from suffocation. The after treatment is to build the animal up by regular exercise and good food. Give him the following tonic mixture :

| | |
|---|---|
| Sulphate of Iron.................................... | ¼ pound. |
| Ground Gentian Root.............................. | ¼ " |

Mix thoroughly and give a teaspoonful three times a day in his feed.

### DISTEMPER (STRANGLES.)

This is very common among Canadian and American horses. It receives the name of strangles from the peculiar suffocating noise he makes in breathing. It generally attacks young animals from the time they are foals until they are five or six years old, but it may be found in older horses.

**Causes.**—It is due to germs in the blood, and all colts are liable to be affected with it. It is seen most in the spring and summer months of the year. These germs float in the air and are taken into the system by the animal breathing.

**Symptoms.**—The animal is dull and languid, and a small amount of work fatigues him. He will probably have a cough and sore throat at first, and there will be saliva running away from the mouth, the bowels are inclined to be a little costive or dry looking, then there will be a little lump begin to form around the throat, sometimes under the throat and sometimes at the sides. This lump gradually gets larger and causes the animal to have difficulty in breathing, he is dull, hangs his head and soon becomes very gaunt. The disease generally runs its course in from six to twelve days, the abscess or lump around the throat generally breaks and runs, and he nearly always runs at the nose, which is a good sign, in some cases there is more than one lump or abscess forms, and if the poison is in the system it is best to have them brought to a head and let the discharge out for this gets the poison out of the blood.

**Treatment.**—This is very satisfactory in most cases: Always allow the disease to run its course, give plenty of pure air, clothe the body according to the season of the year, feed on good food, such as boiled oats or chop stuff with plenty of boiled flaxseed in it to keep the bowels loose. This is a disease that does not require much medicine. Give the following powder:

| | |
|---|---|
| Nitrate of Potash or Saltpetre........................ | ¼ pound. |
| Sulphur........................................ | ¼ " |

Mix thoroughly and give a teaspoonful three times a day on his tongue with a spoon. Rub the throat and lumps well with white liniment three times a day, apply a hot poultice of half linseed meal and bran to the throat every night, this will draw the abscess or swelling to a head and cause it to break, which is better than lancing. In cold weather it is best to apply a mustard plaster to the swelling instead of a poultice, for after the poultice is taken off the animal is apt to catch cold. Never, in any case, burn sulphur under the animal's nose, because it is apt to set up inflammation of the lungs. In case the lump gets so large around the throat that the animal's life is threatened by choking, you may then perform the operation of tracheotomy, which is performed by cutting into the windpipe on the underside about nine inches down the neck from the larynx (Adam's apple). About thi part of the neck you will find that the windpipe is almost bare—just covered with the skin—cut a hole through the skin about two and one-half inches long, then cut across three of the rings of the windpipe and have the regular tracheotomy tube to put in the hole and keep the rings open so the horse can breathe through this opening until the abscess or lump breaks and the swelling goes down so he can breathe through his throat, when this occurs take the tube out and draw the skin back to its place by putting two or three stitches across it, then treat it as a common cut by bathing two or three times a day, and after bathing, each time apply the white lotion, and the hole will soon heal up. During the time the tube is in the throat it is best to take it out and clean it once a day.

### BASTARD STRANGLES.

This generally follows simple strangles or distemper.

**Causes.**—Is from the matter in the abscess or lump being absorbed into the system and poisoning the blood.

**Symptoms.**—Abscesses of the same nature as the one which forms in the throat from distemper are liable to form and break out in any part of the body, sometimes around the shoulder, flank, neck or hips, but the rest of the general symptoms are the same as in simple strangles. These abscesses will keep forming until the poison is out of the blood.

**Treatment.**—The treatment is the same as in a simple case of strangles. Give the same medicine, bathe, poultice and rub the abscesses with white liniment and try and bring them to a head so that they will break themselves, or lance them to get the poison out of the blood, for this is the only way to get rid of the disease. The danger in this disease is that the abscesses may form inside on the lungs or bowels and cause the animal's death. We have known of some cases where the animal would break out forty or fifty times, depending upon the amount of poison in the blood.

### PINK EYE (EPIZOOTIC CELLULITIS).

The disease affects the cellular tissue under the skin, and like the other fevers of the horse, runs a course which takes from six to nine days, and, as a general thing, if treated properly, runs its course favorably.

**Causes.**—This disease is due to germs in the air, and when once it starts it spreads all over the country from one horse to another. There was a great outbreak of this disease between 1875 and 1880, which spread all over the country.

**Symptoms.**—The animal becomes dull and feverish, and his eyes are red and swollen, and afterwards turns to a pinkish color, from which the disease receives its name—pink eye. The animal does not eat well; his temperature runs from a 103 to 105 degrees; his mouth is hot and dry, and he falls off greatly in condition; his manure is hard and dry; there is a discharge from the eyes and a swelling of the limbs and belly, and in some cases he has a cough.

**Treatment.**—This disease is a very simple one to treat, in most cases, if there is not too much medicine used. Allow the disease to run its course; keep the body warm; have the stable clean, and allow the animal to have plenty of fresh air. Feed on soft food with plenty of boiled flax seed to keep the bowels loose, and give the following simple medicine to act on the blood:

Chlorate of Potash.................................... ¼ pound.
Nitrate of Potash or Saltpetre............... .......... ¼ "

Mix thoroughly and give a teaspoonful three times a day on his tongue with a spoon. Never mix chlorate of potash and sulphur together, as it will explode and is very dangerous. Give the animal plenty of cold water to drink in small quantities. In cases where the animal is very weak give as a stimulant a wine glassful of whisky in a pint of gruel three times a day, and never, under any circumstances, give a physic in this disease. It is well to wash out the nostrils and eyes twice a day with luke warm water. After a few days you will see the animal began to take notice of things around him and gradually gets better.

---

## CHAPTER XVI.

# DISEASES OF THE SKIN.

### CRACKED HEELS (SCRATCHES).

This disease is common among horses, and affects the heels, the heels of the hind legs being oftener affected than the front.

**Causes.**—Anything that will irritate the skin tends to produce this disease. In the fall and spring, when there is wet and muddy roads, washing and not properly drying the legs, standing in badly kept stables, or by wearing boots on the legs. Heavy, hairy-legged horses are more subject to it than light horses.

**Symptoms.**—The affected legs have a tendency to swell and are stiff and sore about the heels, the skin becomes cracked and scaly-looking at the back of the heels, and in some cases, when you are driving him, the heels will bleed.

**Treatment.**—The first thing to do is to feed on soft food and give a physic drench consisting of the following:

Bitter Aloes.... ................................ 8 drams.
Ginger ... ........................................ 1 tablespoonful.
Common Soda........... ...................... 1 "

Dissolve in a pint of luke warm water and give as a drench, allowing the animal to stand in the stable the next day. After this give him the following powder:

Nitrate of Potash or Saltpetre........................ ¼ pound.
Sulphur.............................................. ¼ "

Mix thoroughly and give a teaspoonful twice a day in his feed, this will cool his blood and get him in good condition. Bathe the heels with luke warm water and a little castile soap night and morning, after bathing wipe dry with a soft cloth and apply the white lotion. Before you take him out to work rub his heels with vaseline or lard to keep the heels soft and protect them from the dirt and water while you are working him. In cases where the heels are very sore and swollen apply a hot poultice of half linseed meal and bran to them every night for a while until he seems better.

### MUD FEVER.

This is inflammation of the skin of the legs and the under part of the belly.

**Causes** are similar to that of scratches, and the disease is seen mostly in the fall and spring during the wet weather, when the cold, muddy water splashes over the legs and belly, or washing the legs with cold water and not drying them properly.

**Symptoms.**—The legs are swollen and stiff, the skin is hot and tender, and the hair falls off in patches.

**Treatment.** Give the same medicine internally as that given for scratches to cool and clean the blood. If the legs are dirty bathe them off with luke warm water and a little castile soap and dry them with a soft cloth; after this don't bathe, but brush off with a soft brush and apply the white lotion twice a day, which is soothing and healing to the skin. Keep the animal out of the wet and mud as much as possible. If you have to work him, each time before you take him out, rub the parts affected with vaseline or lard, which will soften and protect the affected skin.

### GREASE.

This disease generally follows cases of scratches that are neglected. It affects the glands of the legs, as well as the skin, and is more common in heavy, hairy-legged horses that have round, fleshy legs. It may be caused from bad blood and swollen legs, and is more often seen in the hind legs than in the front ones. It is also brought on by clipping horse's legs in cold, wet weather.

**Symptoms.**—There is a thickening and swelling of the legs, the hair stands out on the legs, and they are hot and tender, and there is an oily discharge from around the heels, which has a bad smell in some cases. This is how the disease got its name, grease. When the animal is working the swelling goes down, but comes back again during the night.

**Treatment.**—It is hard to effect a complete cure, but you can relieve the animal to a certain extent by giving a physic drench, and powders, same as are mentioned in scratches, to cool and clean the blood, feed on soft food, bathe the legs with warm water and castile soap twice a day; after bathing rub dry and apply the white lotion with a few drops of carbolic acid in it to destroy the smell, poultice the legs every night with hot bran and linseed meal to soothe and draw the oily stuff out of them. This will generally fix the legs up in good shape for some time. In cases where it is very bad and in the "grapous" stage, and there is little red growths around the heel, which look like a bunch of grapes, burn them off with caustic potash or chlorate of zinc, which is in the form of little sticks.

### SIMPLE ECZEMA.

This disease is sometimes mistaken for mange, but unlike mange, it is not caused by germs or parasites working in the skin. It is mostly noticed in hot weather, when the animal is fed on very hot food, which heats the blood, such as barley and other hot foods.

**Symptoms.**—First there is a dryness in the skin around the head, neck and tail, then little pimples will form, which will break and run a watery-looking fluid. After the animal is brought in from work he will be very itchy and rub himself against the manger until the skin is almost raw.

**Treatment.**—You can relieve the disease at the time, but when a horse once becomes affected with this disease it will generally break out every summer afterwards during hot weather. Give the animal a physic drench and powders the same as are mentioned for the treatment of scratches to clean and cool the blood, and rub him twice a day with any of the following washes: Corrosive sublimate, one dram to the pint of rainwater, mix together and shake well before using; creolin may be used, two tablespoonfuls to the pint of water, mix together and shake well before using; tincture of iodine may be used, two drams to the pint of water and shake well before using. Of the three receipts mentioned, the cheapest and favorite one, and the one we use mostly is the creolin, which will cost about two or three cents, and acts as well as any. The way to apply the lotion is first to take a fine brush and brush all the dust out of the skin and then apply the wash by rubbing it all over the affected skin with a sponge or cloth. Do this twice a day until he stops rubbing himself.

## NETTLE RASH (SURFEIT).

**Causes.**—This is generally caused from feeding hot and over-ripe food, or giving a drink of cold water when the animal is heated, or from overheating the animal by driving him too fast.

**Symptoms.**—Small pimples will appear in the skin around the head, neck and shoulders, but may affect the skin in other parts of the body. This disease is noticed at all times of the year.

**Treatment.**—Give the animal a physic drench consisting of

| | |
|---|---|
| Bitter Aloes | 8 to 10 drams. |
| Ginger | 1 tablespoonful. |
| Common Soda | 1 " |

Mix in a pint of luke warm water and give as a drench, allow the animal to stand in the stable the next day, feed on soft food, and follow up after this with the following powder:

| | |
|---|---|
| Ground Gentian Root | ¼ pound. |
| Sulphur | ¼ " |
| Nitrate of Potash or Saltpetre | ¼ " |

Mix thoroughly together and give a teaspoonful twice a day on his tongue with a spoon until the pimples disappear.

## WARTS.

Warts are thickenings or growths on the skin.

**Causes.**—It is difficult to say what is the cause, but some animals seem to be more inclined to warts than others, and they may be found on any part of the body.

**Treatment.**—If the warts have a neck to them they are easy got rid of by what is known as cording them, that is tying a small, strong cord on the wart as close to the skin as you can tie it; by tying it very tight it will stop the circulation of the blood in the wart and cause it to die and fall off in a short time. If they have a large neck and you cannot cord them cut them off with a sharp knife or scissors, and burn them a little with caustic potash, which will kill the roots and stop them from growing again. They are also nicely taken off with an ecraseur, if you have one.

### MELANOTIC TUMORS.

These tumors only affect gray horses, and are found mostly to be around the tail, sheath, lips and ears, but may be found on other parts of the body.

**Causes.**—The exact cause is not known.

**Symptoms.**—These tumors are generally small, varying from the size of a bean up to the size of a pigeon's egg, and are in bunches. They are not sore to handle, nor do they break out, but simply grow until they cause trouble to the parts they are in.

**Treatment.**—As long as they do not interfere with the animal in any way it is best to leave them alone, but as soon as you fear they are going to cause trouble it is best to cut them off by using a sharp knife and putting Monsell's solution of iron on them to stop the bleeding; or tie a tight cord around them allowing them to drop off themselves the same as a wart; or take them off with an ecrasure, which is an instrument for that purpose. After they are off treat the part as a common wound by bathing twice a day with luke warm water and applying white lotion each time after bathing.

### MANGE.

**Causes.**—This is caused from a germ or parasite working down into the skin and setting up the disease. One horse will catch the disease from another, also men and other animals will take the disease from the horse, so you see it is best to be careful if you think the animal has mange. It is most seen in animals that are in poor condition, with long, dirty hair, but it will affect animals that are in good condition. The disease is generally carried from one horse to another by the groom's clothes, harness or brushes.

**Symptoms.**—The insects burrow down into the skin and set up an irritation which breaks and runs a watery discharge, the hair falls off in patches, and the animal is extremely itchy. The disease generally commences around the mane and tail, and gradually spreads over the body. To make sure of the disease being mange examine a few of the scales under the microscope, and if you find the small germs or parasites in it you will know that it is a case of mange you have to deal with.

**Treatment.**—Apply something that will destroy the germs or parasites in the skin. If the horse's hair is long, clip him, and wash the body off with luke warm water and a little soap; then apply any of the following washes:

Carbolic Acid........................ ¼ ounce or 1 tablespoonful.
Sweet Oil..............................1 pint.

Mix and shake well together; rub it around his head, neck and shoulders the first day; the second rub it around the chest, belly and over the back; and the third day put it over his hind quarters and legs. It is best not to go over the whole body the same day, as too much of the carbolic acid would be absorbed into the system and might cause poisoning. Another very good wash, and one we think better than the above, is:

Creolin..............................2 ounces or 8 tablespoonfuls.
Rain Water..........................1 quart.

Shake well and rub in thoroughly all over the skin twice a week until the parasites are killed. This is by far the cheapest and best remedy known; you can cure a case for ten cents. To prevent the disease from spreading to other animals, wash with carbolic water, everything that he has come in contact with, that is, a few drops of carbolic acid in warm water; this will kill the germs or parasites and prevent the disease from spreading. Feed on soft food and give a teaspoonful of sulphur in his food twice a day.

### RINGWORM ON HORSES.

**Causes.**—It is caused from germs or parasites working in the skin, but is not so common in horses as in cattle.

**Symptoms.**—It often attacks the animal around the eyes and nose. The germs or parasites work in circles, which causes the hair to fall off and leaves round, bald spots. If not checked, it will soon spread over the body. Men are liable to take this disease from horses and cattle, and horses and cattle are liable to take it from men.

**Treatment.**—The best and cheapest remedy, and never known to fail in our experience, is crude petroleum oil as it comes out of the ground. This can be bought at almost any store in the country or city. The way to use this is to paint it over the spot where the ringworm is working, and let it go a half-inch over the edge of ringworm on to the good skin so as to check the disease from spreading. Apply this once a day until the ringworm is gone. If it blisters the skin, stop using it for a day or so and then continue again. Another very good remedy is to paint the spot where the ringworm is with tincture of iodine every two or three days until the ringworm is gone.

### LICE ON HORSES.

The lice may be common horse lice or hen lice. The hen lice are very small and reddish looking, and travel very fast. They get on the horse from being too near a lousy hen house, or from hens roosting in the stable. The horse lice are larger than the hen lice; they have a long, brownish body and travel very slowly, and they are generally found on horses that are turned out and have long, dirty hair.

**Symptoms.**—The horse's coat looks rough; he does not thrive well, and keeps rubbing his sides, neck and tail until he has the hair worn off, and he seems in perfect misery. By examining the hair closely you will see there is lice on him.

**Treatment.**—Kill the lice by washing the body off well with luke warm water and soap, then dry him off by rubbing him with cloths, and apply the following wash:

Creoline............................2 ounces or 8 tablespoonfuls.
Rain Water.........................1 quart.

Shake well and apply all over the body every third day by rubbing it well into the skin. Keep the treatment up until you have all the lice killed. Another very good remedy is a plug of "black-strap" chewing tobacco well cut up and steeped in a gallon of rain water and rub well in all over the body every third day until the lice are killed.

---

## CHAPTER XVII.

# DISEASES OF THE BRAIN AND NERVOUS SYSTEM IN GENERAL.

### INFLAMMATION OF THE BRAIN AND ITS COVERINGS (PHRENITIS).

This is not a very common disease, but is sometimes met with. Congestion first sets up, which is followed by inflammation.

**Causes.**—From a tumor growing around the brain; or a severe injury to the skull, with or without fracture; or continued exposure to the heat by being out in the hot sun will cause it; it is also sometimes caused after a case of distemper or strangles, where there is an abscess formed in the brain; also from other causes we cannot account for.

**Symptoms.**—The first symptoms are marked dullness. If the animal is standing in the stall, he will rest his head against the manger; the pupil of the eye will be very large; his pulse will be beating quite strong, but will beat slow sometimes—only twenty times per minute—and there will be a peculiar snoring noise made in breathing. After this dullness passes off, then there will be the reverse take place. The animal will be greatly excited—seems perfectly mad, and acts like a mad horse; the pulse, in this stage, becomes a great deal quicker, and the breathing louder; the animal reels around in his box with his head very high, and in some cases will rear up and put his front feet in the manger; and at times he moves like a piece of machinery. Sometimes he will be noticed to be walking around in his box with his head to one side; this is caused from the brain being affected on that side. In some cases, if you try to lead him he will fall down; but at times he will have quiet spells, after which the exciting symptoms will again return and be more alarming than at first. We have seen cases where the animal would twist his head down between his front legs and hold it in that position.

**Treatment.**—It is dangerous treating an animal with this disease, and you have to be careful in going around him. Give him a good physic drench consisting of:

| | |
|---|---|
| Bitter Aloes | 10 to 12 drams. |
| Ginger | 1 tablespoonful. |
| Common Soda | 1 " |

Mix in a pint of luke warm water and give as a drench. One hour after the above drench give the following:

| | |
|---|---|
| Bromide of Potassium | 2 drams or 1 teaspoonful. |
| Fleming's Tincture of Aconite | 10 drops. |

Mix in a pint of cold water and give as a drench every two hours, and apply cold water cloths and ice to his head until he gets relief; cover his body up nice and warm with blankets. In a case where the animal is in good condition, and you notice the disease coming on, bleed him; take a half or three-quarters of a pail of blood from him. If you bleed him, follow up with the above treatment, but only give him about eight to ten drams of bitter aloes and not so much aconite.

## SUNSTROKE.

This disease is common to all animals, and more especially to man. It generally attacks hard-working horses in the hot months of summer. It is a congested state of the blood vessels of the brain, with loss of power and feeling.

**Causes.**—Exposure to the hot sun, as a general thing, and especially so if the horse has been high fed and kept in a poorly ventilated stable. It is often seen in taking a horse out of a pasture field and giving him a hard day's work in the sun when he is not used to it.

**Symptoms.**—The first symptoms are dullness and dryness of the skin, and if the horse is working he will not be sweating as much as he should; he will also be noticed not to take his food very well, and have a staggering gait when he is walking. These symptoms may be noticed two or three days before the disease sets in. If he is working he will finally stagger and fall down, he may then struggle for a short time, but, finally, will lie quiet, with complete loss of power and feeling; if you prick him with a pin or knife he does not feel it, the pupil of the eye is very much enlarged, the pulse quick and weak, he breathes heavy, and when you try to get him up he will not be able to help himself at all.

**Treatment.**—Apply cold, wet cloths and ice bags to the head around the brain, and keep these on until he gets relief, and also give

| | |
|---|---|
| Sweet Spirits of Nitre................ | 1 ounce or 4 tablespoonfuls. |
| Whisky......... .................... | 1 wineglassful. |

Mix in a half pint of cold water and give as a drench. Repeat this every hour and a half or two hours until he gets relief. Be careful in drenching him while in this state for fear of choking him. As well as the above drench give a dose of physic consisting of

| | |
|---|---|
| Bitter Aloes ........... .................... ........ | 8 to 10 drams. |
| Common Soda..... ............................. | 1 teaspoonful. |
| Ginger .............. ........................ | 1 " |

Mix in half pint of luke warm water and give as a drench. If the animal seems to take notice to things around him and wants to eat, give him soft food with boiled flaxseed in it, and give plenty of cold water to drink, in small quantities at a time, but often. If the horse lays for any length of time turn him over two or three times every day, this will prevent his lungs from becoming affected, and he will lie easier. Keep the body very warm with plenty of blankets, this will have a tendency to draw the blood away from his head. In this case, as in others, when the animal is down keep plenty of dry bedding under him. As soon as he wants to get up, and seems able, help him to his feet.

### CONCUSSION OF THE BRAIN.

**Causes.**—It generally occurs in a horse when he is running away and strikes his head against something, or in rearing up and falling back and striking the pole of the head. In some cases he gets better quickly, while in other cases he may die very suddenly.

**Symptoms.**—In a pure case of this the animal looses all motion and feeling, he becomes completely paralyzed, and may lie without much signs of life. The pupils of the eyes will be very large, his pulse will be very weak. If there is no sign of fracture of the skull bones there is hopes of recovery. He will first begin to show signs of consciousness by trying to get up, but will rise on his hind legs first, and it may be some time before he will be able to rise on his front legs. In some cases the animal gets better quicker than in others, according to the amount of injury to the brain.

**Treatment.**—Same as that given for sunstroke.

### STOMACH STAGGERS (MEGRIMS).

**Causes.**—It may be caused in various ways. Anything that will interfere with the flow of blood to the brain, such as heart disease; from indigestion; from horses working in a tight collar; or from a small tumor growing and pressing on the brain. Very nervous animals are more subject to this than others of the opposite temperament.

**Symptoms.**—The horse will be attacked suddenly; he staggers and becomes unmanageable and falls to the ground. These symptoms may pass off in a few minutes, and the animal apparently seems as well as ever. A horse once affected with this disease is unfit to use for single driving, for he may take one of these fits at any time and fall down without showing the slightest sign before it.

**Treatment.**—Dash cold water on his head until he comes to, and afterwards give him a physic drench consisting of

| | |
|---|---|
| Bitter Aloes | 8 to 10 drams. |
| Common Soda | 1 tablespoonful. |
| Ginger | 1 " |

Mix in a pint of luke warm water and give as a drench. If there is anything wrong with his stomach this will generally relieve him, and it might be he would never have another attack of it. If it is from a tight collar, put a large one on him. In some cases it is recommended to give after the physic a teaspoonful of bromide of potassium in his feed twice a day, for a while, to act on his nerves.

### INFLAMMATION OF THE SPINAL CORD AND COVERINGS (SPINITIS).

**Causes.**—It is sometimes caused from the animal injuring its back by falling over a bank while running away, by severe exertion from nervous excitement. We have seen a case where a colt, running in a pasture field, fell and hurt himself. It may also be brought on by throwing an animal.

**Symptoms.**—At first the symptoms may not be so well marked, but they gradually come on. The animal seems very feverish and weak, has a staggering gait, and will sometimes be noticed to strike his hind fetlocks in walking, and, in some cases, if you go to turn him around he will fall down, and, as a general thing, if the disease is allowed to run on, it soon causes paralysis and death.

**Treatment.**—Give a physic drench of

| | |
|---|---|
| Bitter Aloes | 8 drams. |
| Common Soda | 1 teaspoonful. |
| Ginger | 1 " |

Mix in a pint of luke warm water and give as a drench; also, give the following powder:

| | |
|---|---|
| Powdered Nux Vomica | ¼ pound. |
| Ground Gentian Root | ¼ " |
| Nitrate of Potash or Saltpetre | ¼ " |

Mix thoroughly and give a teaspoonful in his feed or on his tongue with a spoon every night and morning. Keep the animal very quiet in a comfortable place and blister him along the back with a mustard plaster—quarter of a pound of mustard and enough vinegar to make it into a paste, put this plaster on every day for a few days. When the horse is getting better and his back very sore where you blistered him, oil his back with sweet oil, lard or goose oil. Feed the animal on soft food with plenty of boiled flaxseed in it to keep his bowels loose. If the animal gets down always assist him in getting up, for when once he gets off of his feet entirely there is very little hopes of recovery.

## PARALYSIS.

Paralysis may be in the complete or partial form. Complete paralysis is when there is loss of motion and feeling all over the body, and in cases of this kind death soon comes on. Partial paralysis is where there will be one part of the body affected, such as the hind quarters, or he may be paralyzed in one side. If the spine is affected all of the parts behind the affected part will be paralyzed.

**Causes.**—From an injury to the brain or from tumors growing around the brain or spinal cord. Nervous excitement and intoxicating liquors cause this disease in man. In stallions it is caused by being put to too many mares, or from a horse getting cast in the stall and hurting his back; throwing an animal for an operation may hurt him; in hunting horses, or other horses, it may be caused from slipping and straining the muscles under the spine, or from a fracture of the spinal bone. Sometimes, when a horse has fractured his ribs he cannot get up. This disease is more often seen in cows than in horses.

**Symptoms.**—If it is a case where the paralysis affects the one side of the body, the animal cannot walk straight, but goes around in a circle, and has not the proper use of that side. In a case where it affects the hind quarters, the animal. when he lies down, cannot get up; he will raise on his front legs, but cannot get up on his hind legs, and if you do not help him onto his feet, he seems very uneasy; but during the time he is down he will eat and drink fairly well. If you prick him with a pin in his hind part he cannot feel it. In cases of complete paralysis, when he cannot move at all, he soon dies.

**Treatment.**—If you think that he could bear his weight on his legs if he was up, raise him with pulleys or slings. Apply a mustard plaster over his back if the weather is warm, but if the weather is very cold, instead of applying mustard, put a half-pail of hot salt in a bag over his kidneys and blanket him so he will be hot, for what you need is heat to the back in these cases. In cases where the animal can stand fairly well, when he is up, keep him on his feet as much as you can, for a horse can stand a couple of weeks without hurting him. If you allow him to get down, and he cannot get up, he will only flounder around and may hurt himself. Give the following drench:

| | |
|---|---|
| Bitter Aloes........................ | 8 drams. |
| Sweet Spirits of Nitre ............ | 1 ounce, or 4 tablespoonfuls. |
| Common Soda........................ | 1 tablespoonful. |
| Ginger ........................... | 1 " |

Mix in a pint of luke warm water and give as a drench; this will get the bowels and kidneys acting; then give him the following powders to strengthen the nerves:

| | |
|---|---|
| Powdered Nux Vomica ............................ | ¼ pound. |
| Nitrate of Potash, or Saltpetre ........................ | ¼ " |
| Ground Gentian Root................................ | ¼ " |

Mix thoroughly and give a teaspoonful three times a day on his tongue with a spoon, or in soft food with plenty of boiled flax seed in it, and if he is able to walk give him gentle exercise every day.

## INFLAMMATION OF THE BRAIN AND SPINAL CORD AND THEIR COVERINGS (CEREBRO-SPINAL MENINGITIS).

This is congestion of the brain and spinal cord and their coverings, and if, in this state, they do not soon get relief, it turns into inflammation. This is, comparatively, a new disease, and is not known in any other country but on this continent.

**Causes.**—From horses being kept in a crowded stable that is poorly ventilated and badly drained, or from eating certain kinds of grasses that contain too much narcotic properties, such as are grown in swamps.

**Symptoms.**—They vary according to the parts most affected. There will be trembling noticed in the different parts of the body; the animal seems very dull and does not feed. As the disease goes on, the animal will have a peculiar jerking in the limbs, and then he will stagger, fall down and be unable to rise; the pulse will be quick and weak; the bowels usually costive, and his water is of a dark brown color. At first he generally lies in a dull, stupid manner, breathing heavy, and sweats freely if it is very warm. After a time the dullness passes off and the animal becomes delirious. If you give him water to drink, he will try, but he cannot, as there is paralysis of the gullet. These symptoms gradually get worse until he dies. Wherever you see one horse affected there is apt to be more affected, for the same cause that brought it on him will bring it on the others.

**Treatment.**—If noticed before the animal is too bad, there is hope of recovery; but if the animal is down and cannot swallow before he is noticed to be sick, the chances are against him. As soon as the disease is noticed, take one half-pail of blood from him and give the following drench:

| | |
|---|---|
| Bitter Aloes | 8 drams. |
| Sweet Spirits of Nitre | 1 ounce or 4 tablespoonfuls. |
| Common Soda | 1 tablespoonful. |
| Ginger | 1 " |

Mix in a pint of luke warm water and give as a drench, then follow up with the following powder:

| | |
|---|---|
| Powdered Nux Vomica | ¼ pound. |
| Nitrate of Potash or Saltpetre | ½ " |
| Hyposulphite of Soda | ½ " |

Mix well together and give a teaspoonful on his tongue every three hours until he gets relief. Apply a mustard plaster and hot cloths along his back and cover the body warm. If the animal gets down turn him over from side to side three times a day, and be very careful while drenching for fear of choking him. Feed on soft food, with plenty of boiled flax seed in it, and give plenty of cold water to drink.

## CHOREA.

This is an affection of the nervous system, where the horse loses the power to back up.

**Causes.**—It is due to some injury to the spinal cord, and sometimes comes on a colt after castration, but we think he must have been injured while throwing him.

**Symptoms.**—The horse will be useful, and as long as he is going ahead he will be all right. He can pull or do anything in the shape of work until you go to back him up; as soon as you try to, you will find that he cannot back; the muscles of the hind quarters will begin to shiver, his tail will rise up, and, no matter how much you force him, he cannot back up. This disease is more common in nervous animals.

**Treatment.**—If you notice it when it is coming on, give:

| | |
|---|---|
| Powdered Nux Vomica.............................. | ½ pound. |
| Nitrate of Potash or Saltpetre.......................... | ½ " |

Mix thoroughly and give a teaspoonful three times a day in soft food. Rub him over the back and hips with the white liniment twice a day. If he is in good condition, give him a physic drench of

| | |
|---|---|
| Bitter Aloes.......................................... | 8 drams. |
| Ginger.......................................... | 1 tablespoonful. |
| Common Soda.......................................... | 1 " |

## STRINGHALT.

This receives its name from the way the animal acts.

**Causes.**—This is caused from some affection of the nerves which go to supply the part affected, but really what parts of the nerves affected has never been found out. This disease is more often seen in highly nervous animals, and is caused from applying severe blisters to the legs, which irritates the nerves, or clipping the legs and having him out in the cold. It is noticed to follow after castration, either from the burning of the rope on the fetlock or the irritation of the cutting.

**Treatment.**—If this disease is once well established it is incurable, but if noticed at the time it is starting give the animal a physic of

| | |
|---|---|
| Bitter Aloes.......................................... | 8 to 10 drams. |
| Common Soda.......................................... | 1 tablespoonful. |
| Ginger.......................................... | 1 " |

Mix in a pint of luke warm water and give as a drench. Leave the animal in the stable the next day, and follow up with the following powder, which acts on the nerves: Bromide of Potassium, two drams or a teaspoonful, twice a day in his food, or on his tongue with a spoon.

## HYDROPHOBIA (RABIES) IN HORSES.

This disease never occurs in a horse unless he has been bitten by a mad dog or cat.

**Symptoms.**—He shows restlessness, will bite and rub where he was bitten. These symptoms are followed by brain disturbance, and the animal will act somewhat as if he was suffering from inflammation of the brain; but in this disease the animal is wicked, and will bite at you in a peculiar way, just like a vicious dog. The animal becomes more excited, turns round and round in his stall until he gets weak and falls down, and gradually keeps getting worse until he is relieved by death.

**Treatment.**—If the horse is showing the above symptoms, and he has been bitten by a dog, have him destroyed at once; but if you were called to see a horse after he was bitten by a mad dog, and before he shows the above symptoms, take a sharp knife and cut away the flesh around the bite, then burn it with a stick of caustic, potash or nitrate of silver. If you have not these, burn it well with a red hot iron or anything to destroy the poison.

## LOCKJAW (TETANUS).

This is purely a disease of the nerves and receives its name by the way it acts on the muscles of the jaw. Sometimes they become so set that you could not pry the animal's mouth open. There are two forms of this disease, one is known as the traumatic form, this is where the disease follows an injury or operation, which can be seen; the diopathic form of the disease comes on the animal without any visible injury or operation. In this case it is thought to be brought on from worms or bots in the stomach or from being exposed to extreme cold.

**Causes.**—The general causes of this disease are when a nail runs in the horse's foot, it may follow a stake or cut just about the time the wound is healed up, and comes on from eight to twenty-one days after being injured. It may also come after wounds which seem to be healing extra fast. It follows operations, such as docking, nicking a horse's tail or castration; it sometimes occurs after very severe blistering. In referring to the disease following castration, it is more apt to come on when a horse is allowed to run through a river or spring creek, or being left out in cold winds. As an example, twenty-four horses were castrated and bathed in cold water a few days after, and sixteen out of the twenty-four died of lockjaw. This shows you the necessity of having the operation done in fine weather and the colt kept away from damp places. The disease is noticed to come on just about the time the colt is healed up, the same as in other wounds.

**Symptoms.**—There is not much difficulty in telling a case of this kind. At first there is a peculiar stiffness of the body, and he walks with his neck high and his head stuck out as if he was suffering from sore throat, he has a peculiarly high action, and if he is on pasture he cannot get his head down to eat, except in a very mild case of the disease. The horse, in most cases, will try to eat if he can, especially at the commencement of the disease. If you walk up to him and suddenly excite him he will almost groan, as if in pain, his head will fly up and the haw of the eye will fly over the eye so that you can hardly see it. When he is excited his tail will raise up and the muscles all over his body seem hard, drawn and set, and the animal will almost fall down. In a case of this kind examine the mouth and you will find that it can only be opened a little. If it is a severe case these symptoms will gradually grow worse, and in a few days he will get off his feet and will be unable to raise them ; he will have fits of severe struggling and be in terrible agony, and the only thing that gives him relief is death, which soon comes.

**Treatment.**—In a very light case of the disease, if in the spring and the animal is able to get down and eat grass, it is best to leave him out in a level pasture field where there is nothing to disturb him. In a case of this kind do not go near him to give him medicine, but watch him to see that nothing turns up that you are not expecting, because catching a wild colt in the pasture to give him medicine would only excite him so much that it would do more harm than good. There has been all kinds of treatment tried for this disease, but the best is to give the animal, as soon as noticed, a dose of physic consisting of

| | |
|---|---|
| Bitter Aloes | 8 drams. |
| Common Soda | 1 tablespoonful. |
| Ginger | 1 " |
| Sweet Spirits of Nitre | 1 ounce or 4 tablespoonfuls. |

Mix in a pint of luke warm water and give as a drench, being careful not to excite the animal much while giving it to him. In case it is caused from a nail in the foot, pare the hole out well where the nail went in and poultice with hot linseed poultices, changing them often to keep them hot. Keep the foot poulticed until the animal seems relieved. If it is from a wound bathe well with luke warm water three or four times a day, and paint the foot over with fluid extract of belladonna after each bathing. Keep the animal in a comfortable stable, free from noise or anything that will excite him, as perfect quietness is what is wanted in treating this disease. Give one dram, or one teaspoonful, of fluid extract of belladonna on his tongue with a spoon three times a day, and feed on soft, easy chewed food, such as gruels of chop stuff with boiled flaxseed in it– make the gruel so he can drink it down—if in the spring give him grass, if in other times of the year scald his hay and make it as soft and easy to eat as you can. It generally takes from three weeks to thirty days for the disease to run its course. It is best not to rely too much on medicine, for it takes time for it to run its course. Good nursing is better than medicine in this disease. As a general thing, before the end of the third week, he begins to get better.

---

CHAPTER XVIII.

# DISEASES OF THE LYMPHATIC SYSTEM.

### WEED IN THE LEG (LYMPHANGITIS).

This disease also gets various other names, such as water farcy, big leg and Monday morning fever. This is a very common disease now in this country, and is liable to be more so, on account of horses being better fed and cared for than they used to be. The disease generally affects the hind legs, but sometimes affects the front legs.

**Causes.**—It is common in hard-worked and highly-fed horses, by letting them stand in the stable for a few days without exercise after being accustomed to working every day and feeding them the same as if they were working; this is why it is seen so often on Monday morning. The direct cause of the disease is from horses getting too much nutriment in the blood, which over stimulates and sets up inflammation in the lymphatic glands in the legs; then when these glands are inflamed they do not absorb the lymph and carry it off as they should when all right, thus the leg becomes very largely swollen with lymph. It is also caused by a horse getting a prick of a nail in the foot, and the soreness extends up the legs and affects the glands and sets up inflammation in them. It is also caused from a horse having bad blood with too much fibrine in it. Heavy horses with sluggish circulation are more liable to it than lighter horses.

**Symptoms.**—The first signs are the animal will quit feeding, tremble and be feverish, which is followed by lameness and swelling along the inside of the leg—usually the hind leg—just inside the thigh. The swelling, at first, will be along the line of the lymphatic vessels in the shape of a hard cord; if you press your hand on this cord the animal will immediately throw his leg out and up and seem very stiff and sore; after this stage the leg becomes greatly swollen all the way down and around it, in some cases as large as a stovepipe, and it will be very sore to handle, and is so painful that the horse will lift it from the floor and look around at it. The symtoms of a case in the front leg are similar, only the swelling starts at the chest and extends down the inside of the front leg. The pulse will be quickened, and the animal breathes heavier than natural. After an animal has once been affected by this disease he is more liable to have it again, and after he has been afflicted with it several times it terminates in what is known as elephantiasis, or big leg.

**Treatment.**—Get rid of the nutriment of the blood as soon as possible, by bleeding; take a half pail of blood from him if the animal is strong and fat, and give him a physic drench:

| | |
|---|---|
| Bitter Aloes | 8 to 10 drams. |
| Common Soda | 1 tablespoonful. |
| Ginger | 1 " |

Mix in a pint of luke warm water and give as a drench; if you bleed, just give 8 drams of aloes, and give the following powders:

| | |
|---|---|
| Nitrate of Potash or Saltpetre | ¼ pound. |
| Sulphur | ¼ " |
| Ground Gentian Root | ¼ " |

Mix thoroughly and give a teaspoonful three times a day on his tongue with a spoon or in his food. Feed on soft food, with plenty of boiled flax seed in it, to keep his bowels loose. Bathe the leg with luke warm water, as warm as you can bear your hand in it, with some saltpetre and vinegar in the water, for nearly an hour at a time, three times a day, and after wiping dry, rub well with white liniment weakened down nearly one-half by adding more water to it. If in very cold weather, bandage the leg to keep him from getting cold in it after the bathing. Allow him to stand quiet for three or four days until you get the inflammation checked in the gland, then commence to exercise him some every day, and gradually bring him back to his natural habits again. The more he is bathed with warm water the sooner he will get relief.

### BIG LEG (ELEPHANTIASIS).

This is a thickened state of the leg from repeated attacks of lymphangitis or weed in the leg. The swelling becomes organized and you cannot effect a complete cure in any case.

**Treatment.**—Give the animal regular exercise, also give him a good physic drench once in a while to keep his bowels right, consisting of

| | |
|---|---|
| Bitter Aloes | 8 to 10 drams. |
| Common Soda | 1 tablespoonful. |
| Ginger | 1 " |

Mix in a pint of luke warm water and give as a drench, then give him the following powder to act on his kidneys and blood:

| | |
|---|---|
| Nitrate of Potash or Saltpetre | ¼ pound. |
| Sulphur | ¼ " |
| Ground Gentian Root | ¼ " |

Mix thoroughly together and give a teaspoonful in his food twice a day, this will help him more than anything you can do for him. Any time you notice the leg swelling repeat the above treatment, and by keeping his blood in good condition he may make a good work horse for a long time.

### SWELLING OF THE LIMBS (ANASARCA).

**Causes.**—This is caused from the lymphatic glands of the legs working sluggish and not carrying the lymph off. It generally occurs in the hind legs, and is commonly called stocking of the legs. The most common cause we have is where a horse is accustomed to standing in the stable during the winter months and then putting him to hard work in the spring before he gets used to it. After resting during the night his legs will be swollen the next morning. Or, it may be caused from giving a horse a long journey when he is not used to it; the next morning his legs will be swollen. Also a horse's blood being in bad condition will cause it.

**Symptoms.**—There is swelling of the legs during the night, and in the morning when he is taken out the swollen legs will be quite stiff. This swelling will go down after he is kept moving for some time, but will return again the next night, perhaps worse than before. If this disease is not attended to, it is apt to terminate in a case of scratches or grease.

**Treatment.**—Keep the animal from work for a few days and give him a physic consisting of

Bitter Aloes .... ............... ...8 drams.
Common Soda................... .... 1 tablespoonful.
Ginger ............................1 "
Sweet Spirits of Nitre......... .....½ ounce or 2 tablespoonfuls.

Mix in a pint of luke warm water and give as a drench. Feed on soft, light food, and allow the animal to stand in a stable a few days after giving the drench. In case the animal is very weak, and you think it is not safe to give him the aloes, give him

Raw Linseed Oil..... .... ..........1 pint.
Sweet Spirits of Nitre................½ ounce or 2 tablespoonfuls.

Mix and give as a drench,, and use him the same as if you gave him the aloes; then give him the following powder to act on his kidneys and blood and build his system up:

Nitrate of Potash or Saltpetre ............................¼ pound.
Sulphur...... ................ .. ........... ....¼ "
Ground Gentian Root.................................¼ "

Mix thoroughly and give a teaspoonful every night and morning in his feed. After he stands a couple days put him at gentle work, not too hard, also hand rub his legs at night and bandage them, which will keep the swelling down and strengthen the leg. Never leave the bandage on longer than three hours, for if you do it will do more harm than good.

## CHAPTER XIX.

# DISEASES OF THE HEART, ARTERIES AND BLOOD.

### I.—HEART DISEASES.

These diseases are very uncommon in the horse, but sometimes we have what is known as rupture of the valves and palpitation of the heart.

#### RUPTURE OF THE VALVES OF THE HEART.

**Causes.**—It is hard to tell just what the cause is, but the valves are diseased in some form before the rupture takes place.

**Symptoms.**—When rupture takes place the horse dies almost instantly, for it stops the circulation in the system. In some cases the horse may be subjected to weak spells before rupture takes place, while other times this symptom will not be noticed.

**Treatment.**—There is none; and the only way to tell if this caused death is to examine the heart, which will be clotted with blood, caused from the rupture of one of the valves in the heart, and also you will see the ruptured valve.

### PALPITATION OF THE HEART.

This is weakness of the heart.

**Causes.**—It generally comes after weakening diseases, such as the fevers of the horse, lung troubles, or dropsy of the heart.

**Symptoms.**—The animal will be very weak, and you can hear the heart thumping in its cavity; the symptoms will be greatly increased when the animal is excited. His pulse will be very quick and weak, and in this state the animal will be unable to do any work.

**Treatment.**—Keep the animal quiet and strengthen the system as much as you can by good food, and give him the following tonic medicine which acts as a tonic to the heart and system in general :

| | | |
|---|---|---|
| Pulverized Digitalis | ¼ | pound. |
| Ground Gentian Root | ½ | " |
| Powdered Nux Vomica | ¼ | " |
| Ground Sulphate of Iron | ¼ | " |

Mix thoroughly and give a teaspoonful three times a day in his feed or on his tongue with a spoon.

---

## II.—DISEASES OF THE ARTERIES.

### TUMOR (ANEURISM).

This is a diseased state of the walls of an artery.

**Causes.**—The causes are unknown.

**Symptoms.**—This disease often goes on in the artery unknown until the diseased part of the artery will give way; if it is internally, and a large artery, he will bleed to death very quickly; if it is a small artery in the muscles he will not bleed to death, but you will notice a large swelling appear suddenly around the diseased part of the artery, and the artery will bleed until it is stopped by clotting and pressure from the muscles and skin. The animal will be a little weak, but the swelling will not be sore, and pressing on it you can tell there is a fluid in it.

**Treatment.**—Allow the fluid to remain in the swelling two or three days, until you are sure the artery has stopped bleeding, then, with a sharp knife, open into the lower part of the swelling; make a big cut into it, and remove all the clotted blood, then take a large syringe and wash out all the blood with luke warm water with a few drops of carbolic acid in it. After this, bathe

the parts well with luke warm water twice a day, and inject in the hole each time, after bathing, with the white lotion. Keep the animal quiet and it will heal up in the course of two or three weeks all right.

### RUPTURE OF AN ARTERY.

**Causes.**—It is sometimes caused from straining while the mare is foaling; drawing heavy; severe exertion of any kind; or a severe bruise.

**Symptoms.**—Same as a tumor or aneurism of an artery, only you will not find the artery diseased.

**Treatment**—This disease is treated the same as tumor or aneurism of an artery.

---

## III.—DISEASES OF THE VEINS.

### INFLAMMATION OF A VEIN (PHLEBITIS).

This disease is mostly noticed in connection with the jugular veins.

**Causes.**—Usually from bleeding a horse with a rusty fleam; using a rusty pin or your hands having dirt on them while putting in the pin (so you see the necessity of having everything clean when bleeding); sometimes from a horse rubbing his neck against something after bleeding; from turning the horse out to grass right after bleeding and allowing him to have his head down, this interferes with the circulation of the blood, causes swelling and clotting of the blood, which sets up inflammation.

**Symptoms.**—There is swelling along that side of the neck you bled him from, and if he is allowed to have his head down that side of his head will also be swollen. The swelling will be hard and painful when you touch it, and, in a few days, the clotted blood formed in the swelling will start to fester and break out in little boils or abscesses along the side of the neck.

**Treatment.**—Keep the animal's head well tied up and bathe the sides of the neck with warm water and vinegar four times a day for half an hour at a time, and each time, after bathing apply white liniment. If the neck beals or festers, open the places up with a knife and allow the matter to escape. When you do this, change the treatment to white lotion instead of liniment. Do this until you get all the swelling and inflammation out and the bealing places all healed up, then blister, using the following:

Vaseline, or lard ........................................ 1 ounce.
Pulverized Catharides, or Spanish Fly .................. 1½ drams.

Mix well together and rub the blister along the swollen part of the neck. Tie the horse's head short so he can not rub it. Rub the blister in well and grease the place where you blistered in three days after; allow it to remain two or three weeks, or until it heals up, and if the swelling is not down, blister again and follow out the same directions in this as in the first blister. In after treatment do not turn the animal out to pasture for a year or so, for his head will swell up on account of his not having the use of this vein; keep him in the stable and feed him out of a high manger, and by doing this he will be just as useful as ever, for after a while the other jugular vein will enlarge so it will do the work of both the veins. In buying a horse look at both sides of his neck to see that the veins are all right.

---

## IV.—DISEASES OF THE BLOOD.

### BAD BLOOD.

This is when the blood gets very impure and the horse does not thrive well.

**Causes.**—From working very hard and feeding very high, or from turning the animal out to a straw stack and allowing the system to run down.

**Symptoms.**—The animal will get weak, and will not thrive well, and gets hide bound. The hair will be rough, dry and scruffy; and, if the blood is hot, there will be pimples form over the body; and when he stands in the stable over night his legs will swell; and if he is out in wet weather scratches will come on; and the animal is dull and unfit for work, and no matter how much you feed him he will not thrive.

**Treatment.**—Get his blood in good shape by giving a physic drench to start on, consisting of:

Bitter Aloes ............................ 8 drams.
Ginger.................................... 1 tablespoonful.
Common Soda.... ...................... 1 "

Mix in a pint of luke warm water and give as a drench. Allow the animal to stand in the stable a couple of days, and feed on soft food with plenty of flaxseed in it to keep the bowels loose, then give the following powder:

Nitrate of Potash or Saltpetre ........................ ¼ pound.
Sulphur ..................................................... ¼ "

Mix and give a large teaspoonful twice a day in his feed, after you have given him this, follow up with the following tonic powder to make him strong:

Ground Gentian Root .................. ..... .........¼ pound.
Sulphate of Iron.........................................¼ "

Mix well together and give a teaspoonful twice a day in his feed; if in the spring of the year and you can turn the horse on grass, bleed him, taking half a pail of blood, then turn him out, and this will cure him as quick as anything. When you are treating a horse for bad blood, if you have him in the stable, give him a little exercise every day.

### AZOTURIA.

This disease, at one time, was not very common, but is getting more so all the time.

**Causes.**—From allowing a horse to stand in the stable and feeding him well for a few days without any exercise, the blood gets too full of albumen, and then, by taking him out and driving him, brings on the disease. When you drive the horse the blood gets heated and he takes in a great deal of oxygen from the air into the blood, which unites with the albumen in the blood and changes it into acids, which are known as hippuric and urea acids; these acids stop the action of the kidneys and then act on the muscular system, and cause the muscles of the back and hips to become swollen and paralyzed. This is the only disease that is noticed to come on a horse very soon after starting on a drive. It is more common in the winter months, on account of the animal standing in the stable more, but is liable to come on at any time of the year.

**Symptoms.**—On taking a horse out of the stable, he will go off full of life for a quarter of a mile to a mile, or even a longer distance in some cases, then you will notice him break out into a sweat; he becomes stiff in the hind quarters and not able to trot. If you examine him he will be breathing heavy, his pulse quick and weak, and will be trembling about the flanks. Look over the back and hips and you will notice the muscles swollen and as hard as a board. If you attempt to drive him still further, he will get so stiff he will not be able to go, and will fall down and not be able to rise, and all the symptoms mentioned above will be increased greatly. His urine will be of a dark red color and very little at a time, for the kidneys are not acting much. If you catch some in a dish, and allow it to stand, the acids will come to the top. In severe cases the legs and ears are very cold.

**Treatment.**—As soon as the symptoms are noticed, stop driving him and take him to the nearest place, for if you keep on driving him he will only fall down and be a great deal more bother to you. When you get him into the stable, cover him well with blankets and let him have a good sweating; this will relieve the kidneys, give him a good rubbing all over the back and hips with white liniment; if you have not got the liniment along with you apply a coat of mustard and vinegar over the back, or even a half pail of hot salt in a bag would be good, for you must get heat to the kidneys to start them to act. Give the following:

Sweet Spirits of Nitre................1 ounce, or 4 tablespoonfuls.
Bitter Aloes.................. ......8 drams.
Common Soda................ .....1 teaspoonful.
Ginger...................... .......1 "
Fleming's Tincture of Aconite........10 drops.

Mix in a pint of luke warm water and give as a drench; this is to get the bowels and kidneys acting. As a general thing, if this treatment is given as soon as noticed, he will be able to work in three or four hours. If it is a severe case, and the horse does not get relief in three or four hours, follow up with the following drench:

Sweet Spirits of Nitre................1 ounce, or 4 tablespoonfuls.
Common Soda ..... .................1 tablespoonful.
Fleming's Tincture of Aconite.........10 drops.

Mix in a pint of luke warm water and give every three hours until the animal gets relief; also, keep the heat applied to the back. If the animal is so bad he gets down, turn him over from side to side, twice a day, and as soon as he able to get up, help him to his feet. It is well to use slings for a short time every day after he begins to get strong. We saw one case where a horse had lain nine days and afterwards got up and was all right again by means of the above treatment without the slings. Feed on soft food and give all the luke warm water he can drink; tend to his general comfort, such as a comfortable stall with good bedding. Watch if he makes water, and if he does not, take it away with a catheter. When the animal is getting better give the following powders to get his kidneys working:

Nitrate of Potash or Saltpetre..........................¼ pound.
Ground Gentian Root...................... ...........¼ "

Mix and give a teaspoonful twice a day in his feed. After an animal has suffered from this disease once he is more apt to be troubled again. Watch that his kidneys are in good shape, and have the animal exercised every day. If you have fear of a horse becoming attacked with this disease when you are going to drive him let him walk easy for the first mile or so, and then he will be all right.

---

CHAPTER XX.

# DISEASES OF THE BONES.

## BIG HEAD (OSTEO-POROSIS).

This disease is more common in some localities than others; it is not met with much in Canada or Great Britain, but is frequently seen in the United States and Mexico. It is a disease which attacks horses from one to four years old, and rarely attacks old horses, and it gradually comes on as the animal grows, without any signs of soreness, only the bones of the head and legs get larger and lighter and become very brittle, that is, easy to break. Thus it gets the name "big." It more often affects the bones of the head than those of any other part of the body.

**Causes.**—It is difficult to say what is the real cause of this disease, but it is supposed to be caused from an animal feeding on pasture land deficient in the salts of lime. Some think it is wholly caused from animals grazing on low-lying, swampy land, where the grass grows long and does not contain the full amount of nourishing substances.

**Symptoms.**—At first the symptoms are not very well marked, but the animal is noticed to be dull, fall off in condition, and his muscles get very soft. The animal may run along this way for four to six months, then the true nature of the disease will show itself. The animal will seem stiff in traveling, his belly becomes very gaunt, and the bones of the head will be noticed to be getting larger than natural; then, after this, the bones of the legs may also become enlarged, and as the disease goes on the bones will continue getting larger, and the animal keeps gradually going down in condition until he dies. In some cases the bones become so brittle that while he his walking along one of his legs would break and he would have to be destroyed.

**Treatment.**—The treatment is not very successful, and if the disease has been running on for any length of time it is best to destroy the animal. If in the early stages, and he is on low pasture, move him to a good, high pasture field, and give a teaspoonful of sulphate of iron in some oats twice a day to build the system up and overcome the disease.

### SPLINTS.

This is a bony enlargement on the inside of the leg, between the knee and fetlock, and is sometimes noticed to affect the outside of the leg, and it is noticed to affect the bones of the hind legs, below the hock joint, either on the inside or outside. Any enlargement of the bone along the places mentioned comes under the name of splint.

**Causes.**—Certain breeds of animals are more liable to splints than others. For instance, horses with small, weak bones below the knees, or colts that are very fat, and heavy on their legs, are the most liable to splints. Driving or riding colts on hard roads, or working them on hard roads. Horses that are driven on the pavements of large cities are very apt to have splints, from shoeing too heavy; or from the animal striking his foot against the other leg in traveling. In all of these cases, whether it is due to hard roads, or from striking the leg with anything, inflammation will set in between the bone and the covering of the bone, then there is a deposit of bony matter, this is what causes the enlargement and soreness. When the splint affects both sides of one leg it is called a double splint.

**Symptoms.**—At first it is a little hard to detect; but when the splint gets any size it is very easy to tell what is wrong, and when a young horse goes lame on the hard road it is well to examine for splints. The lameness has a peculiarity about it, when the animal is walking he walks perfectly sound, and he also stands on the leg as if nothing was wrong, but when you come to trot him he will be very lame, and he will drop and raise his head to a great extent. Always bear in mind that when a lame horse is trotting his head goes down when he strikes his weight on the sound leg, he does this in trying to favor the lame leg, and in all cases, no matter how sure you are about the part the horse is lame in, examine the foot to see there is no nail in it.

**Treatment.**—As a general thing it is successful. Keep the horse from work as much as possible, and if in summer time bathe the leg in cold water with a little salt in it; do this a couple of times a day, and after rubbing dry, apply the white liniment; keep at this treatment until you get the inflammation and soreness out of the splint, after this blister to remove the enlargement. Use the following blister:

| | |
|---|---|
| Vaseline or lard | 1 ounce. |
| Biniodid of Mercury or Red Precipitate | 1 dram. |
| Powdered Cantharides or Spanish Fly | 1 dram. |

Mix thoroughly together and there will be enough to blister an ordinary splint three times. If there is much hair on the splint clip it off and apply one-third of the blister; the more you rub it in the better the blister works; tie his head a little short for a few hours so he cannot get down to bite the blistered part; let this stand for three days, then grease the blistered part with vaseline or lard, then let it go for two or three weeks after the first blister and blister again, and follow out the same directions as in the first blister, and, if it is not all gone, blister the third time in two or three weeks more.

### SORE SHINS.

This is inflammation of the covering of the shin or metacarpal bones and nearly always affects the front legs, but sometimes affects the shin or metatarsus bones in the hind legs. This is more common in some parts of the country, and is generally seen in young race or trotting horses that are put to severe exertion.

**Causes.**—It is from hard and continual driving in training young horses for races; from a continual steady strain on the legs, it sets up an inflammation in the covering of the shin bones; or from taking a colt out of the pasture field and putting him to too hard work when he is not used to it; or from running and striking the front of the shins against anything.

**Symptoms.**—They are very plain. The colt is very lame at first, and in examining him, if you run your hand down over the shins, you will find them very sore and hot. He will flinch and jump away from you when you press on the parts affected; and in trotting he drops his head similar to splint lameness. If it is allowed to run without being treated, a thickness will soon appear in front of the shin bones where the soreness was.

**Treatment.**—Give the animal as much rest as you can. If in the summer, bathe with cold water and salt. If in the winter, bathe with warm water and salt. After bathing twice a day, rub dry and apply white liniment. After you get the soreness and inflammation out by bathing and using the liniment, blister him to take down the enlargement. Use the same blister and same directions as is given in treatment of splints.

## SIDE BONE.

This is ossification, or turning into bone, of the lateral cartilages of the foot; these are two cartilages, one situated on each side of the foot, and by pressing on them at each side of the foot, just above the hoof, you can move them in and out, that is when they are in their healthy state, but when they become diseased or changed into bone, they become enlarged and you cannot move them at all. This disease is more often met with in heavy breeds of horses, but it is sometimes met with in light horses, when it is harder to treat and more of a detriment to them on account of being used for fast work.

**Causes.**—From hard work, as a general thing.

**Symptoms.**—In heavy horses, they are not lame in some cases, just the enlargement at each side of the foot, just above the hoof, but in severe cases there may be lameness. In light horses, used for drawing, the first symptom noticed is lameness, afterwards followed with the enlargement at the sides of the foot, just above the hoof.

**Treatment.**—Rest the animal as much as you can, and, if in the summer, bathe the foot well with cold water and salt twice a a day; after bathing each time, wipe dry and apply the white liniment. If it is in the winter, bathe with warm water and salt, and also poultice with half linseed meal and bran; put the poultice on as warm as you can without burning the animal, and each time after bathing and poulticing, rub with white liniment, the same as mentioned above. After you get the soreness and lameness out by the above treatment, blister with the following receipt:

**Biniodid of Mercury or Red Precipitate..................2 drams.**
**Vaseline or Lard.......................................1 ounce.**

Mix well together and there is enough in this receipt to blister a small side bone four times. Apply quarter of the blister and rub in thoroughly, leave it for three days and then grease with lard, and allow it to go for two or three weeks, then wash the parts clean with luke warm water and soap, and blister again, just the same as the first time; repeat the blisters until the lameness is entirely gone and the side bone stops growing. In buying a horse always examine him closely for side bones, especially if it is a heavy horse. In cases where you want to work the animal shoe him with a bar shoe.

## RINGBONE.

There are two kinds of ringbones—the high-up ringbone and the low-down ringbone. The high-up ringbone affects the pastern joint, the low-down ringbone affects the coffin joint. A ringbone is a bony growth around the pastern or coffin joints. They may affect the front legs, but are more often found on the hind ones, and there are cases where the whole four legs were affected at the same time.

**Causes.**—Like most other bone diseases, it runs in some breeds of horses to be affected with ringbones, that is to say it is hereditary, so you see the necessity of breeding sound animals. But there are other well-marked causes, such as hard or fast work; or an injury or severe sprain of the joint will bring it on. A common cause in colts is allowing their feet to grow too long; or, in foals, by allowing them to follow their mother when she is working, which is very hard on foals; sometimes by a horse running a nail in his foot, and if it runs up far enough to wound the coffin joint, it will cause the joint to become diseased and throw out a ringbone; or it may come on by a horse standing on one leg while he is very lame in the other.

**Symptoms.**—They are very plain. There is lameness, followed by an enlargement around whichever of the joints are affected. Now, if it is the pastern joint, the enlargement will be about half way between the fetlock joint and the hoof; this is the form known as high-up ringbone. If it affects the coffin joint, the enlargement will be noticed bulging out around the top of the hoof, and this form of the disease is known as a low-down ringbone. The parts around the ringbone will be hot, and the lameness will increase as the disease goes on. The peculiarity of ringbone lameness is that the animal takes a longer step on the affected leg and puts his heel down first; also, that he is more lame when starting off than after he is driven a piece and gets warmed up.

**Treatment.**—The treatment, in some cases, is not attended with very great success, although, in others, it is very successful, depending, of course, on how much disease there is going on in the joint. The treatment is similar to spavin—the main object being to set up what is known as anchylosis of the diseased joint, that is, to have the bones forming the joint become united solid to each other. As soon as this takes place, the lameness and soreness leaves, and this is what is called a cure. Of course, after it is cured, the motion of that joint is gone, and the anmial does not have quite as free action as before the leg was affected, but will be very useful for years after. In order to get a real good idea about this, examine some ringbone, after a horse dies, that was cured, and it will give you an idea how the bones unite. In all cases first have the foot pared down to its natural shape, and have the toe cut off very short and keep it cut short afterwards; this throws the strain off the joint. Cut the hair off the enlargement, if it is long, and blister with the following:

Biniodid of Mercury or Red Precipitate .... ...... ......2 drams.
Vaseline or Lard.......... ..........................1 ounce.

Mix well together. There will be enough in this receipt to blister an ordinary ringbone two or three times, according to the size of it. Apply one-third of the blister and rub in thoroughly—the more you rub the better it will work—and grease the third day after blistering, and keep on blistering every three weeks until it is cured and follow out the same directions as the first blister; each time before you blister wash off the parts with warm water and soap. If, in the course of a few months, this does not help him, "fire" him. Use the same kind of a firing iron as that used in firing a spavin. Place a twitch on his nose and have one of his front legs held up, and when your irons are red hot burn all around the ringbone in streaks—running up and down—three-quarters of an inch apart; don't burn too severe, just enough to leave white seam where you run the iron. In doing this take your time and do not lean heavy on the iron. After you fire, leave it six days and blister the same as above mentioned anc follow out the same directions. If the horse will not stand, throw him the same as you would if you were going to castrate him. Don't get discouraged if he does not get better right away, for it generally takes from six months to a year, and even longer in bad cases. In colts the treatment is just the same, only not quite so severe, and will vary according to the size of the animal.

### BONE SPAVIN.

A spavin is a disease affecting the bones of the hock joint, and generally throws out a bony enlargement on the inside of the hock joint. When the spavin is inside of joint and does not show itself outside in the form of an enlargement it is then called an occult spavin. They are divided into high-up spavins and low-own, or what is co n n only called jack spavins.

**Causes.**—The same as in ringbone, certain breeds of horses are nearly all spavined, for it comes on in a hereditary form, that is where the dam or sire is affected with spavin and their colts become affected with spavin, and so on for generations of horses. A great many horses have spavins in this way, therefore you will see the necessity of breeding good, sound horses, although when a spavin comes on from some well marked cause, such as from hard work, or from a severe sprain of the joint or an injury, such as a kick from another horse, it is all right to breed from these.

**Symptoms.**—Are very plain and easily noticed as a general rule. Inquire into the history of the case; how long the animal has been lame and how he acts when travelling. If the horse steps shorter than natural and strikes the toe first in putting down the foot. If he is noticed very lame in starting off after standing for a while, or from being kept in the stable over night, and gets better after he goes a mile or so, and the further he goes the less he shows the lameness, then look for a spavin; if there is an enlargement you can easily see it, but if it is an occult spavin there will be no enlargement, just heat and soreness in the parts. When the above symptoms are present you may be sure it is a spavin. After a time the muscles of the hip on the side he is lame on will waste away from not being properly used. You must not let this symptom mislead you as to where the lameness is. As the enlargement grows the lameness will increase.

**Treatment** is the same as in ringbone. Try and get the diseased joint to become united and form what is called anchylosis of the joint, then the soreness and lameness will disappear; but on account of there being no movement in the joint, the animal will not have as free use of the leg as he had before he got the spavin, but may be very serviceable for a number of years after being cured. High-up spavins are very much harder to cure than the low-down or jack spavin, for they affect the largest articulation in the hock joint, but they are treated just the same. It is always best to first try a blister on spavins before firing, for in some cases a blister will cure them all right. Blister with the following receipt:

Biniodid of Mercury or Red Precipitate. ................2 drams.
Vaseline or Lard.......................................1 ounce.

Mix thoroughly together. There will be enough in this receipt to blister an ordinary sized spavin twice. Apply half of it over the inside of the hock where the enlargement is; rub in thoroughly and grease the parts three days afterwards. If it is an occult spavin, or if the enlargement goes right through the joint, put all the blister on at once, that is, half on the inside and half on the outside of the joint; then let it alone for three weeks and wash off the parts with warm water and soap, and blister again just the same as the first time. Repeat the blister a few times, and, if not somewhat better, fire it with the feather iron, the same as shown in the cut. Have three of these irons and get

FEATHER IRON.

them red hot, then place a twitch on the horse's nose; also have one of his front legs held up, and proceed to fire. Draw the lines the same as seen in the cut, and always have them a good half-inch apart each way. Run the iron lightly over the lines (same as shown in cut) until there is a white line appears, then that is plenty deep enough. The best place to heat your irons is at a blacksmith shop, and if the horse is very ugly you may have to throw him in the same way as if you were going to castrate him, and then fire. Keep on firing and blistering until you effect a cure; it may take from six months to a year or two to effect a complete cure. After firing do not blister for six or seven days. It is all right to work the animal a little between times, it will help on with the cure, but when you are working the horse, shoe him with a high healed shoe, this helps to throw the strain off the hock joint. After firing and blistering, always keep the animal out of the water, for getting the leg wet scalds the hair and skin and makes it very sore.

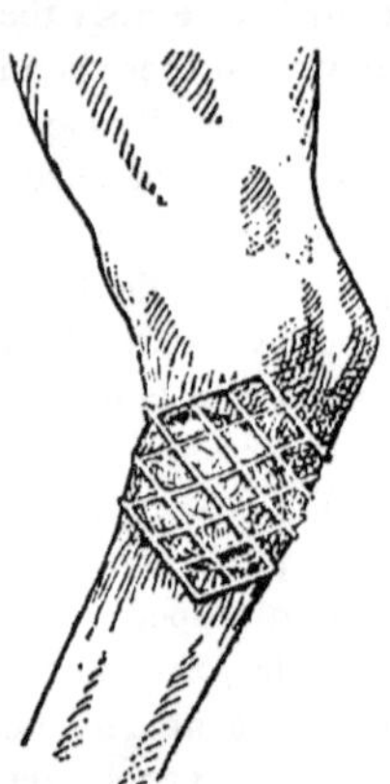

CHAPTER XXI.

# WOUNDS AND TREATMENT.

Wounds are of various kinds, as follows:

(1) **An Incised Wound.**—This is a wound caused by a clean cutting instrument where its length is greater than its depth.

(2) **A Punctured Wound.**—This is a wound caused by some sharp or blunt instrument, and its depth exceeds its length.

(3) **A Lacerated Wound.**—This is a wound where the flesh is both torn and cut, as in a case where one horse kicks another.

(4) **A Contused Wound.**—This is a wound where the skin is not broken, but the tissues under the skin are very much bruised, as, for instance, a man with a black eye. This is one of the best examples of this kind of a wound.

## GENERAL TREATMENT FOR ALL KINDS OF WOUNDS.

If the wound is bleeding, stop it in any of the following ways: By applying cotton batting over the wound and a tight bandage over the cotton batting, and leave it on for twenty-four hours; or, by putting on astringent medicine. The best medicine for this is Monsell's solution of iron; apply it with a feather. Another very good way is to tie the end of the artery tightly with a small string, that is, if you can get hold of the artery. A common sewing needle that is used for sewing up wounds comes in very handy in some cases, by just running the needle in under the artery or vein that is bleeding, and tie up some of the muscle along with the artery, and leave it tied until it comes off of its own accord. If the wound is deep, and you cannot catch the artery, plug the hole with cotton batting and leave it in for twenty-four hours, then examine the wound to see if there is any pieces of stick or anything left in it. If it is a lacerated wound, sew it up after washing the wound out with a little luke warm water to make it nice and clean. The best way to sew the wound is to place a twitch on the horse's nose and have one of his front legs held up, and put in the stitches with a common sewing needle used for sewing wounds, which can be got at any wholesale drug store, and use the small carriage trimmers' twine to sew it up with, which can be got at any hardware store. Before starting to sew up a bad cut, examine as to where you should put in your first stitch, so as to bring the edges of the wound evenly together. Put in the stitches one-half inch apart, tying each stitch separately until you have the whole wound sewed up. Just draw the stitches moderately tight, they will stay in longer than if they were drawn too tight. In case it is a wild animal, throw him down the same as if you were going to castrate him, then sew it up. After stitching up the wound, bathe well with luke warm water twice a day until the stitches come out ; after bathing each time apply the white lotion, this is to keep down the inflammation and swelling. Allow the stitches to gradually rot out of their own accord, then the wound will open up again ; still continue to bathe and apply the white lotion. After applying the white lotion each time after the stitches come out and the wound is opened up some, paint the wound with compound tincture of benzoin or Friar's balsam ; this will keep any proud flesh from growing in the wound. When the wound is just about healed up, and is hard, then change the treatment. Quit bathing and apply the lotion and benzoin. After this, use the green salve twice a day, every night and morning. The best way to apply this salve is to melt a little of it in a spoon by holding it over the lantern blaze until it is thin enough, and then apply it with a feather all over the wound. The idea of using this salve is that it not only heals, but it keeps the skin soft around the wound, and allows the edges to draw well together, and makes the wound look nicer when it is healed. If the wound is deep, and there is not a very large hole in the skin, do not sew it up, but treat it just the same way as you would after the stitches have come out of a wound you have sewed up, only work the medicine up into the wound by using a feather or a syringe. The main thing in this kind of a wound is to get it to heal from the inside first. If you have a wound where the skin is not much torn, mostly bruised, give it lots of bathing with luke warm water two or three times a day. After bathing each time, wipe dry and apply the white lotion to take down the swelling. If you have a case where the wound is of long standing and does not seem to heal, burn it a little with caustic potash, and then treat it the same as an ordinary wound. Sometimes, in treating a wound, the matter forms a pocket and does not run out. In this case you must cut a hole at the bottom of the pocket or sack, where the matter is lodged, and allow it to run out; after this treat, and you will find it will heal up all right. The above treatment will do for wounds on any part of the body except around the eyes or feet. You will find separate treatment for these wounds given along with the diseases of the parts. Look in the index.

### MODES OF HEALING WOUNDS.

There are two ways of healing: 1st, "Healing by the first intention;" 2d, "Healing by the second intention." Healing by the first intention is the way which is mostly used by the medical doctors. This is when the wound is nicely brought together and kept in an antiseptic state and heals from the start without suppurating or running matter. Healing by the second intention is the way that wounds are mostly treated in veterinary practice, for you cannot keep the horse quiet, and the wound soon becomes irritated and begins to run matter, which is a good sign as long as the discharge is of a nice white color. After the wound begins to run matter it gradually becomes knit together. In all cases of wounds keep the horse quiet and feed on soft, light food with plenty of boiled flaxseed in it, this will keep his bowels regular while he is standing so quiet. If there is a thickening left after you have the wound healed up give it a light blistering with the following:

Pulverized Catharides or Spanish Fly.................. 1½ drams.
Lard or Vaseline.................................... 1 ounce.

Mix thoroughly and apply a little of it, according to the size of the enlargement, rub it in well and let it go for three days and then grease the blistered part. In the course of two or three weeks, if it is not gone completely down, wash the parts with luke warm water and soap, then dry and apply more blister. After blistering treat the same as above mentioned; keep on blistering till you get the thickening all out.

### INFLAMMATION OF A WOUND.

This generally follows deep punctured wounds, but might follow other kinds of wounds.

**Symptoms.**—The wound becomes very much swollen and tender, the animal seems feverish and in pain, and the cut discharges watery-looking blood.

**Treatment.**—Keep the animal quiet and warm, and give a teaspoonful of nitrate of potash or saltpetre in his feed twice a day, also feed on soft food to keep his bowels loose. Bathe the wound three times a day with hot vinegar and water—half and half—bathe for half an hour or more, rub dry and apply the white lotion. Swab out the wound by means of a sponge or syringe with the following: Carbolic acid, 25 drops to 1 pint of water. Swab out each time after bathing.

### BLOOD POISONING.

**Causes.**—From a wound being handled with dirty hands or dirty or rusty instruments, or anything that will poison the wound.

**Symptoms.**—The wound becomes very sore, and commences swelling and festering, and he is apt to break out and run matter in different parts of the body.

**Treatment.**—Give the animal

Hyposulphite of Soda........................... 1 teaspoonful.

Give three times a day on the tongue with a spoon or in his feed. This is to kill the poison in the blood. Bathe the wound with hot water three or four times a day and poultice between the times of bathing with hot linseed meal. Swab the wound out by means of a sponge or syringe with carbolic acid and water—25 drops to 1 pint of luke warm water. This is to clean the wound. Any place that the animal breaks out treat it same as the wound.

---

## CHAPTER XXII.

# INJURIES AND DISEASES OF THE HEAD, NECK, SHOULDERS, BODY AND HIPS.

---

### I.—INJURIES TO THE HEAD AND NECK.

### POLL EVIL.

It gets its name from affecting the poll of the head.

**Causes.**—It is caused by an injury of some kind, as a horse striking his head against the top of a low doorway when going in or out of it, or from pulling back on his halter. It is also caused by putting on a heavy poke to keep the animal from jumping.

**Symptoms.**—Are swelling and soreness of the parts, and the horse carries his head stiff, for it hurts him to move that part of his neck. Now, if he does not get relief soon, matter will form and work down around the bones, and then the bones themselves become diseased.

**Treatment.**—If it is treated as soon as it is noticed it is easily done. Just remove the cause and bathe the parts well with warm water and a little vinegar twice a day, and after bathing rub dry, and each time apply the white liniment; also, keep the animal very quiet and feed him out of a high manger. This is to keep the sore parts as quiet as possible until the soreness and swelling is all gone. In a case where matter forms open it with a knife at the lowest part so as to allow the matter to run out. After opening bathe well twice a day and melt green salve and pour it into the cut each time after bathing. In a case where the bones become diseased you will have to throw the animal and cut into it and scrape out the dead bone before it will heal, then treat the same as above mentioned. In case where poll evil gets very bad it generally leaves the horse's neck stiff, and he will not be able to eat off of the ground. In cases of this kind, if the animal is ugly to handle, always put a twitch on his nose while you are dressing it. After you get the parts healed up all right, and if it is thickened, sweat it down by rubbing the parts with the acid liniment, apply every third day after bathing with warm water and salt until the thickening has gone down.

## INJURIES TO THE MUSCLES OF THE NECK.

**Causes.**—Are from pulling back on the halter, or from getting cast in the stall. In some cases it is caused from having been bit by another horse.

**Symptoms.**—Are stiffness of the neck, also swelling and soreness of the parts. In some cases he will carry his neck to one side.

**Treatment.**—Is to bathe well with warm water and salt two or three times a day. After bathing wipe dry and apply the white liniment until the soreness and swelling is all gone. In cases where matter forms open it at the bottom of the lump and allow it to run out, and afterwards treat it as an ordinary wound by applying the white lotion.

## FRACTURE OF THE BONES OF THE NECK.

**Causes.**—Are generally from falling on his head while running away, or in stumbling. It may be done in other ways, as jumping over anything that is high and falling on his head. It may also occur while throwing an animal and allowing him to fall on his neck.

**Symptoms.**—If the fracture causes the bones to press on the spinal cord it causes death immediately by stopping the heart's action. If it is only a piece off of one of the bones the horse will be very sore around the part, and will carry his head to the side the fracture is on. If you go to straighten his neck he will almost fall down.

**Treatment.**—Keep the animal as quiet as you can and feed him out of a high manger. Bathe the parts well with warm water twice a day and apply the white liniment each time after bathing, and the bones, after a time, will unite together. If it does not heal together the part broken will fester and form matter in it, then open up the parts with a sharp knife and remove the piece of broken bone, then treat as a common wound and you will find it will soon heal up.

### SORE ON THE TOP OF THE NECK FROM THE COLLAR.

**Causes.**—Are from a bad fitting collar irritating the neck, or from sometimes taking a horse that is not used to a collar and working him too hard before his neck is hardened to the collar, causing his neck to become scalded and sore.

**Symptoms.**—Are very plain, the top of the neck becomes sore and raw looking, and the horse is afraid to let you handle that part of his neck.

**Treatment.**—Is to wear a nice fitting collar on him and have a regular pad for the top of the neck; dry and clean it every time after using it; clip off the mane around the sore and wash it every night with cold water and salt, then dry it well and apply the white lotion, also apply the lotion every morning, without washing, before you take him out to work; sometimes it is necessary to lay him off work for a few days to get it healed.

---

## II.—INJURIES AND DISEASES AROUND THE SHOULDERS AND WITHERS.

### FISTULOUS WITHERS.

This disease resembles poll evil, and also gets its name from the part it affects.

**Causes.**—Are from riding a horse with a poor-fitting saddle, which bruises the withers; from a horse rolling or getting cast on his back, and in this way bruise the withers and bring on this disease.

**Symptoms.**—Are swelling and soreness of the withers, and if allowed to run on without being treated soon forms matter, which will burrow down around the bones; if it is not attended to soon after it affects the bones the matter that comes from it will have a very bad smell, and little pipes will form which will run down to the bones.

**Treatment.**—If you notice it as soon as it begins to form matter, bathe well two or three times a day, and each time after bathing rub the parts with white liniment; if matter forms, open the swelling up at the very lowest part so as to let the matter run out, then bathe, as above mentioned, and pour hot green salve right into the hole. In a case where the discharge of matter smells very bad, and you think the bone is diseased, throw the animal, cut down to the diseased bone and scrape it, then treat with green salve by pouring it into the cut twice a day after bathing. Both this disease and poll evil, in bad cases, are tedious to treat. In dressing the wound always put a twitch on the horse's nose if he is very ugly. If there is a thickening after you have the parts healed up, rub the parts with acid liniment; apply the liniment every third day after bathing with luke warm water and salt until the thickening is all gone down.

### SWEENY (SHOULDER SLIP).

This is a common occurrence in young horses.

**Causes.**—It is more likely to occur in young horses than in old ones. It may be caused in ploughing, when the plough strikes solid against anything, causing a sudden jerk; or in horses that are used on a tongue and the wheel passes over a stone or rough piece of ground, swinging the end of the tongue around, causing the neckyoke to jerk and bruise the shoulder. Anything that will bruise the muscle of the shoulder will bring it on.

**Symptoms.**—They are well marked. As soon as the horse receives the injury there will be a slight swelling and soreness of the part, afterwards there is a wasting away of the muscles that are injured. It is generally the muscles lying over the shoulder blade that become affected with this disease, and in some cases the muscles fall away until there is a large hollow over the shoulder blade. In some cases the horse is not very lame, but goes a little stiff.

**Treatment.**—If the animal is at very heavy work, change it to light work, and if you can allow him to run without work so much the better. Bathe the parts well with cold water and salt every night if in the summer; after bathing, wipe dry and hand rub and pull the skin out on the hollow; this is to get the skin loose from the muscle. Every third day after bathing apply the acid liniment in and around the hollow part of the shoulder; this is the best remedy known. Continue this treatment until the muscle becomes its natural size and the shoulder is filled out. It sometimes takes the muscles a long time to fill out to their natural size, but, as a general thing, it gets all right in the course of time. If you can, turn him out to pasture, then blister in and around the hollow place with the following:

| | |
|---|---|
| Pulverized Cantharides or Spanish Fly.................. | 1½ drams. |
| Vaseline or Lard............. ......................... | 1 ounce. |

Mix well together and rub it all in to the parts, and grease three days after with lard, and repeat blister every three weeks until cured.

## SCALDED OR GALLED SHOULDERS.

**Causes.**—Generally from the collar getting hard from the sweat drying on it; or from a bad-fitting collar—either too big or too small. Sometimes, in the spring of the year, when the horse's shoulders are tender and not used to work; by putting the horse to hard work scalds and makes the shoulders sore.

**Treatment.**—Use a good-fitting collar, and keep it clean by brushing or scraping the dirt off it before using. While you are working your horses, if you are going to allow them to stand a few minutes, raise the collars so as to allow the air to get to the shoulders, also keep the mane from getting under the collar. Every night, after working, bathe with cold water and salt, and after wiping dry, if they are sore, apply the white lotion; this will heal and harden the skin.

## BRUISES OF THE SHOULDER.

**Causes.**—This is caused from a bruise of any kind.

**Symptoms.**—A swelling appears around the bruised part soon after it is done, and when you come to examine it, it is found to contain water, or serum.

**Treatment.**—Cut into the swelling at the bottom part of it and allow the water to run out; make a hole large enough for you to put in your finger; after opening, bathe well with luke warm water twice a day; after bathing, wipe dry, and apply the white liniment around the swelling, and, as well as doing this, dress the inside of the swelling, to get it to heal from the inside, with compound tincture of benzoin or Friar's balsam, or you might use green salve instead of the benzoin to put in the wound; insert it up into the hole with a feather.

### TUMORS OR LUMPS ON THE SHOULDER.

**Causes.**—From the animal receiving a bruise of any kind on the shoulder, and not being properly treated afterwards, it turns into a hard, callous lump.

**Treatment.**—The best treatment is to take a sharp knife and skin the lumps right out; after this, sew up the skin with a needle and twine, and treat them as a lacerated wound. There is not much danger in cutting around the shoulder, for there are no large vessels where these lumps are generally found.

### SHOULDER JOINT LAMENESS.

This is a very bad lameness, and the seat of the trouble is generally situated where the large muscle of the shoulder passes down through the pulley-shaped part on the bones on the front of the shoulder joint.

**Causes.**—From a severe sprain of the joint, and is generally caused by a horse becoming cast in his stall; or from going through deep snow; or from falling on his shoulder; or from a kick from another horse; and sometimes it is caused from a kind of rheumatic inflammation settling in the joint.

**Symptoms.**—When the animal is trotting he will step short on the affected leg, and will often strike the toe and stumble, on account of not being able to lift the leg up properly. In standing, he will bring the foot well under him, resting on the toe, allowing the knee to bend forward; this is to give the shoulder a rest. If you pinch him around the shoulder, he will flinch from soreness.

**Treatment.**—If the case is allowed to run on for a long time, and the bone becomes diseased, it is incurable; but if taken in time, you may effect a cure by bathing the shoulder with luke warm water and salt twice a day. After bathing, wipe dry and rub well with white liniment. Keep the animal in the stable and keep the shoulder as quiet as you can. After the soreness is pretty well out, blister with the following blister:

| | |
|---|---|
| Powdered Cantharides or Spanish Fly.................. | 1½ drams. |
| Vaseline or Lard. .................................... | 1 ounce. |

Mix and apply half of this mixture around the front and side of the shoulder joint; rub in well and allow it to stand for three days, then grease with lard. After two or three weeks, if the animal does not seem better, wash the shoulder off and blister again, following the same instructions as given before.

## SORE BACK.

**Causes.**—From a bad-fitting harness; or from the backing of the harness rubbing on it.

**Treatment.**—In all cases remove the cause, and, if in the summer time, bathe with cold water and soap; if in the winter, use warm water and soap; do this twice a day. After bathing each time, wipe dry and apply the white lotion.

## CHRONIC SORES ON BACK OR SHOULDERS (SIT-FAST).

**Causes.**—From working and neglecting a sore shoulder or back.

**Symptoms.**—The sore spot has no tendency to heal and keeps raw all the time while working.

**Treatment.**—Burn the sore with caustic potash and afterward treat by keeping the animal from work, bathe with luke warm water, and, after bathing, apply the white lotion twice a day until it is healed.

## BROKEN BACK.

**Causes.**—From an animal receiving a fall of some kind.

**Symptoms.**—If the break is near the front of the bones of the back it will cause instant death, but if it occurs further back it will cause paralysis of the hind quarters.

**Treatment.**—There is none but to destroy the animal.

## INJURIES TO THE MUSCLES OF THE BELLY.

**Causes.**—Sometimes when a horse steps on a sharp piece of stick it turns up and catches the muscles of the belly, or from the hook of a cow, or from the kick of another horse.

**Treatment.** If it is a lacerated wound, stitch it up, and treat it same as given for lacerated wounds; if it is a punctured wound, and only has a small hole in the skin, examine to see that there is nothing in it, and then use the same treatment as that given for punctured wounds. If the wound is severe enough to allow the bowels to come out, then it is a more serious case. If the bowels are out to any great extent, and are injured by the horse tramping on them, it is best to destroy him at once, but if the bowels are only out a little, and are not injured or blackened, throw and secure the animal, and wash the bowels off with a little luke warm water and shove them back into their place, then stitch up the wound good and tight, afterward bathe the wound twice a day with luke warm water and apply the white lotion each time after bathing. If the opening is very large it is well to wrap a bandage of factory cotton around the body to help to support the stitches. Keep the animal quiet and feed on soft food until the wound heals up.

### STAKED.

This is a very common occurrence among horses.

**Causes.**—The animal steps upon a piece of stick and it turns up and catches him, or from jumping a fence and getting caught, or running against a stake sticking up in a field, or a plow handle or anything that the animal will run against.

**Treatment.**—Pull the stake out if it is still stuck in the horse, and examine it to see that there is no pieces of the wood left in the wound. There is never much bleeding in a case of this kind, but if it does, stuff the hole with cotton batting and leave it in for twenty-four hours; after this time it will be safe to take it out, then treat same as for punctured wound. Keep the animal quiet until it begins to heal, and feed on soft food with plenty of flaxseed in it to keep his bowels loose.

### HIPPED.

This is when the point of the hip bone is knocked down.

**Causes.**—From running through a narrow doorway and striking the side of it, or from falling on hard ground, or anything that will strike the point of the hip hard enough to break a piece off the bone.

**Symptoms.**—At the time it happens there will be swelling and soreness around the point of the hip. After it gets well you will notice that the injured hip is not as large as the other, and it is a nasty eye-sore on a horse.

**Treatment.**—The only thing to be done after it is knocked down is to bathe and apply the white liniment after bathing until you get the swelling and soreness out, then leave it alone. If it is a case where the hip swells up and begins to fester around the broken piece of bone, cut into it and take the piece out, then treat same as lacerated wound.

### FRACTURE OF THE HIP BONES.

This is generally caused from a horse slipping and falling on ice, or may be done in any other way when the animal receives injury enough to fracture the bones.

**Symptoms.**—There is severe lameness, and if you examine closely by twisting on the bones you will hear the broken ends of the bones grind on each other. In a short time the parts around the fracture become greatly swollen.

**Treatment.**—If the fracture is very bad it is best to destroy the animal, but if you want to try to save it, all that can be done is to keep the animal very quiet and bathe the parts well with luke warm water and vinegar, after rubbing dry, rub well with white liniment twice a day, and in some cases this will affect a cure in the course of a few weeks.

### FALLING AWAY OF THE MUSCLES OF THE HIP AFTER FOALING.

**Causes.**—The muscles become injured, while the mare is foaling, by throwing herself down or striking herself against anything and bruising the muscles on either hip.

**Symptoms.**—The mare will be noticed to be stiff and sore after foaling, and after a few days there will be a falling away of the muscles of whichever hip was injured. The lameness varies, in some cases it is worse than others.

**Treatment.**—The best and only treatment for this is to let the mare run out and apply a fly-blister to the wasted muscles. Use the following blister :

| | |
|---|---|
| Pulverized Cantharides or Spanish Fly | 1½ drams. |
| Vaseline or Lard | 1 ounce. |

Mix well and apply all over the wasted muscles, rub in well and tie the mare so she cannot bite or rub the blistered parts for a few hours, then turn her out and grease in three days afterward, allow it to go for a month, and if she is not better blister again, and keep on blistering every month until she is better.

### SORE TAIL FROM THE CRUPPER.

**Causes.**—From reining the horse too high, which draws the crupper too tight, or from the horse sweating and allowing it to get dry and hard on the crupper, or anything that will irritate the tail.

**Treatment.**—By not using the crupper under his tail for a few days you will have removed the cause, and to heal it wash the parts off with luke warm water and soap ; after bathing it once apply the white lotion twice a day, without bathing, until it is healed up.

### FRACTURES OF THE BONES OF THE TAIL.

**Causes.**—Usually from a horse rearing up and falling back on his haunches.

**Treatment.**—Keep the animal quiet, and if it is the season for flies, cover the animal so it will not have to use its tail in trying to keep them off; if the tail is much swollen rub with white liniment twice a day until the swelling is gone down, then leave the animal quiet until the bones unite. If there is not much swelling bandage moderately tight and leave the bandage on a few days at a time, while the bones are uniting.

### HIP JOINT LAMENESS.

This is a sprain of the round ligament in the hip joint.

**Causes.**—The way this disease generally occurs is from a horse stepping on a small, round stone, and the stone turns and throws the leg forward, or from slipping on ice, or falling in other ways and striking the hip.

**Symptoms.**—The horse steps short on the affected side, and in trotting he goes kind of three-cornered, and every time the horse raises his leg the hip raises with it, showing that he is trying to keep the hip quiet; he will also flinch when you press around the hip joint. If the disease is allowed to run on for some time there is a wasting of the muscles around the part.

**Treatment.**—If it is an old standing case, and the joint is diseased, it is incurable, but if taken in time the animal may be cured by keeping him quiet and bathing the parts well with luke warm water and salt twice a day; after bathing, each time, wipe dry and apply the white liniment. Keep on with this treatment until you get the soreness and swelling out, then blister with the following:

Pulverized Cantharides or Spanish Fly.................1½ drams.
Vaseline or Lard.........................................1 ounce.

Mix well and there will be enough to blister the hip three times, each time rub the blister in well, allow it to go for three days, then grease with lard; let it go for three or four weeks, then blister again; repeat the blister this way until the animal is over the lameness.

### SPRAIN OF THE MUSCLES OF THE HIP.

This is more common than hip joint lameness, and is usually caused in heavy horses by slipping while pulling a heavy load; or while driving light horses, you pull them up suddenly, causing them to slip, and in this way they sometimes sprain the muscles of the hip.

**Symptoms.**—There is swelling and soreness of the muscles, which will be best noticed while standing behind the animal, and there will be great difficulty in bringing the leg forward.

**Treatment.**—Keep the animal quiet; bathe well with luke warm water and salt two or three times a day; after bathing, wipe dry and rub the parts well with white liniment. Follow up this treatment until the animal is better.

---

## CHAPTER XXIII.

# DISEASES OF THE FRONT AND HIND LEGS.

### CAPPED ELBOW.

This is a thickening at the back of the elbow joint.

**Causes.**—From shoeing and leaving a long heel on the shoe, which catches the animal in that part while he is lying down; or from wearing a heavy belly band on the harness, which rubs against and irritates the elbow.

**Treatment.**—If it is caused by the shoe, have him shod with a short-heeled shoe, and, for a time, tie an old bag around the foot, so when he is lying down the foot will not irritate the elbow. Treat with the following as con as noticed: Bathe well with luke warm water and vinegar twice a day. If it is not raw, only swollen, use the white liniment after bathing; but if it is raw, use the white lotion; this will generally bring it all right. In some cases it festers and matter forms in it; in this case open it up with a knife at the lowest part of the swelling, and treat it the same as above mentioned, only melt a little green salve and put it in the hole each time after dressing it until it heals. If it is of long standing, and there is a large, hard lump formed on the cap of the elbow, the only treatment is to throw the animal and secure him, and skin the lump out carefully, which can be done without much bleeding. After you cut the lump out, draw the skin together with a few stitches and treat the same as a lacerated wound, by bathing with luke warm water and applying the white lotion twice a day until it heals up.

### FRACTURE OF THE BONES IN THE SHOULDER.

**Causes.**—Generally from falling, or from running against somethi g.

**Symptoms.**—There will be swelling and soreness of the parts, and the animal will be very lame, and, on moving the shoulder, you will hear the ends of the bones grating on each other.

**Treatment.**—In most cases it is best to destroy the animal, but if you attempt to treat it, keep the animal quiet and bathe with luke warm water twice a day. After bathing, apply the white liniment; this is to keep down the swelling and inflammation. The treatment in this case is of very little success. The parts become badly swollen and the animal is in severe pain, and it is best to have him destroyed.

### INJURIES AND SPRAINS OF THE KNEE.

**Causes.**—Generally from the animal falling and hurting the knees.

**Symptoms.**—The animal cannot bend the knee in walking, and when you handle it, it seems very sore.

**Treatment.**—Keep the animal quiet, and bathe twice a day with luke warm water and salt, and after bathing apply the white liniment; but if the knee is cut, use the white lotion after bathing. If the cut is large, put a few stitches in it, and treat as above mentioned.

### STRIKING THE KNEE (SPEEDY CUT).

This is where the animal strikes the knee with the opposite foot.

**Causes.**—This is caused generally by bad shoeing, or using too heavy a shoe on colts; some horses with high action have a tendency to strike the knees.

**Symptoms.**—There is swelling on the inside of the knee; it is very painful, and as soon as you go to handle it the animal will jerk his knee away from you. In some cases it becomes very much swollen and forms matter in it. When the horse is walking, in bringing his leg forward he will swing it out and can hardly get along, it being so sore; while in other cases, where he does not strike it so severe, the inside of the knee becomes thickened and hard.

**Treatment.**—Leave the animal off work, and bathe the parts well with luke warm water and salt two or three times a day; the more bathing the better. Each time after bathing, rub dry and apply the white liniment. If there is matter formed, open it up with a sharp knife to allow the matter to run out, then treat as above mentioned. After it is healed up, if there is a thickening, blister the thickening with the following:

Pulverized Cantharides or Spanish Fly ..................2 drams.
Vaseline or Lard..........................................1 dram.

Mix well and rub what you can nicely get on over the swelling; rub in well and grease the third day; let it go for a couple of weeks, then blister again, and repeat the blistering until the thickening is gone. After this, be careful how you have the animal shod, and if he is inclined to strike, wear a knee boot on him.

### FRACTURE OF THE BONES OF THE KNEE.

**Causes.**—From falling on the knees.

**Symptoms.**—The animal will be very lame, the knees become swollen and is hot and tender.

**Treatment.**—Keep the animal quiet and bathe well with luke warm water—the more the better—after bathing, each time apply white liniment. It generally takes a case of this kind four or five weeks to get better. After the soreness and swelling has gone down it is well to blister a few times with the fly blister to get the soreness and thickening out of the knee.

### PUFFY ENLARGEMENTS AROUND THE KNEES AND FETLOCKS (BURSAL ENLARGEMENTS).

**Causes.**—From striking the knee against the manger while pawing, or from lying on the hard floor, or anything that will bruise the knee or fetlock.

**Symptoms.**--There is a puffy enlargement, but not sore to handle, and it feels as if it was full of oil.

**Treatment.**—Bathe well once a day with cold water and salt, rub dry and bandage for three hours. After you remove the bandage—every third day—apply acid liniment, which will make, as a general thing, a complete cure. Never attempt to open a Bursal, for it will allow the oil that is in it to run out. In all cases, if you find the cause, remove it.

### SPLINT AFFECTING THE KNEE.

This is where the splint is very high up on the bones and affects the knee.

**Treatment.**—Use the same treatment that is given for splint, but in some cases it is very hard to effect a cure.

### KNEE SPRUNG.

This is when the knee is bent forward.

**Causes.**—From hard and fast work, or from standing in a stall that has a big slant, high in front and low behind, or from feeding out of a very high rack, or from a sprain and contraction of the back tendons, or a horse being shod with high heels will cause it. Horses with weak, small knees are more subject to this than horses with good, strong straight knees. An animal may be very badly knee sprung and still be a good work horse.

**Treatment.**—If you want to work the animal and treat it, first fix the floor, see that it is level and also feed him off the floor. Bathe the legs with cold water and salt twice a day, after bathing rub dry and apply the white liniment every night, after the liniment is dried in, bandage for a few hours, shoe the animal with a flat shoe, if the roads are not slippery. If you do not want to use the horse turn him out and blister the back cords once a month with the following:

| | |
|---|---|
| Pulverized Cantharides or Spanish Fly | 2 drams. |
| Vaseline or Lard | 1 ounce. |

Mix and there will be enough in this to blister both legs once, rub in well and tie his head short so he can he can not bite it for a few hours, then in three days, grease it and let him run for a month, and repeat the blister until he is better.

### CALF-KNEED.

This is not a disease but a fault in the formation of the knee, this is where the animal stands with his knees bent back.

**Treatment.**—There is none; only in buying a horse watch that he is not calf-kneed for they are bad stumblers.

### SPRAIN OF THE BACK TENDONS.

There is more or less swelling around the sprained tendons, the animal will be lame in travelling, and if you press on the cords he will flinch.

**Treatment.**—Bathe with cold water and salt if in summer, but if in winter bathe with luke warm water and salt twice a day, rub dry and apply the white liniment, and an hour or so after applying the liniment, bandage, leaving the bandage on for three hours. After he is pretty well over the lameness, and if there is a thickening left, blister with the following:

| | |
|---|---|
| Pulverized Cantharides or Spanish Fly | 2 drams. |
| Vaseline or Lard | 1 ounce. |

Mix and rub enough in to blister him well, and grease the third day. Keep the animal quiet while treating him.

### SPRAIN OF THE FETLOCK JOINT.

**Causes.**—From the horse stepping crooked or turning over on his ankle.

**Symptoms.**—The animal will be very lame, the joint will be swollen and sore to handle.

**Treatment.**—Use the same treatment as given in sprain of the back tendons.

## KNUCKLING.

**Causes.**—From hard and fast work. Horses that stand straight on the fetlock are more apt to knuckle.

**Symptoms.**—There will be a jerking forward of the fetlock every time the animal steps, and in standing the animal stands with his fetlock joint bent forward instead of back.

**Treatment.**—If it is a case of long standing there can be nothing done, but if it is only coming on blister him around the fetlock and turn him out. Blister with

| | |
|---|---|
| Powdered Cantharides or Spanish Fly | 2 drams. |
| Vaseline or Lard | 1 ounce. |

Mix and apply around the fetlock, rub the blister in well and allow it to stand for three days, then grease and let it go for a few weeks and blister again ; repeat the blistering until the joint gets strong. There will be enough in the above prescription to blister two fetlocks.

## KNUCKLING IN COLTS.

**Causes.**—From being left in the stable without getting much exercise during the winter and in the spring he will be knuckled in the fetlocks.

**Treatment.**—Before turning him out blister around the fetlocks well with the above blister, and in three days afterwards grease and turn him out to pasture.

## INTERFERING, OR STRIKING THE FETLOCKS.

This trouble is mostly met with in colts when they are first shod.

**Causes.**—Sometimes from shoeing with too heavy a shoe and putting the horse on a long trip before he is used to it. Some horses naturally travel so close behind that they brush the legs together.

**Symptoms.**—Soreness on the inside of the fetlock, and sometimes the hair and skin will be knocked off, and in severe cases will swell badly When the horse, in travelling, strikes himself he will hop on three legs with pain for a little piece, then be all right until he strikes again.

**Treatment.**—Bathe the leg well with lukewarm water and a little castile soap twice a day, if it is swoolen, after bathing rub dry and apply the white lotion. Have the shoes changed and made light. If the animal is in poor condition, feed him well and do not work him too hard until he gets strong and in good shape for his work. While you are driving, it is well to wear an interfering boot on the leg, and by careful shoeing and getting him strong and able for his work, he generally gets over it all right.

### WIND GALLS.

These are little puffy swellings at the back part of the fetlock joint.

**Causes.**—From hard driving, in some horses it comes on easier than others.

**Symptoms.**—It does not generally lame the animal, but it is an eye sore. These little puffy swellings are full of oil, which comes from the bursal that secretes the oil which lubricates the back tendons where they work over the back of the fetlock.

**Treatment.**—If in the summer, bathe the legs well every night with cold water and salt, then rub dry and put on a bandage ; leave this on for a couple of hours every night after bathing, and every third night give the leg a good rubbing with acid liniment after taking the bandage off. If this does not cure him in a few weeks, blister with the fly blister used in knuckling, and treat him the same after blistering.

### THICKENINGS AROUND THE FETLOCK.

**Causes.**—This is generally from hard work and from sprains of the joint, and is often seen in livery horses.

**Treatment.**—Blister the same and use the same treatment afterwards as is given for knuckling, only blister heavier.

### FRACTURES OF THE BONES BELOW THE KNEE.

**Causes.**—Driving a horse fast along a very hard road when he is feeling good, will sometimes fracture these bones, or in jumping a fence, or in running away, or getting the leg caught, or anything that will give the leg a blow hard enough to fracture the bones.

**Symptoms.**—The horse will be very lame and will hold the leg up with pain, and by taking hold of the leg and twisting it, you will hear the ends of the fractured bones grate together.

**Treatment.**—If it is in an old animal destroy him at once, but if it is a colt or young horse, keep him quiet in slings, after you get the leg straight, then apply a starch bandage, which is a bandage wrung out of starch, put a good lot of the bandage on and hold the leg straight until the bandage hardens, after that it will hold the broken leg straight ; leave it on for four or five weeks until you are sure the bones are well knit together.

### BREAK DOWN.

This is where the tendons and ligaments at the back of the fetlock give way and allows the fetlock to drop down almost on to the ground. This is mostly seen in running or racing horses where they are put to very severe exertion.

**Treatment.**—You cannot restore the fetlock back to its natural state, but blister once a month with the following:

| | |
|---|---|
| Powdered Canthardies or Spanish Fly.................. | 2 drams. |
| Vaseline or Lard..... ........... .................... | 1 ounce. |

Mix and rub about half of this in along the back of the fetlock, allow it to stand for three days and let the animal run out to pasture. Repeat the blister until you get him pretty well over the lameness. If he is very lame, better keep him in the stable for a while, and bathe with warm water and salt twice a day after bathing, apply the white liniment until he is pretty well over the lameness, then blister and turn him out.

### SPRAIN OF THE STIFFLE JOINT.

**Symptoms.**—The animal cannot bring the leg forward, and it is very sore when you press around the joint.

**Treatment.**—Bathe twice a day with hot water and vinegar, with a little salt in it, after bathing wipe dry and apply the white liniment. Keep the animal quiet to give the sprained parts rest.

### DISLOCATION OF THE PETELLA OR STIFFLE OUT.

This is where the petella or stiffle bone slips out of the pulley-like process of bone in front of the stiffle joint, and as soon as this slips out it locks the joint so the animal cannot move it.

**Causes.**—It generally occurs in young colts or young horses that are worked hard and get down thin; it comes on by the animal slipping off the end of a plank, or slipping while in the act of getting up.

**Symptoms.**—The horse will act like an animal with its foot nailed to the floor; it cannot get it forward or back. When you feel around the joint it will be drawn and hard, and you can see the joint bone is out too far, and when it is left out for a long time the joint becomes swollen.

**Treatment.**—It is very simple to treat in most cases. Have an assistant to hold the animal's head and another to pull the affected leg well forward while you place your hand against the joint and shove it in toward the animal as hard as you can, until the bone goes into its place, after which the animal can move his leg; as soon as it is in walk the animal on a level piece of ground until the affected parts regain their strength. If it slips out the second time it is easier put in, by pressing it the same way. Bathe the parts with warm water and salt, after bathing rub dry and apply the white liniment twice a day until the joint gets strong and the soreness is all out. If the animal is in poor condition feed well and try to get his strength up.

### PARTIAL DISLOCATION OF THE STIFFLE.

This generally occurs in young foals when they are running over a rough pasture field, or in colts in the spring of the year when they are very weak. At first, when noticed, the stiffle will slip out and the leg will be locked for a hop or two, then it will slip in and he will go on all right again. The stiffle keeps slipping in and out for some time until the bones gets diseased and weakened, and then the bone gets about half way out and stays there. After this the joint will be larger than natural and the animal will never have proper action again.

**Treatment.**—As soon as it is noticed is the time to treat it. Place the animal in a level place where he can not run around much and feed well to get him strong as soon as you can; apply a light blister around the stiffle joint made of the following:

Pulverized Cantharides or Spanish Fly..................1½ drams.
Vaseline or Lard.....................................1 ounce.

Mix and apply a light coat of the blister over the joint, rub in well and grease the third day. Let it go for a month then blister again and repeat this every month until the stiffle is good and strong.

### FRACTURE OF THE THIGH BONE.

**Causes.**—Generally from a kick of another horse, and, although the bone itself is fractured, the coverings will hold the bone to its place in some cases.

**Symptoms.**—For a few days after the kick the animal will seem lame and you may not suspect a fracture of the bone, but all at once the covering of the bone gives way and the horse will almost fall, and when you come to examine it you will see the leg is fractured, for you can swing the leg around.

**Treatment.**—When the bones give right away, destroy the horse, but if the horse gets a severe kick and is a little lame and you are afraid of fracture, keep the animal perfectly quiet, bathe with warm water and salt and after bathing apply white liniment, to get soreness and inflammation out. By doing this, the covering of the bone is so strong that it will hold the bones together until it knits together and he gets all right.

### SPRAIN OF THE MUSCLES ON THE FRONT OF THE HIND LEG, BETWEEN THE HOCK AND STIFFLE JOINT.

**Causes.**—Similar to other sprains.

**Symptoms.**—The muscle will be swollen and sore in moving the animal forward, when he lifts the leg and goes to bring it forward—instead of it going forward it will fly backwards and upwards. The muscle that is affected is called the flexor metatarsi muscle.

**Treatment.**—Keep the animal very quiet, bathe well with luke warm water and salt three times a day, and after bathing rub dry and apply the white liniment.

### SPRAIN OF THE MUSCLE ON THE INSIDE OF THE HIND LEG RUNNING UP FROM THE HOCK.

**Causes.**—From a severe sprain of the leg by slipping in drawing.

**Symptoms.**—There is thickening of the muscles above the hock, and the animal will be stiff for a few days, the stiffness will soon disappear, but it will leave a thickening if not treated.

**Treatment.**—Leave the animal off work and bathe the parts twice a day with luke warm water and salt, if in the winter, but, if in summer use cold water and salt ; after bathing rub dry and apply the white liniment, after the animal gets over the stiffness and soreness, blister to get the thickening out, using the following:

Pulverized Canthardies ................................2 drams.
Vaseline or Lard ........................ ............. 1 ounce.

Mix and apply about half the amount, rub it in well and let it go for three days, then grease. Repeat the blister in three weeks and repeat it every three weeks until the thickening is all gone down. Turn the animal out while you are treating it.

### BOG SPAVIN.

This is a puffy enlargement partly on the inside and partly on the front of the hock joint. There is an over amount of joint oil secreted in the joint, and this bulges out the capsular ligament at this part of the joint.

**Causes.**—Horses with round, meaty joints are most subjected to this. Keeping young colts in the stable and feeding them high without much exercise, or fast work or strain of the joint will cause it.

**Symptoms.**—There is a puffy enlargement on the inside and front of the hock joint sometimes larger than others, by feeling it you can tell it is full of oil. If it comes on from a sprain the animal will be lame for a few days, but this passes off and leaves an enlargement.

**Treatment.**—If the animal is lame, bathe with luke warm water and salt twice a day, in warm weather use cold water, and in cold weather use warm water, after bathing rub dry and apply the white liniment. Keep up this treatment for a few days until he is over the soreness and lameness, then blister with the following, and turn the animal out.

| | |
|---|---|
| Pulverized Canthardies or Spanish Fly | 2 drams. |
| Vaseline or Lard | 1 ounce. |

Mix and there will be enough to blister two or three times according to the size of the bog, rub in well and in three days grease. Repeat the blister in a few weeks as soon as the skin is nicely healing up. It needs to be blistered several times before you get it all down. Often where there is a bog spavin there is a thoroughpin in connection with it. For further particulars refer to last paragraph under the heading of capped hock.

### THOROUGHPIN.

**Causes.**—Similar to those given for bog spavin and the enlargements are on each side of the hock near the back.

**Treatment.**—The treatment is the same as given for bog spavin. Never, under any circumstance open either a bog spavin or a thoroughpin, for if you do, you will have a case of open joint to deal with.

### CAPPED HOCK.

This is a common disease. It is a swelling or thickening on the cap of the hock.

**Causes.**—From an injury of some kind, such as from kicking in the stable and striking the hock against the stall, or some horses will do it in laying down.

**Symptoms.**—Swelling on the cap of the hock, which is generally soft but not sore to handle, and is of a puffy nature, for it is mostly oil that causes the enlargement on account of the little bursa being injured, it secretes too much oil and that is what causes the enlargement. It does not lame the horse or interfere with his usefulness, but it is a bad eye-sore.

**Treatment.**—If it is caused from kicking the stall, place the horse so he cannot strike it. If being treated as soon as it is done, bathe with luke warm water and salt, rub dry and apply the white liniment. Keep this treatment up until the soreness and swelling is all out, then blister the thickening with the following:

Pulverized Cantharides or Spanish Fly..................1½ drams.
Vaseline or Lard..........................................1 ounce

Mix thoroughly and there will be enough to blister one cap four or five times; blister light and blister about every two weeks; grease the third day after each blister. In case you want to use the animal bathe the parts once a day with cold water and salt, rub dry and apply the acid liniment every third day after bathing. The acid liniment is also good when used in this way for bog spavins and thoroughpins when you want to work the horse.

## SPRAIN OF THE HOCK JOINT.

**Symptoms.**—The horse is very lame and does not use the hock right in bringing the leg forward. There is swelling and heat around the joint.

**Treatment.**—Bathe well with hot or cold water, according to the season of the year, after bathing, each time wipe dry and apply the white liniment; bathe twice a day and keep the horse quiet.

## FRACTURE OF THE BONES OF THE HOCK.

Fractures of these bones do not occur very often.

**Symptoms.**—The animal will be very lame, and will hold the leg up with pain; the joint will become swollen and very sore to touch, and by working the joint you can hear the grating sound made by the broken bones.

**Treatment.**—If it is a very bad fracture destroy the horse, but if not, and a young animal, keep him quiet and bathe well with luke warm water and salt three or four times a day after bathing, rub dry and apply the white liniment, this is to keep down the inflammation and swelling. It is always best in treating any kind of a fracture to keep the animal in slings, but do not sling him entirely off his feet, just enough to ease the weight off his legs. After a few weeks the bones will become united, but there will be lameness and soreness in the joint, which is best relieved by blistering with

Pulverized Canthardies or Spanish Fly.....................1 dram.
Biniodide of Mercury or Red Precipitate...................1 dram.
Vaseline or Lard.............................................1 ounce.

Mix thoroughly and there will be enough to blister twice, rub a little in on each side of the joint, rubbing it in well, grease the third day and let it alone for three or four weeks, then blister as before. Let the animal have a good rest by turning him out to pasture.

### BLOOD SPAVIN.

This is an enlargement of the vein which runs down in front of the hock joint.

**Causes.**—From a weakened state of the wall of the vein.

**Symptoms.**—There is an enlargement of the vein, but the animal is not lame, nor it does not hurt the animal for any kind of work, but it is an eye sore.

**Treatment.**—There is no treatment for it.

### OPEN JOINT.

This may affect any of the joints of the legs or body, and from some injury the joint becomes punctured, allowing the joint oil to leak out.

**Causes.**—Generally from a kick, or prod with a fork, or running against some sharp obstacle.

**Symptoms.**—There will be lameness and swelling around the injured joint and leaking from the hole, of an oily looking substance which is the synovia or joint oil. If the leaking is not stopped, inflammation gets into the joint and destroys the joint, and if the horse does get better, it leaves him with a stiff joint.

**Treatment.**—The larger the joint the more troublesome it is to treat, and the more danger there is of loosing the animal. Do not bathe or apply any liniment, for it only increases the flow of the joint oil. If the case is taken in time by applying Monsell's solution of iron in and around the hole every couple of hours with a feather, in most cases it will completely check the running of the oil, and the animal will recover all right. This is the best treatment known for open joint. In cases where the hole is very large and the bone injured, it is best to destroy the animal. Keep the animal quiet and feed on soft food with plenty of boiled flaxseed in it to keep the bowels open. Sometimes after the joint seems healed up for a couple of weeks it will break out again; treat this the same as at the first time until it closes the hole up. After the joint is healed up and it is still swollen, bathe with cold water and salt once a day, and bandage for a couple of hours after bathing, and every third day, after taking off the bandage, rub the joint well with acid liniment.

### TUMORS AND CANCERS.

They are very rare in the horse but are fully described in tumors and cancers in cattle.

### CURB.

This is rupture and enlargement of the ligament that runs down the back part of the hock joint—this ligament receives the name of calcaneocuboid ligament.

**Causes.**—Horses that have crooked or curby legs are more liable to this than horses with straight legs. It generally comes on from a severe sprain by slipping while drawing or driving, or, from rearing up or backing the animal forcibly with a heavy load or in deep snow.

**Symptoms.**—It is easily detected by looking at the hind leg at the side, you will see the enlargement at the back of the hock, or by running the hand down over the back of the hock joint you can feel it. When the curb is first sprung on, the animal will be lame, and, in travelling, he will step long, somewhat similar to ringbone lameness. After the animal rests a day or two, and is driven, at first, he will not be very lame, but after driving a while he becomes very lame, and when allowed to stand he rests the leg by standing on his toe and throwing the fetlock forward. After a time the animal will get over the lameness if not treated, but the enlargement will remain.

**Treatment.**—Get the soreness and lameness out of the ligament by bathing with luke warm water and salt twice a day, if in winter, but if in summer cold water and salt ; after bathing rub dry and apply the white liniment. When the soreness and swelling is all out of the ligament, blister with the following to reduce the thickening.

| | |
|---|---|
| Pulverized Cantharides or Spanish Fly.................. | 2 drams. |
| Vaseline or Lard...................................... | 1 ounce. |

Mix, and there will be enough in this to blister three times. Cut the hair off around the curb and rub one-third of the mixture in well, in three days grease and let it go for a couple of weeks, then wash the leg off with luke warm water and soap and blister again, then grease as before, repeat this blistering every three weeks until the enlargement is all gone. During the treatment do not work the animal, but if you must work him keep him well shod so he will not slip.

### RHEUMATISM.

This is a kind of chronic inflammation in the tendons and ligaments around the joints, and may affect any joint of the body.

**Causes.**—It often follows weakening diseases when the animal is allowed to run out and lay on the cold, damp ground or from bad blood, when there is too much acid in it.

**Symptoms.**—There is a slight soreness, swelling and lameness in the joints, changing from one joint to the other, and is noticed to be worse in damp weather.

**Treatment.**—Rub the affected joints well two or three times a day with white liniment ; also give the following :

| | |
|---|---|
| Salicylic Acid.................................. | ¼ pound. |
| Nitrate of Potash or Saltpetre...................... | ¼ " |
| Common Soda .................................. | ¼ " |

Mix thoroughly and give a large teaspoonful three times a day. This medicine acts like a charm in rheumatism.

---

## CHAPTER XXIV.

# DISEASES OF THE FEET.

### ACUTE FOUNDER (LAMINITIS).

This is inflammation of the sensitive part of the foot, or what is commonly called the quick of the foot.

**Causes.**—This disease is very easily brought on by driving or working a horse hard and then giving him a cold drink of water, or allowing him to stand in a draft while he is warm. This checks the perspiration suddenly and drives the blood to the feet, which sets up inflammation in them. Sometimes by giving an animal a small feed of wheat it will first cause acute indigestion, and then turn to a bad case of founder. It may affect mares a few days after foaling, especially if they do not have their natural flow of milk, or if by catching cold after foaling it should settle in their feet. Hard driving and bad shoeing will also cause it. Lung troubles sometimes terminate in founder.

**Symptoms.**—Founder generally affect the front feet, but may affect the hind feet and the symptoms are plain. The pulse beats strong and runs up to from 50 to 75 beats per minute. The animal sweats freely and breathes heavy and quick ; he generally stands on his feet for a few days at the first of the disease ; he stands in a peculiar way ; his front feet will be stretched out as far as he can put them, and standing on the heels trying to relieve the feet, while he has the hind feet drawn forward and well under him to throw as much weight as he can on his hind legs to relieve the front feet His feet are very hot and feverish, and the horse can hardly move forwards or backwards. On account of the inflammation being inside the hard resisting hoof where there is no room for swelling ; it is one of the most painful diseases that the horse is liable to, and if he does not get relief in a few days the quick of the foot becomes destroyed, and the bone will get loose from the inside of the wall and drop down on the sole, and when it does it will cause the sole to become bulged out in an unnatural way, and then it is known as a club (pumiced) foot.

**Treatment.**—Give a dose of physic consisting of

| | |
|---|---|
| Bitter Aloes.............................................. | 8 to 10 drams. |
| Common Soda.............................................. | 1 teaspoonful. |
| Ginger.............................................. | 1 " |
| Fleming's Tincture of Aconite.............................................. | 10 drops. |

Mix in a pint of luke warm water and give as a drench. If he is in high condition take half a pail of blood away from him, then take off his shoes and place his front feet in a tub and bathe them for two or three hours at a time with luke warm water—hot as you can bear your hand in it—two or three times a day ; after each bath rub dry and apply white liniment around the feet and legs, poultice the feet with hot linseed meal and bran and leave the poultice on till you are ready to bathe again, repeat this treatment until the inflammation is checked ; give the animal plenty of cold water to drink—a little at a time—and if he wants to eat give soft feed and the following drench :

| | |
|---|---|
| Nitrate of Potash or Saltpetre.............................................. | 1 teaspoonful. |
| Fleming's Tincture of Aconite.............................................. | 10 drops. |
| Water.............................................. | 1 pint. |

Mix and give as a drench three times a day, continuing the drenches until the animal gets relief. Clothe the body well and have the stall well bedded to induce him to lie down for it will help to relieve his feet. The after treatment is, blister with the following and turn him out to pasture :

| | |
|---|---|
| Pulverized Caneharides or Spanish Fly.............................................. | 1½ drams. |
| Lard or Vaseline.............................................. | 1 ounce. |

Mix well and apply all of it around the tops of both the feet, rub in well and grease three days after, then turn him out to pasture.

### SORE FEET (CHRONIC FOUNDER).

This is a soreness or a chronic inflammation of the feet.

**Causes.**—From hard work, especially driving on hard roads. Horses with small or flat feet are more subject to this than other horses, but any of them are liable to it; bad shoeing, letting the shoe rest too much on the sole will cause it; standing on a dry floor and the feet becoming dry, hard and contracted.

**Symptoms.**—It generally affects the front feet; they become hot, dry and very hard, and, in some cases, become contracted at the heels, which is caused from the fever and soreness in the foot. The horse has a peculiar stumbling action, and he tries to step on the heels first; if you press around the feet with a pincers or your hands the animal will flinch. After a time, from trying to favor his feet, the muscles of the chest will gradually waste away and leave the chest hollow. This must not mislead you and make you think there is anything wrong with the chest. Sometimes when the chest falls away some call it chest founder, but this is a mistaken idea, for there is no such thing as chest founder, the whole trouble arises in the feet.

**Treatment.**—The treatment is not very satisfactory in some cases. If it a valuable animal soak the feet well in warm water and salt, in a tub containing six or eight inches of water; leave the feet in the water two or three hours at a time, twice a day; every night put on a hot poultice of half linseed and bran, leave it on all night. After the soreness is pretty well out blister around the top of the hoof with the following:

Powdered Cantharides or Spanish Fly....... ...........1 dram.
Vaseline or Lard......................................1 ounce.

Mix and apply all of it around the tops of both front feet and turn the horse out to pasture for a long while, grease the blister the third day. After the horse is all right and he is brought in to work again let him stand on a ground floor, for an animal once affected with this disease is more liable to be affected again. Be careful in shoeing; we recommend the bar shoe. If it is a horse you want to keep shod, and he is not of much value, keep him on a ground floor and pack the feet every night with cow manure, blue clay, or anything that will keep the moisture in the foot, and in very bad cases by blistering and turning out to pasture for a while will help it.

### CLUB (PUMICED) FOOT.

This is when the foot bone becomes separated from the inside of the walls of the foot and drops down on the sole and frog of the foot. This disease is generally the result of acute founder when it is allowed to run on too long. This disease spoils a horse for road work, but he may be fixed up so he will work pretty well on the farm at slow work by blistering him around the tops of the hoofs, same as is done for chronic founder, and turn him out for some time and shoe with a heavy shoe well corked up, and have it well beveled out so it will not bear any weight on the sole of the foot; have the whole weight to come on the wall, also have the shoes set regular about once a month.

### CORNS.

Corns are generally found affecting the front feet, but may be found in the hind feet, and are also more frequently found on the inside heel of the front foot, but may be found in either or both heels.

**Causes.**—Horses with weak, flat heels are more subject to it, and it is generally brought on from bad shoeing, and by the shoe resting too heavy on the heel and bruising the horn between the bar and quarter of the wall; driving on hard roads has a tendency to bring on corns; steady driving will also cause them.

**Symptoms.**—The horse is more or less lame in most cases, and is more so just after he bruises it. When standing, if it is in one foot, he will be noticed to point that foot out, if in both feet, he will first point one out and then the other, changing from one to the other every little while. The animal will go pretty well on soft ground, but will be noticed to get lame as soon as he strikes a hard piece of road. On raising the foot and tapping or pressing on the affected quarter, he will flinch. In taking the shoe off and paring down the quarter there will be a red spot in the corner of the sole. In case the corn becomes bruised and festers, the symptoms will be more severe, he will hold up the foot and in walking he will step long, and will step on the toe, then hop on the other foot to get along. The affected quarter will be hot and tender when pressed on, and if he does not soon get relief it will fester and break out at the top of the hoof.

**Treatment.**—In many cases of corns it is not necessary to lay the animal off, but shoe him so the shoe will not press on the heels and apply a poultice of hot linseed and bran to the foot for a few nights. A bar shoe is a very good one to put on, for it throws the weight on the frog and relieves the heels and quarter. In case it is a festering corn remove the shoe and pare down into the corn until you strike the matter, allow it to escape, this will give him relief, then poultice as above mentioned and allow the animal to rest a few days, and when you are going to use him again shoe him with a bar shoe, and see that the shoes do not press on his heels and quarters, and if the soreness continues, blister around the affected quarter with the same blister that is used for chronic founder. The after treatment is to keep the animal shod regular and see that the shoes do not press too much on the heels.

### THRUSH IN THE FEET.

This is a disease that affects the frog of the foot, and is mostly seen in the hind feet, but often affects the front feet.

**Causes.**—From standing in wet and filth, or anything that will rot the frog. Heavy horses are more subject to this disease than light horses.

**Symptoms.**—They are very plain, the animal may be just a little lame, but if he steps on anything very hard he will flinch, and by examining the foot you will find that the centre of the frog is eaten out by the disease, and there is a discharge that comes from it which has a very bad smell.

**Treatment.**—Keep him out of the wet and dirt, and keep the stable very clean. Cut off the dead horn from around the frog and wash out the diseased part of the frog with warm water and a little soap, after it is cleaned well poultice with a hot poultice of half linseed meal and bran for a few days until the foot is nice and soft, after this clean the poultice out of the frog and dust in some dry calomel about twice a week until it is better; another cheap remedy is to pack the foot full of common salt a couple of times a week; another is to pour a few drops of butter of antimony into the diseased foot once a week. Do this until all the discharge and smell is gone from the frog, after that leave the frog alone until it goes down itself.

### NAILS IN THE FEET (PUNCTURES).

This is a very common thing, especially in large cities. This is where a horse steps on a nail and it runs in to the bottom of his foot. When a horse is lame always examine the bottom of the foot to see that there is no nail or anything in it.

**Symptoms.**—If it is in the hind foot the animal knuckles over and becomes lame very suddenly. If in the front foot he points it out while standing, and when he steps on it will put as little weight on it as possible and hop on the sound leg. If you examine the foot you may find the nail itself; if not, by tapping around the foot with a small hammer you will find where the tender spot is, then pare around it and you will find a small, black spot where the nail went in.

**Treatment.**—If you find the nail pull it out and pare out the hole where the nail went in almost down to the quick, after this drop in a few drops of butter of antimony, which will kill any poison or rust that may be left in. After this keep the animal quiet and apply a hot poultice of half linseed and bran for a few days until he is entirely over the lameness before you work him. The danger of these punctures are that they may start to fester, which is a very painful thing, and also very dangerous. If it starts to fester, the animal will hold the foot up with pain ; it will be swollen around the top and very hot. Cut down well into the hole and allow the matter to escape ; bathe with hot water and poultice to relieve the pain and draw all the matter out. In cases where you cut a large hole and the quick bulges out, burn it with butter of antimony once a day. After the animal is able to walk nicely blister the foot around the top of the hoof with the same blister used in chronic founder and turn him out to pasture till the soreness is all out.

## PRICKS IN THE FOOT.

These are injuries which are caused in shoeing by driving nails too close to the quick.

**Symptoms.**—At the time of driving the nail the horse will give a sudden jerk, showing that the nail struck the quick.

**Treatment.**—Remove the shoe and keep the horse quiet for a day or two ; if he is lame poultice the foot with hot linseed. If he continues to be lame, pare around the nail hole and give it the same treatment that is given for punctures of other nails. In any of these cases, if you want to work the animal when he is getting better, plug the hole with tar and cotton batting and put a leather under the shoe to keep the dirt out.

## QUARTER CRACK.

This receives its name on account of the part of the foot it affects; it generally affects the front feet, and is more often seen to affect the inside quarter of the foot.

**Causes.**—Certain breeds of animals are more subject to this than others, especially if the feet are of a brittle nature, and it is often seen in road horses, more especially if they have high knee action and strike the ground heavy.

**Symptoms.**—At first a small crack appears in the quarter of the foot just below the hair; if the animal is kept to work he will get lame, the crack will get larger and longer, and will sometimes bleed. If the animal is kept working, it will sometimes fester on account of the irritation being kept up.

**Treatment.**—Remove the shoe, pare out the bottom of the affected quarter so the shoe will not rest upon it, then shoe with a bar shoe; as well as this, pare out the crack all the way down on each side—almost to the quick—from the top to the bottom of it so it will not be pressing on itself. If the animal is a little lame poultice the foot for a few days until the soreness is all out. After this apply a little of the same blister that is used for chronic founder just above the crack in the hair. This is to stimulate the growth of the horn and make the hoof grow and draw out all the soreness.

## CAULKS.

This is common in the fall and spring when horses are sharp shod.

**Causes.**—From the horse or his mate stepping on his foot and cutting around the top of the hoof.

**Treatment.**—If it bleeds freely apply Monsell's solution of iron, and if you have not this apply a pad of cotton batting and bandage tightly over it, leave it on for twenty-four hours. The danger in caulks is that sometimes hair and dirt gets worked down into the caulk and it begins to fester and works down into the foot. In all cases where the animal shows any signs of lameness, pare out the hoof around the caulk to the bottom of it, then poultice for a few days, change the poultice twice a day to keep it hot and each time the poultice is changed, pour a little hot green salve into it.

## OVERREACH.

This generally occurs in fast horses where they are speeded without having on quarter boots, by stepping too far with their hind foot and catching the heel of the front foot.

**Treatment.**—Treat the same as a lacerated wound, and in all cases where you are speeding fast horses, wear boots on them to prevent them from catching their quarters.

## BRUISES OF THE SOLE OF THE FOOT.

**Causes.**—From stepping on a stone or any hard substance or from the shoe pressing on the sole.

**Symptoms.**—The animal in travelling steps very long on the affected foot, and by tapping the sole of the foot with a hammer he will flinch when the bruised spot is struck. If the bruise is allowed to run on without being treated it will commence to fester and the symptoms will be more severe ; the horse will hardly be able to put his foot to the ground, and it will be hot and swollen around the top.

**Treatment.**—Remove the shoe and find the affected spot by tapping on the sole with a hammer, and if you do not think it is festering poultice with a hot poultice of half linseed meal and bran ; change the poultice twice a day to keep it hot until the soreness is all out, but if you think it is festering pare a small hole in through the sole to the festering part, to allow the matter to escape, after this poultice as above mentioned until the animal is better. Each time, when changing the poultice, melt a little green salve and dop it into the hole you cut.

## CUTS OF ANY KIND AROUND THE FEET.

Refer to the treatment of wounds.

## FALSE QUARTER.

**Causes.**—From a cut around the top of the hoof which sometimes causes a false growth of horn, in the form of a ridge, to grow down the hoof as it grows out.

**Treatment.**—File the ridge down level with the wall of the foot with a rasp.

## COFFIN JOINT LAMENESS (NAVICULAR DISEASE).

This is getting to be a very common disease, and is seen mostly in driving horses.

**Causes.**—From hard and fast work. Animals with short, upright pastern joints, short, stubby action, or horses with high pounding action are more often affected with this disease; allowing the toes to grow too long and cutting down the heels when shoeing them will cause it.

**Symptoms.**—There is usually more or less lameness ; in some cases it comes on sudden and severe, while in others it gradually comes on for weeks, and sometimes for months, before it is much noticed. While standing the horse will point the feet out, and in some cases this is the first symptom that will be noticed. If both feet are affected the horse suffers pain and while standing will first throw the weight on one foot and then on the other. In travelling he has what is known as a groggy action. Another well marked symptom is a wasting of the muscles of the chest. On examining the feet the heels will be found to be contracted and hard, and by raising the foot up and pressing with your thumb on the back part of the heel the horse will flinch. If you notice the shoe the horse has been wearing it will be found to be worn most at the toe.

**Treatment.**—If it is a bad case of long standing it is incurable, but if taken in time, by resting the horse and by bathing the feet twice a day for an hour or two at a time (if in the winter bathe with warm water, if in summer use cold water). After bathing apply white liniment around the top of the hoof and every night poultice the foot with hot linseed meal and bran, half and half. Continue this treatment until he is pretty well over the lameness, then blister the foot same as in chronic founder and let him out to pasture. If it is of long standing and you want to use the horse, have him stand on a ground floor and pack his feet every night with cow manure or blue clay or anything that has a tendency to soften the foot. Shoe him with high heeled shoes and have it low in front. In some cases neurotomy is performed, that is, where the operation of nerving the foot is performed, this is done with a view of taking the feeling away from the foot. The way this is performed is by throwing the animal and securing him, then make a cut along the inside and the outside of the leg between the knee and fetlock about half ways, make the cut about two inches long lengthwise in the groove between the shin bone and the back tendons, in this groove you will find the nerve, artery and vein which runs down the leg, then cut about an inch out of the nerve so it will not unite together again, do this on both sides of the affected leg or legs, then stitch the cut up, bandage the leg and treat as a common wound, bathe and apply the white lotion twice a day. After this is performed the horse will get up and go off as though there was nothing wrong, but after this operation be careful in shoeing him and examine the foot every night, for if anything went wrong with the feet they would rot off before the animal would show any lameness. This operation at one time was performed to a great extent, but is not so much done now. In examining a horse if you think that he has been nerved in this way, prick him with a pin around the feet and if he has been operated on he will not feel anything, but if he has not he will show signs of feeling.

---

## CHAPTER XXV.

# MISCELLANEOUS INFORMATION.

Horses that are wide between the eyes with a nice tapering muzzle and a nice bright eye shows a good disposition and a kind horse. Nice large ears, and in travelling they should be carried forward, this also is a good indication. The neck should be good and long and bowed upwards, and well cut out under the jaws. The shoulders should be long and well slanted forwards with nice high withers, also a short strong back with long round well developed hips and rump. Also a good round deep chest with a tidy belly not too small; it is also a good point to have him well ribbed up, that is, not to have too much space between the last rib and the hip bones; also look to the legs that the bone is flat and clean with good shaped feet.

### HOW TO EXAMINE FOR SOUNDNESS.

Give the horse a quick trot or gallop to see that he is not lame, and as soon as he stops put your ear to his nose to hear that his wind is not affected. Then examine his nostrils and mouth, at the same time look to see the age of the animal, which is told by his teeth; then look closely into the eyes to see that there is nothing wrong with them, pass the hand up around his ears and the pole of the head to see that they are all right; then examine one side of him first by starting at the neck, running the hand over it to the withers, then over the shoulder, down the outside and inside of the front leg, and watch carefully for splints, sidebones, ringbones and such like; then raise the foot and see if it is a well formed one, with good, strong heels; look back along the belly for warts and running sores caused from castration, then pass the 'hand down the back of the hips and see that the hip bones are both the same size, then follow the hind leg over the hock and hind fetlock and look for spavins, windgalls, curbs, splints, side-bones and ringbones, also look at the stiffles and see that they are all right; examine the foot as you did in the front, then examine the other side same as side mentioned; after this stand back and take a look at him to see how he stands on his legs, and, also, how he holds his head and neck; notice if he stands with his front feet well under him, for this is a good sign; at the same time see if he is inclined to be weak in the knees and fetlocks. Beware of calf-kneed horses for they are always stumblers; see that he stands neither too straight nor too crooked on his hind legs. After this take a walk around to the front and see that he has a well formed breast, and that he does not toe in nor toe out too much; then make an effort as if you were going to strike him with a stick or whip over the side, if he grunts examine close to see that he is not a roarer, then give him another good, sharp trot or gallop to see that he carries his front and hind legs nice and straight and that he has good action, also notice whether he carries his tail straight or not. Step up quick and place your ear to his nostril to see if he makes any noise, and be sure that he is all right in his wind. To make sure that the horse is all right put him in a stable for an hour or two, giving him a pail of water and a feed, for in some cases of lameness the animal will not show it until he has stood for a while. After this go into the stall and take the horse out yourself, watching how he steps over and how he backs out of the stall for fear of string halt and corea. After this trot him again and test him for lameness and soundness in his wind.

## HOW TO TELL A HORSE'S AGE BY HIS TEETH.

Commencing at the time the colt is foaled: At nine days old he gets two centre nippers, or front teeth, two above and two below; at nine weeks old he gets four more front teeth, one on each side of the other two pairs, these are called the lateral teeth; at nine months old he gets four more front teeth, called the corner teeth, one on each side of the two pair below, and one on each side of the two pair above. So, at nine months old, the colt has a full mouth of milk, or temporary front teeth. At the age ol two years you have to judge from the general appearance of the colt as to his age, as there is no change in the front teeth. At three years old he sheds the four centre nippers or front teeth, two above and two below, and gets in permanent ones; at four years old he sheds the four lateral teeth, two above and two below, and gets in permanent teeth in their places; at five years old he sheds the four corner, or outside teeth, and gets in permanent ones. So, at this age, the horse has a full mouth of permanent front teeth.

Each one of the front teeth has a black ring on it at five years old; at six years old the rings on the two centre teeth in the bottom row disappears and only a black spot is left on each; at seven years old the rings on the lateral teeth, or the ones next to the centre in the lower row, disappears and only a black spot is left on each tooth; at eight years old the rings on the corner teeth of the lower row disappears and only a black spot is left on each tooth; at nine years old the rings on the two centre front teeth on the upper row disappears and a black spot remains on each tooth; at ten years of age the rings on the lateral teeth, or the ones next to the centre pair in the upper row, disappears and a black spot is left on each tooth; at eleven years old the rings on the corner teeth of the upper jaw disappears and a black spot is left on each tooth. At twelve years old we come again to the bottom row, and looking at the front of the teeth the two centre ones will be found to be much narrower and longer than the others; at thirteen years the lateral, or teeth next to the centre pair—one on each side—become much longer and narrower; at fourteen years old the corner teeth in the lower row become much longer and narrower; at fifteen years old the two centre teeth on the upper row become long and narrow; at sixteen years old the lateral teeth on the upper row become long and narrow; at seventeen years old the corner teeth of the upper row become longer and narrower. At eighteen years of age we look again to the bottom row and the two centre teeth will be much shorter than the others; at nineteen years old the lateral teeth of the bottom row have become much shorter; at twenty years old the corner teeth have become much shorter; at twenty-one years of age look again to the teeth in the upper row and the two centre ones will be much shorter; at twenty-two years old the lateral teeth will be shorter, and at twenty-three years old the corner teeth of the upper row will be short. After twenty-three years old the age can not be told accurately.

### HOW TO DRENCH A HORSE.

Have a halter on the horse and take him some place where there is room for his head to be held up high, then place a twitch on his nose. A twitch is a handle of some kind, about three feet long, at the end is a hole through which is a small rope tied to form a ring. This rope is slipped over the horse's nose and twisted up pretty tight, then hold the horse's head up high by lifting on the twitch. This is to keep the medicine from running out when poured into his mouth. Have an assistant to hold the twitch while you, with a black bottle which contains the drench, slip the neck of the bottle into the mouth from the side where there is no teeth ; only pour a little out of the bottle at a time, as you might choke the horse. If he goes to cough let his head down immediately until through coughing, then continue the drenching, but don't be in too big a hurry to empty the bottle.

### HOW TO BLEED A HORSE.

Place a small rope, or plow cord, in the form of a slip-knot, over the horse's neck and draw it tight enough to swell the jugular vein on the left side of the neck, moisten the hair over the swelling and hold the fleames—which must be clean—lengthwise with the vein ; have an assistant to hold the rope, and also cover the horse's left eye so he cannot see you. When you are satisfied the fleames set over the vein strike them a quick, sharp blow with a stick of some kind, which, as soon as the vein is cut, will send a stream spurting out of the hole. Continue holding the rope tight until it has bled enough. Always catch the blood in a pail so you know how much you take away. When it has bled enough let the rope slack, which will stop the bleeding. Then take a bright new pin and run it through the two edges of the skin bringing them together, then wind a string around the pin so it cannot get off ; leave the pin in for twenty-four hours after bleeding ; tie the horse's head up and do not let him out to grass or eat anything that will cause him to keep his head down to the ground until the vein is healed. Be sure your hands, the fleams and the pin are clean, for dirt or rust will set up inflammation of the vein.

### A CHILL.

**Causes.**—It is often noticed after a horse has been working hard and takes a cold drink of water, or is allowed to stand in the cold afterwards, or anything like that, will bring on a chill.

**Symptoms.**—The horse will be shivering, looks very dull, his back will be humped up a little, he breathes heavy and refuses his feed. This is not looked on as a disease itself, but a symptom of some other disease, and if allowed to run on without being checked is liable to set up inflammation of any of the organs of the body, so you see the necessity of checking a chill in time.

**Treatment.**—Put the animal in a warm stall and put an extra blanket or two on, to get him heated up, if his legs are cold rub them to get the circulation started. Give as a drench:

Sweet Spirits of Nitre ............... 1 ounce or 4 tablespoonfuls.
Fleming Tincture of Aconite ...................... 5 to 10 drops.
Ginger ........ ...................... .......... 2 tablespoonfuls.

Mix in a pint of luke warm water and give as a drench, repeat the dose every hour until he is better. If you have not the sweet spirits of nitre give a wine glass full of whisky. As soon as the horse will eat give him a hot drink or a bran mash.

## HOW TO FEED AND TAKE CARE OF A HORSE.

Always water the horse before feeding if he is not too warm. Feed regular, that is, set certain times for feeding him and feed as near that time as possible. Give him a certain amount of exercise every day that the weather is fit to take him out, if not working him allow him to run out around the yard, and always remember that good cleaning and a clean stable is half the feed. A good plan to cleanse the stable is to throw a little lime around the stalls every week or so, also in the summer when the horses are all out of the stable, close up the doors and burn sulphur in the stable, this will kill all the germs of distemper and all other diseases that are in the stable. Always have a little box of salt kept in front of the horse so when he wants it he can have it, and by doing it this way he will never take more than is good for him; rock salt is the best if you can get it. Whitewashing the stable about once a year is a good plan to keep it clean.

---

## NOTICE.

All the doses of medicine which are mentioned in this book in the diseases and treatment of the horse, if not specially mentioned for the age of the animal, is intended for the average size horse, so in giving doses to colts or very small horses you must regulate the dose to the age and size of the animal. Yearling colts would take one-third of the dose mentioned for a horse. A two year old could stand one-half the dose mentioned for a horse. A three year old can stand nearly as large a dose as is mentioned for a horse. Four year old and upwards take the full dose.

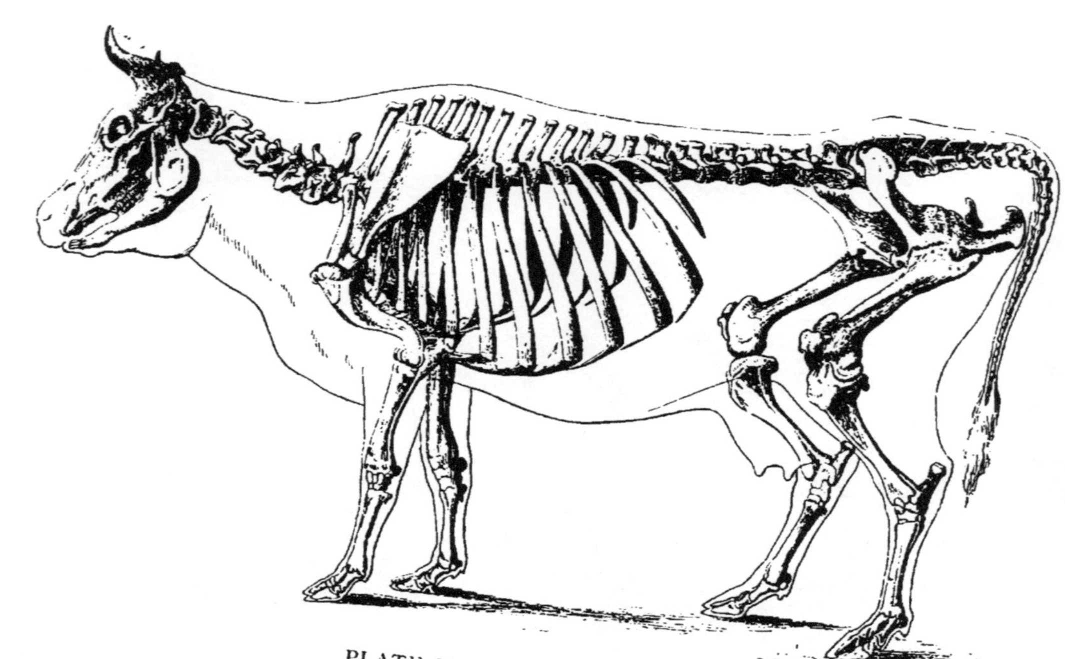

PLATE V.—SKELETON OF THE OX.

# EXPLANATION OF PLATE V.

## SKELETON OF THE OX.

This cut shows the position of each of the bones and joints in the frame of the ox. For particulars in connection with this cut refer to the anatomy of the ox—the part which deals with the bones.

SPECIAL NOTICE.—Every place where Tablespoonful is mentioned in this book should read: SMALL, or DESERT TABLESPOONFUL, which is equal to two Teaspoonfuls.

# PART III.

# ANATOMY, DISEASES AND TREATMENT OF CATTLE.

## CHAPTER I.

## ANATOMY OF THE OX.

Each of the bones and joints of the ox receive the same name as those of the horse. There are a few important points of difference in the structure of the skeleton of the ox and horse, viz.: There are only thirteen pairs of ribs in the ox while the horse has eighteen pairs. The breast bone or sternum is much larger and flatter in the ox than it is in the horse. This is important to remember because when there is anything wrong with the lungs or chest of cattle they always lie down, for the breast bone and the ribs are so formed that when they lie on the breast bone it expands the chest, giving the animal more relief than if it was standing, while the horse always stands in lung trouble, for his breast bone and ribs are so formed that it contracts the chest and gives him more pain while lying down. The bones of the head of the ox differ from those of the horse by being very heavy and wide on the upper part of the skull and has a projection of bone at each side on the upper part of the skull called the core of the horn. This is important on account of dehorning, and every one interested should become familiar with it. This projection or the core of the horn is hollow inside which is a continuation of the sinuses or cavities in the bones of the head. Attached to the core of the horn and covering it, is the horn itself. Another point of difference between the ox and the horse, is that there are two small bones found in the heart of the ox called the cardiac or heart bones, while in the horse's heart there are no bones at all to be found.

Another point of difference between the ox and horse is that the bones in the leg of the ox are divided from fetlock down into two parts, while in the horse they are not divided.

## DIGESTIVE ORGANS OF THE OX.

**The Lips** of the ox are thick and hard, the upper one in front has no hair on it and varies in color with the color of the animal, and when cattle are in good health this space is always moist.

**The Cheeks** on the inside are covered by numerous little rough processes which give the cheeks a very rough appearance.

**The Tongue** of the ox is stronger than that of the horse and is more movable, it is very thick and heavy at the back; it is pointed at the front end and the upper part of it is very rough; it is by means of the tongue the ox takes most of the food into his mouth.

**The Glands** which secrete the saliva are similar to those of the horse.

**The Teeth** differ very much from those of the horse, the ox has no front teeth in the upper part of his mouth, their place being taken by a pad of cartilage or gristle. This pad takes the place of the upper row of front teeth for the lower row of teeth presses against it when the animal is cropping grass; on account of this pad it can be seen why cattle will not do as well on short grass as horses. The front teeth in the lower row also differs from those of the horse, they are eight in number, chisel-shaped and are loosely set in the gum. The molars or back teeth are similar to those of the horse only they are smaller and not so smooth on their upper surface. The ox has twenty-four molars or back teeth, and eight incisors or front teeth, making thirty-two in all.

**The Pharynx,** or gullet in the ox is much larger than that of the horse.

**The Œsophagus,** or tube, which carries the food down from the mouth to the stomach is well developed, the fibres in it are very strong and have a double action. When the animal is eating they carry the food from the mouth down to the stomach, and during the time the animal is chewing its cud they act the very opposite, they carry the food from the stomach back up into the mouth.

**The Stomach** is a very important organ to understand, for cattle suffer a great deal from diseases of the stomach. The stomach of the ox has a capacity of sixty gallons, and is divided into four divisions. The first department is the rumen or paunch; the second is the reticulum, or honey comb; the third is the omasum, or manyplies, and the fourth is the abomasum, or true digestive part of the stomach. The first three departments of the stomach prepares the food for digestion, while the fourth digests it. The rumen or paunch is very large, and in an aged animal it fills up three-quarters of the belly cavity, it lies up against the left side of the belly, where it is attached and held to its place by ligaments; its situation is important to remember, for in many diseases of the rumen, or paunch, it is first noticed on the left side, and in tapping for bloating it is always done on the left side because the paunch lies right up against the left side. The walls of the paunch of an ox resembles the stomach of the horse, but is not nearly so sensitive, and will stand a great deal of abuse before inflammation will set in. The paunch has two openings, both of which are at the front, one is where the food enters the stomach, while the other is where the food passes out into the next division of the stomach called the reticulum, or honey comb, which is the smallest division of the stomach, and resembles a honey comb in appearance. This part of the stomach has not very much to do in preparing the food; it has two openings, one in front, where the food enters into it, the other at the back, where the food passes through into the third part of the stomach called the omasum, or manyplies, which is the second largest division of the stomach. When this is full it is ovoid in shape and placed just behind the second division of the stomach and at the right side of the paunch, and if you examine the inside it will be found to be full of folds, or layers of membrane. The use of this part of the stomach is, while the food is passing through it to draw into its folds all the coarse parts of the food and roll it about in the layers until it gets it fine and well prepared to pass into the last part of the stomach, where it becomes digested. When this part of the stomach becomes deranged and the food becomes dry and hard between the folds it then sets up the disease called impaction of the manyplies, or dry murne. When the food passes out of this third division it is emptied into the fourth part of the stomach called the abomasum, or the true digestive part of the stomach. This is where the food is digested. The walls of this part of the stomach is redder in color than the three first divisions and has the glands which secrete the acids and gastric juices of the stomach which assist greatly in digestion. This stomach, also, has two openings, one where the food enters and the other where the food goes out of it and enters into the small bowels.

**The Bowels,** or intestines, of the ox are divided into large and small bowels, same as that of the horse, the structure and action of them resemble those of the horse. The small bowels are only half the size of those of the horse, they being about one-half inch in diameter, and about 150 feet in length. The large bowels are not nearly so large as those of the horse, and are 36 feet in length.

**The Liver** of the ox resembles that of the horse only it has a gall bladder which resembles a pear in shape and it acts as a vessel to store up the gall during the time there is no digestion going on. But during the time digestion is going on the walls of the vessel contracts and forces the gall down on to the food. The other two glands, the pancreas and spleen resemble those of the horse. These juices have the same action in cattle as they have in the horse.

## HOW RUMINATION OR CHEWING OF THE CUD IS PERFORMED IN THE OX.

The ox takes the food into its mouth principally with its tongue, it is then roughly chewed and swallowed and passes into the rumen or paunch where it is rolled about by the action of the walls of the paunch and mixed with the juices which are secreted in the paunch. After the animal has finished eating, it then, as a general thing, lays down (but may stand up) and commence chewing its cud, the way this is performed is, the walls of the rumen contracts and forces the food up into the bottom part of the œsophagus, which then takes on a reverse action and forces the food up into the mouth where the food is all thoroughly chewed over again, and in swallowing it this time the end of the œsophagus, instead of opening out and allowing the food to drop into the paunch, keeps closed and passes the food right into the second stomach, thus you see the ox first roughly swallows its food and is held in the paunch until the animal has time to bring it back to the mouth and chew it over again. After it enters into the second stomach or reticulum, which is the honey comb part, the food is shifted around for a short time and mixed with the juices secreted in that part, then it passes back into the manyplies or third division of the stomach where the fine parts of the food pass right along to the abomasum or fourth stomach, while the coarser parts of food are drawn between the folds of membrane in the third stomach, here it is worked about between the folds until it is fine and ready to pass into the fourth stomach, where it becomes fully digested by the action of acids and gastric juices which are secreted in this part of the stomach, then it passes into the small bowels and is acted upon by the bile from the liver and the pancreatic juice from the pancreas, these juices are emptied into the first part of the small bowels by little tubes or ducts. which lead from the glands down to the bowels, just on the same principle as that of the horse. After this, throughout the rest of the bowels, the nourishment of the food is taken up into the system by means of little glands or villi which are situated in the coats of the bowels, and the nourishment when once in the blood goes to supply the different parts of the body, while the part of the food there is no nourishment in passes off through the back bowels in the form of manure.

**The Feet** of the ox and bones below the fetlock are divided and receive the name of trotters.

### THE RESPIRATORY OR BREATHING ORGANS.

These organs in the ox work on the same principle, and resemble in structure those of the horse, only they are not, as a general thing, so liable to diseases as the breathing organs of the horse.

### THE URINARY ORGANS.

The chief point of difference in these organs are the kidneys. In the ox they are larger, and instead of being smooth, like those of the horse, they are rough, and resemble a bunch of grapes. The bladder and the rest of the urinary organs resemble those of the horse.

### THE GENITAL ORGANS OF THE COW.

**The Ovaries** in a cow are smaller than those of the mare, but resemble them in structure.

**The Womb** in the cow somewhat resembles that of the mare, but the inside lining is different, for it is covered with button-like processes about the size of a pigeon egg, only more flattened out. These processes receive the name of cotyledons ; this is what the cleaning or afterbirth is attached to and is a very important point which every person interested should become familiar with by examining a womb after calving or by opening a cow that dies soon after calving. The passage out of the womb of the cow is shorter than that of the mare, but is formed on the same principle.

**The Bag,** mammary glands, or udder, is very important to understand in the cow. It is first divided into two halves by a partition or division in the center of the bag. Each one of these halves are again divided into two parts, each part is known as a quarter of the bag, each quarter having a mammary, or milk gland and a sinus, or pouch to hold the milk where it is secreted until the cow is milked. This sinus or pouch is situated just above the passage of the teat.

### THE GENITAL ORGANS OF THE BULL.

**The Testicles** are ovoid in shape and well developed. The spermatic cord and artery are small compared with those of the horse; the penis is long and pointed, and has an S shaped curve in it just below the pubis, or hip bones; this curve can be felt by feeling carefully just behind the bag; the sheath is long and runs further forward on the belly and has a tuft of hair on the point of it. During the time the bull is serving a cow the S shaped part of the penis straightens out.

---

## CHAPTER II.

# DISEASES AND TREATMENT OF CATTLE.

### CATARRH OR COLD IN THE HEAD.

Cattle do not suffer so much from this disease as horses. This is an inflammation set up in the lining membrane or the sinuses of the head.

**Causes.**—It is generally brought on from exposure or a sudden change in the weather.

**Symptoms.**—The nose is rough and dry and has a mattery discharge from it; the animal has a slight cough and makes a rattling in the head when it is breathing.

**Treatment.**—Give a mild dose of physic consisting of

Epsom Salts........................½ pound.
Sweet Spirits of Nitre................1 ounce, or 4 tablespoonfuls.

Mix in a pint of luke warm water and give as a drench; keep the animal dry and warm and feed on mashes and good hay until it seems better. After the first drench follow up with the following medicine:

Nitrate of Potash or Saltpetre..........................½ pound.
Ground Gentian Root..................................½ "

Mix together and give a teaspoonful three times a day until the animal is better. In cases where the discharge continues from the nose and becomes chronic, it is then called nasal gleet, same as in horses; the treatment then is to give a teaspoonful of ground sulphate of iron three times a day in its feed or on its tongue with a spoon, and this will soon stop the discharge.

### SORE THROAT.

This is a kind of inflammation affecting the larynx, or Adam's apple; it may also affect the pharynx, or gullet.

**Causes.**—Exposure to cold and sudden changes of the weather, or from choking, when something lodges in and irritates the throat, or from roughly passing a probang down the throat and bruising it, or from giving medicines that are not properly diluted with water.

**Symptoms.**—Difficult breathing, the throat will be swollen, and it hurts the animal when you press on it, the nose will be dry, and the animal will be noticed to keep swallowing, it refuses its feed and does not chew its cud, it also holds its head poked out trying to ease its throat.

**Treatment.**—Give the following drench:

Epsom Salts.......................................... ¾ pound.
Sweet Spirits of Nitre................1 ounce or 4 tablespoonfuls.

Mix in a pint of luke warm water and give as a drench, taking care not to choke the animal, as well as this follow up with

Nitrate of Potash or Saltpetre..........................¼ pound.
Sulphur..................................................¼ "
Pulverized Alum.............. .......................⅛ "

Mix and give a large teaspoonful three times a day on his tongue, this is to gargle the throat, give until the animal is better. Rub the throat three times a day with white liniment and in bad cases apply a mustard plaster to the throat or you might use a hot poultice every night on his throat.

### FILARIA BRONCHITIS.

This disease affects young cattle and sheep and is caused from small germs or parasites getting down into the lining of the bronchial tubes; these germs receive the name of strongylus micrurus and they get into the system by being taken into the stomach in the water or food, then they pass from the stomach into the bowels, from there they get into the blood and pass around until they get to the bronchial tubes where they lodge and set up this disease. This disease is most common where animals are grazing on low pasture fields.

**Symptoms.**—The animal makes a wheezy noise while breathing and has a dry husky cough with a slight discharge from the nose and by examining the discharge with a microscope you will find those little germs or parasites in it. The animal gradually runs down in condition and if the affected one is not soon separated from the rest the other young cattle will become affected.

**Treatment.**—Kill the germs or parasites by giving

Raw Linseed Oil........................................½ pint.
Spirits of Turpentine..................½ ounce or 2 tablespoonfuls.

Mix and give as a drench once a week. The turpentine has a special action in cases of this kind for it gets into the blood and therefore comes in contact with the germs or parasites and kills them. If this should fail, drive the affected cattle into a stable, then take a pan of sulphur, throw some live coals in it and let the animal's breath the fumes of it, stand in the stable with the cattle and just give them as much of the fumes as you can stand yourself and then turn them out, this is the best way to tell how much to give them. Do this every day for a week or so until they are better. By breathing the fumes of the burning sulphur it comes in contact with the germs or parasites in the bronchial tubes and destroys them and stops the disease. The main thing in burning sulphur is not to strangle the cattle by letting them inhale too much of the fumes, as well as this feed them well to get them up in condition.

### INFLAMMATION OF THE LUNGS IN CATTLE (PNEUMONIA).

**Causes.**—They are similiar to the causes of inflammation of the lungs in horses; catching cold in some way, such as being out in cold rains or standing in a cold draft or catching cold after calving.

**Symptoms.**—The animal refuses to eat, has a slight cough and the nose is sometimes dry, then wet, changing frequently; he breathes heavy and quick and in breathing makes a groaning noise similar to impaction of the stomach, and care must be taken that you do not mistake one for the other. By placing your ear over the side of the chest a sound is heard similar to that made by rubbing some hair of your head just over your ear between your thumb and finger. If it is a milch cow she will be noticed to drop off some in her milk. By pressing on the ribs over the lungs with your hand the animal is noticed to be sore, also feverish and very thirsty and from the fever of the lungs the bowels are often a little costive, the pulse is quick and strong at first but after a couple of days, if not better, the pulse gets quicker and weaker, sometimes running as high as one hundred beats per minute. In this disease cattle lie almost all the time, and lie well upon the breast bone for this position seems to give them relief.

**Treatment.**—If the bowels are not very free and the animal is in fair condition, give

| | |
|---|---|
| Epsom Salts | ½ pound. |
| Sweet Spirits of Nitre | 1 ounce, or 4 tablespoonfuls. |
| Fleming's Tincture of Aconite | 15 drops. |

Mix in a pint of luke warm water and give this drench three times, leaving out the salts after the first drench. Rub the sides with white liniment three times a day, and keep a half pail of hot salt over the lungs, changing it about every hour during the day, and at night apply a mustard plaster; take half a pound of mustard with enough vinegar to make it into a paste and rub well in over the sides and cover the animal up warm. Keep this treatment up until relief comes. When the animal seems to be getting better quit the above drenches and give the following:

| | |
|---|---|
| Nitrate of Potash or Saltpetre | ¼ pound. |
| Ground Gentian Root | ¼ " |
| Ground Aniseed | ¼ " |

Mix together and give a teaspoonful three times a day on his tongue. Keep the bowels regulated by giving one-half pound drenches of epsom salts once or twice a week and feed on soft food with plenty of boiled flaxseed in it.

### BRONCHITIS IN CATTLE.

This disease, when it is the result of a cold, comes from causes similar to that of inflammation of the lungs; the symptoms are same as those of inflammation of the lungs, only by listening with your ear at the windpipe you will hear more of a wheezing noise.

**Treatment.**—Is the same as is given for inflammation of the lungs.

### PLEURISY IN CATTLE.

This is inflammation of the coverings of the lungs.

**Causes.**—They are similar to those of inflammation of the lungs, and this disease is very often found in connection with it.

**Symptoms.**--They are similar to those in inflammation of the lungs, only that when there is pleurisy there is more of a grating sound heard when you place your ear to his side and the animal's sides seem sore and he flinches more when you press on it than in inflammation of the lungs.

**Treatment.**— The treatment is the same as for inflammation of the lungs. In this disease, as well as in all other lung troubles, be very careful in drenching so as not to choke the animal, and give him plenty of fresh air without being in a draft and keep him as comfortable as possible.

---

## CHAPTER III.

# DISEASES OF THE DIGESTIVE ORGANS OF CATTLE.

### SLAVERING.

This is a dribbling of saliva from the mouth.

**Causes.**—From a wound, or something being caught between the teeth, or from eating wild mustard or poisonous grasses.

**Treatment.**—Give the following:

| | |
|---|---|
| Epsom Salts | 1 pound. |
| Common Soda | 1 tablespoonful. |
| Ginger | 1 " |

Dissolve in a quart of luke warm water and give as a drench. Sponge the mouth out with the following lotion once a day:

| | |
|---|---|
| Powdered Alum | 1 tablespoonful. |
| Water | 1 pint. |

After the medicine has operated and you have washed the mouth out two or three times the slavering generally disappears. In all cases of this kind examine the mouth thoroughly by looking into it, and if you find anything caught between the teeth remove it at once.

### SHARP MOLARS OR BACK TEETH.

This is where the edges of the teeth are sharp and cuts the tongue and cheeks. When the animal is feeding it will sometimes stop and spit the food out of its mouth and does not thrive well. The best way to fix this is to run a tooth rasp a few times over the outside edges of the upper teeth and the inside edges of the lower teeth.

### DECAYED TEETH (CARIES) IN CATTLE.

**Symptoms.**—The animal does not thrive well, will stop eating and spit the food out of its mouth; on examining the animal's mouth you will find the breath smells very bad; if you put a clevis in its mouth and run your hand back you will find the decayed tooth.

**Treatment.**—Tie the animal up short and have the tongue held out of the mouth by an assistant; if the animal seems ugly place a small sized clevis crossways in its mouth, then pass your hand in along the teeth until you find the affected tooth. Then remove the tooth with a large pincers or forceps for that purpose. In drawing the tooth, if the animal is ugly, throw it down and secure it by the same process as is given in castrating a bull.

### CAPS ON THE TEETH.

This occurs in cattle from two to four years old when they are shedding their milk grinders; instead of the milk teeth dropping out, as they should, caps hang on the new teeth and cause them to fester at the roots, causing a lump on the jaw bone.

**Symptoms.**—The animal will be noticed to hold its head to one side, have difficulty in eating and sometimes spit the food out, will fall off greatly in condition and if allowed to run on for some time a lump will form on the jaw opposite the festered tooth.

**Treatment.**—Examine the mouth carefully till you find the tooth which has the cap on it which can be easily told by its being stuck up higher than the other teeth, as soon as you find it remove the cap with a pincers or anything that will knock the cap off the tooth, a cold chisel and hammer will do, by tapping it gently until you knock the cap off. If this is done in time it will save a lump from forming on the jaw.

### LUMPS ON THE JAW BONES FROM THE TEETH OR FROM AN INJURY.

These lumps are hard and immovable.

**Treatment.**—If it is caused from a cap staying on the tooth too long, causing the roots of the tooth to fester, remove the cap but do not pull the tooth at first, try a blister on the lump of the following:

| | |
|---|---|
| Biniodid of Mercury | 1½ drams. |
| Powdered Cantharides or Spanish Fly | 1 " |
| Vaseline or Lard. | 1 ounce. |

Mix and apply one-third to one-half of this amount, according to the size of lump, rub in well and grease in three days; between two and three weeks after, wash off the lump with soap and water and blister again, greasing in three days after. If this treatment does not stop the lump from growing throw the animal and secure it and pull the tooth with a large forceps used for that purpose. If the cause is from an injury blister same as mentioned above. For other information of lumpy jaw see actinomycosis.

### INFLAMMATION OF THE PAROTID GLANDS.

This is inflammation of the glands situated at the side of the throat.

**Causes.**—From an injury of some kind such as another animal hooking it; or by striking an animal with a stick or stone while driving them.

**Symptoms.**—The affected side of the throat will be swollen and very sore, and the animal will walk about with its head stuck out, and will fall off in condition, for on account of the throat being so sore it is unable to bend its neck down to eat, and it also hurts it to swallow.

**Treatment.**—Apply hot poultices to the throat of half linseed meal and bran ; change the poultice every three or four hours to keep it hot, and each time—when you are changing it—rub the gland well with white liniment, this will either check the inflammation and bring down the swelling, or if it festers, will bring it to a head, and then if it does not break of its own accord lance it in the softest part. After you have lanced it and let the matter out, poultice every night and apply white lotion before putting the poultice on and after taking it off. While you are treating the animal keep it in the stable and feed it out of a high manger so it will not have to lower its head to eat ; give it plenty of slops and boiled feed, or anything that is easily chewed and swallowed so as to keep its strength up until it gets better.

### CHOKING.

This is a very common thing among cattle.

**Causes.**—From the animal attempting to swallow something that is too large for its throat, such as an apple, potato, slice of turnip, mangel, or carrot, and sometimes from a bone when the animal has a fashion of licking or chewing them, and by mistake it slips down the throat and chokes it.

**Symptoms.**—The animal will stop eating, slaver at the mouth, cough, breath heavy, and after a time will become bloated in the paunch, which is noticed at the left side, and if the animal does not soon get relief it becomes so bloated it will drop down from suffocation and will soon die. Bloating is generally worse when it is an apple or potato, for they act just like a cork in a tube.

**Treatment.**—Examine the mouth and throat, or gullet, and see if you can tell what is choking the animal ; also examine along the left side of the neck to try and find out where it is lodged in the tube. If you can feel or see whatever is choking the animal, and you think you can reach it, put a clevis crossways in its mouth and run your hand down the throat and bring it up. If you cannot reach it pour half a pint of linseed oil down as a drench, then move the obstruction by working it with your hand from the outside until the oil gets worked around it and it may slip down when you get it started. If you have no raw linseed oil any other kind of oil will do. If this fails pass down a probang, which is an instrument for that purpose; there is a small wooden gag goes with the probang; the wooden gag is placed in the mouth crossways; have a man to stand on each side and hold a horn and one end of the gag, keeping the cow's head in line with the body; oil the probang and pass it through the hole in the gag, on through the center of the mouth into the gullet or throat, down into the tube leading to the stomach, pass it gently down until you strike the obstruction, then press on it until you force the obstruction down into the stomach. In cases where the animal is badly bloated, and you cannot press the obstruction down into the stomach, tap her on the left side to let the gas out with a trocar and cannula, or with a penknife, then you can easily shove the obstruction down and the bloating will soon disappear. Do not, under any circumstances, go to shove down a broom-handle, or anything of that kind, to remove the obstruction, for you are apt to burst the tube which leads to the stomach, and you will then lose the animal. Probangs cost from $1.00 to $1.50, and can be got at a wholesale drug store or veterinary instrument store, and is something that every stock owner should have.

### BLOATING (TYMPANITES).

This is gas forming in the paunch, or rumen, and is a common occurrence among cattle.

**Causes.**—From choking, sudden change in the food, wet clover, or eating frozen roots of any kind.

**Symptoms.**—The left side will be greatly swollen with gas, and in severe cases the whole belly will be distended; by tapping with your fingers on the left side, over the paunch, it will make a hollow, drum-like sound, and on account of the stomach being so much swollen it presses on the lungs, causing the animal to breath very heavy. In severe cases the eyes will be bloodshot, and if it does not soon get relief will stagger, fall and die, actually smothering to death.

**Treatment.**—The treatment must be quick in severe cases. Give the following drench:

Spirits of Turpentine.. ..............2 ounces, or 8 tablespoonfuls.
Raw Linseed Oil..... ..............1 pint.
Common Soda.......................1 tablespoonful.

Mix and give as a drench, and if the animal does not get relief in one hour and a half give

Epsom Salts.........................1 pound.
Sweet Spirits of Nitre......... .....2 ounces, or 8 tablespoonfuls.
Common Soda..................... 2 tablespoonfuls.

Dissolve in a quart of luke warm water and give as a drench, after this give the following drench every hour until the animal gets relief:

Sweet Spirits of Nitre................1 ounce or 4 tablespoonfuls.
Common Soda........................2 tablespoonfuls.

Apply plenty of heat to the body by means of blankets, and hot salt over the kidneys. The after treatment is to feed light, give luke warm water to drink and principally mashes to eat for a few days, so as to allow the stomach to have a rest. Where the animal is very much bloated and in danger of its life, perform the operation of tapping. The way this is done is to cut a small hole in the skin on the left side, midway between the point of the hip bone and the last rib, and about four inches down from the back bone. After the hole is cut in the skin take a trocar and cannula and run them downwards and inwards, then pull out the trocar and leave the cannula in, which lets the gas come spouting out of the cannula ; leave the cannula in for an hour or so until the bloating is all gone down, then pull it out and let the hole heal up itself. A trocar and cannula can be got at a wholesale drug store from 50 cents to 75 cents, and is a very valuable thing to have on hand. If you have not one of these instruments take a sharp penknife and stick it in the same place as you did the trocar and cannula.

### IMPACTION OF THE RUMEN, OR PAUNCH, WITH FOOD.

This is when the rumen, or paunch, gets full of food and is unable to work it out.

**Causes.**—From feeding on poor food, such as straw and such like, and the animal takes a large feed of it, and on account of the walls of the stomach being weak it is unable to work the food out of it as it should, or from a large feed of over-ripe grass, or from the animal getting loose and getting a large feed of corn, bran, oats, wheat or other grain, or from a large feed of any kind of food the animal is not used to. It is also caused by feeding fattening cattle too heavy.

**Symptoms.**—In some cases the animal is slightly bloated, while in other cases there is no bloating at all; the animal seems uneasy, and makes a peculiar grunt or groan nearly every time it breathes. If it is a cow, and milking, the flow of milk will fall away in one night, the nose will be dry, breathing and pulse will be quicker than natural, and the animal will keep getting up and down once in a while, and will not take much food or water. By pressing on the left side, over the paunch, or rumen, it will be found hard and full of food, and where you make the dent in it with your finger the dent will stay in it some time. By tapping over the stomach there will be a dull, heavy sound. The bowels are costive, and the passage is dry, slimy looking, and has a bad smell.

**Treatment.**—Give a dose of physic consisting of

| | |
|---|---|
| Epsom Salts | 1½ pounds. |
| Bitter Aloes | 1 ounce. |
| Common Soda | 2 tablespoonfuls. |
| Ginger | 1 tablespoonful. |

Dissolve in a quart of luke warm water and give as a drench. After this follow up with a stimulant to help the physic to work through the bowels.

| | |
|---|---|
| Whisky | 2 wineglassfuls. |
| Ginger | 2 tablespoonfuls. |
| Common Soda | 2 " |

Mix in a pint of luke warm water and give as a drench three times a day until the animal is better. If the physic has not operated in twenty-four hours after giving it, walk the animal a quarter of a mile, the exercise often starts the physic to work, but if it does not operate wait twenty-four hours more before giving any more physic, at the end of this time, if the physic has not operated, give a stronger physic consisting of

| | |
|---|---|
| Bitter Aloes | 2 ounces. |
| Gamboge | 2 drams. |

Mix in a pint of luke warm water and give as a drench, then at the end of another twenty-four hours give the animal a short walk, and this will generally work the physic; if this has not operated, after walking it, give

| | |
|---|---|
| Raw Linseed Oil | 1 pint. |

Give this drench every day until the physic does operate. Keep the animal well blanketed and warm, and place a half pail of hot salt over the back, for heat helps the stomach to act. Give the animal all the luke warm water it will drink; give it sloppy feed, but no hay. In the course of a week or so after, if the physic has operated without carrying off the load of food out of the rumen or paunch, then, as the last resort, perform the operation of rumenotomy.

### HOW TO PERFORM THE OPERATION OF RUMENOTOMY.

The animal, when sick, is generally very easy held; shove it up against the side of the stall or partition, then fasten it there by means of a long rope, any way at all so you get it solid, and tie the animal so the left side will be out, then, with a sharp knife, make a cut four or five inches long, right through the skin and muscles to the stomach, commencing the cut two inches below the back bone, and half way between the last rib and the front of the hip bone, then pass your hand into the cut and take hold of the stomach, drawing it well out, then make a large enough cut in the stomach to pass your hand in and out free, then have some person to hold the stomach still out through the cut in the side so the food will not fall down between the wall of the stomach and the belly; remove all the hard food in the paunch or rumen, then sew the cut you made in the stomach up with a needle used for sewing wounds; use carriage trimmers' twine, and put the stitches half an inch apart, then wash off the cut which was made in the stomach with a quart of luke warm water and 10 drops of carbolic acid in it; after it is washed off nice and clean shove it back to its natural place and sew up the outside cut, putting the stitches in half an inch apart. The treatment for the outside wound is to bathe the cut twice a day with luke warm water, and, after bathing, apply the white lotion until it is healed up. The after treatment is to give the animal a few drenches of stimulants, such as are mentioned in the treatment of impaction of the rumen, or paunch. Feed the animal on soft food made into the form of soft gruels for a few days and keep it quiet until the cut is healed up.

### VOMITING.

This is sometimes met with in cattle, but never in the horse.

**Causes.**—Generally from some irritation of the fourth, or true digestive part of the stomach, or it may be caused from eating some bones, old clothes, or a boot or something of that kind. Sometimes it is caused from nails being taken into the rumen with food. Often after an animal is slaughtered there are nails and rubbish found in the stomach.

**Treatment.**—Give a physic consisting of

| | |
|---|---|
| Epsom Salts | 1 pound. |
| Brown Sugar | ½ pound. |
| Salt | 2 tablespoonfuls. |

Mix in a quart of luke warm water and give as a drench. This will generally give relief, but if it does not, give:

| | |
|---|---|
| Raw Linseed Oil | 1 pint. |
| Sweet Spirits of Nitre | 1 ounce, or 4 tablespoonfuls. |
| Common Soda | 2 tablespoonfuls. |

Mix and give as a drench every second day until the animal gets relief. Feed on soft food and give luke warm water to drink for a while. In cases where this does not give relief and you are sure there is something in the stomach, perform the operation of rumenotomy and take it out.

### HAIR BALLS IN THE RUMEN OR PAUNCH.

**Causes.**—They are caused from animals licking each other in the spring when the hair is loose, and this hair collects in a ball in the stomach.

**Symptoms.**—The animal soon stops chewing its cud, and has slight spells of bloating; it does not feed well and soon falls off in condition.

**Treatment.**—Give a physic of

| | |
|---|---|
| Epsom Salts | 1 pound. |
| Common Soda | 2 tablespoonfuls. |
| Ginger | 2 " |

Mix in a quart of luke warm water and give as a drench. Give this drench once a week, and if it does not help the animal, and you think there is a hair ball in the stomach, then, as a last resort, perform the operation of rumenotomy. There has been cases known where snakes, from three to four feet long, have been found in the stomach of the ox; there are also cases known where nails have worked their way through the stomach into the heart and killed the animal.

### BINDER TWINE IN THE RUMEN OR PAUNCH.

**Causes.**—From eating straw that has been bound with binder twine, and the twine collects and rolls up in the form of a ball in the rumen or paunch.

**Symptoms.**—Same as that given for hair balls in the rumen or paunch.

**Treatment.**—Same as that given for hair balls in the rumen or paunch.

### IMPACTION OF THE MANYPLIES OR THIRD PART OF THE STOMACH.

This also receives the name of fardelbound, or dry murrain. This is when the food in the third part of the stomach gets hard and dry between the folds and cannot work out. In some cases it gets packed in between the folds as hard as a board.

**Causes.**—From the animal eating dry and over-ripe food that does not contain much nourishment. A common cause is when the cattle are turned out too early in the spring, before the new grass has grown much, and in trying to get at the new grass they fill themselves with old, dry grass, which lodges in third part of the stomach and sets up impaction of it.

**Symptoms.**—First there will be diarrhœa, followed by costiveness and stoppage in the bowels, and anything that does come away is hard and slimy looking. The animal will fall off in condition, and if it is a milch cow she will nearly go dry. The nose will be dry and hot, the pulse quickened and breathing increased, and sometimes there will be slight moaning and grating of the teeth; the belly has a tucked up appearance, and on account of the animal not eating much it does not chew its cud. If the disease is allowed to run on without being relieved it affects the brain and the animal becomes delirious, which is followed by convulsions and death.

**Treatment.**—Give the following drench :

| | |
|---|---|
| Epsom Salts | 1 pound. |
| Ginger | 2 tablespoonfuls. |
| Common Soda | 2 " |
| Salt | 2 " |

Mix in a quart of luke warm water and give as a drench. After this give the following stimulating drenches :

| | |
|---|---|
| Whisky | 2 wineglassfuls. |
| Ginger | 1 tablespoonful. |
| Common Soda | 1 " |
| Powdered Nux Vomica | 1 teaspoonful. |
| Salt | 1 tablespoonful. |

Mix in a pint of luke warm water and give as a drench ; repeat this drench three times a day until the animal is better; keep the bowels open by giving pint doses of raw linseed oil every three days ; by keeping the bowels open and giving these stimulants it will generally work the food out of the stomach in a few days. Give the animal plenty of luke warm water to drink, and feed on soft food. Salt is given to get the animal to drink, which helps to work the food out.

### INFLAMMATION OF THE FOURTH PART OF THE STOMACH.

This is inflammation of the fourth, or the true digestive part of the stomach, and is more often met with in calves than in older cattle.

**Causes.**—In cattle it is caused from eating frozen roots or grass, also over-ripe food. In calves it is caused from changing too suddenly from sweet to sour milk, especially when the calf is young this irritates the stomach and sets up the disease.

**Symptoms.**—First diarrhœa, then constipation, and keeps changing from one to the other every day or so; the animal breathes quick and groans with pain; the nose is hot and dry, and the belly has a tucked up appearance and is sore when you press on it; the legs and ears will be cold.

**Treatment for Calves.**—Give the following drench:

Raw Linseed Oil........................½ pint.
Tincture of Laudanum...................1 dram, or 1 teaspoonful.

Mix and give as a drench. Give its belly a good rubbing with mustard and vinegar; blanket to keep it warm, also place some hot salt in a bag over the back, after this give a teaspoonful of laudanum in half a cup of milk three times a day until it gets relief, and if it will drink give it small quantities of new milk, and in its drinking milk put one teaspoonful of common soda every time you feed it.

**Treatment for Larger Cattle.**—Give the following drench:

Raw Linseed Oil....................1 pint.
Tincture of Laudanum...............1 ounce, or 4 tablespoonfuls.

Mix and give as a drench, afterward give the following:

Tincture of Laudanum...............1 ounce, or 4 tablespoonfuls.
Fleming's Tincture of Aconite........15 drops.

Mix in a pint of water and give as a drench three times a day until it gets relief. Clothe the body well and keep hot salt in a bag to its back. Give luke warm water to drink, and feed on soft food.

### DIARRHŒA IN CATTLE.

**Causes.**—From drinking impure, or stagnant water, eating frozen roots; a sudden change of food, or from excitement by being chased.

**Symptoms.**—There are large passages from the bowels of a fluid nature, and the animal soon gets gaunt and falls off in condition, and in bad cases will not eat, but seems very thirsty.

**Treatment.**—Give the following drench:

Raw Linseed Oil.............. ......½ pint.
Tincture of Catechu..................1 ounce, or 4 tablespoonfuls.

Mix and give as a drench. Clothe the body well and place hot salt to the back. Keep the animal very quiet; give very little water to drink, and take the chill off it; feed on soft, hot mashes and very little other food. In severe cases also apply a mustard plaster over the belly to heat the bowels. If the animal does not get relief in five hours after the first drench then give the following:

Tincture of Catechu..................1 ounce, or 4 tablespoonfuls.
Ginger..............................2 tablespoonfuls.
Common Soda.......................2 "

Mix in a pint of luke warm water and give as a drench every five hours until the animal gets relief. This disease, after it runs on for some time, is apt to terminate in bloody flux (dysentery).

### BLOODY FLUX (DYSENTERY) IN CATTLE.

This comes after diarrhœa, when the manure is streaked with blood.

**Treatment.**—The treatment is the same as given for diarrhœa, only in each of the drenches put a teaspoonful of ground chalk.

### COLIC IN CATTLE.

**Causes.**—This is generally caused from taking a cold drink of water or a change of food especially if it is green or frozen.

**Symptoms.**—This is very painful while it lasts, the animal is very uneasy, lies down, gets up, stretches out, strikes the feet against the belly and moans, and looks around at the side with pain and in some cases is slightly bloated on the left side.

**Treatment.**—Give the following drench:

Epsom Salts.........................1 pound.
Tincture of Laudanum................1 ounce or 4 tablespoonfuls.
Sweet Spirits of Nitre................1 " 4 "
Fleming's Tincture of Aconite.........10 drops.

Mix in a quart of luke warm water and give as a drench, keep the animal warm by blanketing and repeat this drench every hour until the animal gets relief, but after the first dose leave out the epsom salts. Another good drench is

Raw Linseed Oil.......... ........1 pint.
Spirits of Turpentine................1½ ounces or 6 tablespoonfuls.

Mix and give as a drench, after this follow up with the above drench every hour. Another good drench is

Whisky.................................... ½ pint.
Black Pepper............ .... ................ 1 tablespoonful.

Mix in a pint of luke warm water and give as a drench. The danger of this disease is that it may terminate in inflammation of the bowels.

### INFLAMMATION OF THE BOWELS (ENTERITIS)

This is not so common in cattle as it is in horses. It generally affects the small bowels, in severe cases the animal dies in four or five hours.

**Causes.**—It sometimes follows cases of colic or from the bowels getting twisted and stopping the passage, from being out in cold rains; or a sudden change in the temperature and the animal gets a chill which settles in the bowels or from eating musty or frozen food, or from drinking ice cold water when it is hot or anything that will chill the body.

**Symptoms**—There is dryness of the muzzle, loss of appetite and on account of not eating does not chew its cud, it seems very restless and is in severe pain, pawing and getting up and down and does not seem to have a minute's ease. Its urine or water is of a red color and the manure that it passes is covered with slime, the legs and ears are cold, the animal keeps gritting its teeth, and on listening at the side there is no movement to be heard in the bowels, the pulse is very quick but after a short time gets very weak, so weak you can hardly feel it.

**Treatment.**—Bleed the animal as soon as noticed, take three-quarters of a pail of blood from it, if it is in fair condition and a medium sized animal; after this give

Tincture of Laudanum................1 ounce or 4 tablespoonfuls.
Linseed Tea..... ............... .... 1 pint.

Mix and give as a drench every four hours, clothe the body well, place hot salt over the back and a mustard plaster to the belly.

### CONSTIPATION OF THE BOWELS IN CATTLE.

This is not so common in cattle as it is in horses and is more likely to occur in cattle that are feeding high on strong feed such as corn, shorts and mill sweepings or any other rich food is apt to cause it especially if the animal is not getting a few roots along with it to keep the bowels loose.

**Symptoms.**—The animal seems dull, does not care to eat or drink, the muzzle is dry and there is no passage from the bowels.

**Treatment.**—Give the following:

Epsom Salts.................................1 pound.
Bitter Aloes ............................. ...1 ounce.
Ginger.......................................2 tablespoonfuls.
Common Soda...............................2 "

Mix in a quart of luke warm water and give as a drench. If this has not operated in twenty-four hours walk the animal for a quarter of a mile and if it has not operated in twenty-four hours after the walk give it the following:

Gamboge.......................... 2 drams or 1 teaspoonful.
Bitter Aloes.......................... 2 ounces.

Mix in a quart of luke warm water and give as a drench; keep exercising the animal every day, and if the last drench has not operated in twenty-four hours give

Raw Linseed Oil.......................... 1 pint.
Whisky.......................... ½ "
Powdered Nux Vomica.......................... 1 teaspoonful.

Mix and give as a drench, and give this drench every day until there is a passage; keep the body warm with blankets and hot salt in a bag over the back.

### INFLAMMATION OF THE LINING OF THE BELLY CAVITY (PERITONITIS).

The causes, symptoms and treatment are similar to those in this disease in the horse. It is rarely met with in cattle.

### DROPSY OF THE BELLY.

This follows cases of peritonitis, and the symptoms and treatment are similar to that in the horse. This disease is rarely met with in cattle.

### TAPEWORMS IN CATTLE.

This is about the only kind of worms the bowels of the ox are subject to. This complaint is rarely met with in cattle, but in cases where it is there may be from twenty-five to one hundred feet of the worm found in the bowels.

**Symptoms.**—The animal runs down in condition, but still keeps feeding and seems always hungry. The only way to be sure that it is a tape worm is to watch the manure and you will find joints of the worm coming away with the manure.

**Treatment.**—Get rid of the worm by starving the animal for four days, that is, just give it enough to keep it from starving to death. Give the following drench:

Oil of Male Fern.......................... ½ ounce, or 2 tablespoonfuls.
New Milk.......................... 1 pint.

Mix and give as a drench. Give this drench three times a day during the four days you are starving the animal, and at the end of that time give one pint of castor oil, which will bring the worm away all right. Young calves are more often affected than cattle; treat them the same, only give quarter of the dose. As soon as the worm passes away bring the animal back to its regular feed and habits again.

### JAUNDICE OR YELLOWS.

**Causes.**—This may be caused from congestion or inflammation of the liver, or from bile stones forming in the duct of the liver, damming back the bile. It is most often seen in stall-fed cattle.

**Symptoms.**—By pressing on the right side of the belly it causes the animal pain, its appetite is poor and it does not want to drink much; the white of the eyes and the lining of the mouth and nose is of a yellow color. If it is a milk cow the milk falls off in quantity, and has a bitter taste like bile; the animal will sometimes cough a very painful cough, and will soon run down in condition and have a very dull appearance.

**Treatment.**—Give the following:

Epsom Salts.................................1 pound.
Salt.........................................2 tablespoonfuls.

Mix in a quart of luke warm water and give as a drench, but before giving it to the animal put one dram of dry calomel (which acts on the liver) on the tongue with a spoon and wash it down with the drench. Repeat this drench once or twice a week until the animal is better. Feed on soft food, give plenty of water to drink and gentle exercise every day.

### FLUKE DISEASE IN CATTLE AND SHEEP.

This occurs in cattle and sheep pasturing on low-lying lands, and is more often met with in rainy seasons.

**Causes.**—Animals drink the eggs of the flukeworm out of pools of water, or take them in along with the grass; after they get into the stomach in this way they get into the blood along with the nourishment, and pass around in the blood until they come to the liver, where they lodge and form into flukeworms, after this they lay eggs, which pass down out of the liver along with the bile, then out of the system along with the manure; they become dry and are blown into pools of water and over the grass, where the animals again take them up. This is the way they generate.

**Symptoms.**—At first, when the eggs are taken into the liver, they seem to stimulate the action of the liver, and the animal seems to thrive better than ever for a time, but after the worms become full grown the liver becomes diseased, which stops the secretion of the bile, and the animal soon falls off in condition, becomes very dull and weak, and has dropsical swellings under the jaws, throat, chest and belly, and these symptoms are soon followed by death.

**Treatment.**—There is nothing that can be be done but to destroy the diseased animal to keep the disease from spreading, and move the unaffected cattle to a higher and dryer pasture. This disease does not affect horses, so horses can be turned on the pasture the cattle are taken from. On examining the liver after death you will find it diseased, and you will also find worms which are from one-half to one inch long, and have round bodies. In some cases you will find them in great numbers.

## WHITE SCOWERS IN CALVES.

This is a form of diarrhœa in calves.

**Causes.**—Is from an inflammation of the lining of the fourth part, or true digestive stomach, and is generally caused from changing the calf's milk by taking the milk of another cow to feed it on, or from giving it cold or skimmed milk when it is not used to it.

**Symptoms.**—The manure it passes is very thin, and is of a yellowish white color; the calf is in pain, breathes heavy, and groans in spells, keeps gritting its teeth and looking around at its sides.

**Treatment.**—Try and find out the cause of the trouble, and if caused from a change in the milk, or from giving it too cold, give good, warm milk to drink, and follow with the following:

| | |
|---|---|
| Raw Linseed Oil.................. | 2 ounces, or 8 tablespoonfuls. |
| Lime Water........................ | 2 " 8 " |
| Tincture of Laudanum.............. | 1 dram, or 1 teaspoonful. |

Mix and give as a drench, and if this does not give relief give the following:

| | |
|---|---|
| Tincture of Laudanum.............. | 1 dram or 1 teaspoonful. |
| Lime Water........................ | 2 ounces, or 8 tablespoonfuls. |

Mix and give this three times a day in a little milk as a drench. Keep this treatment up, and see that the animal is kept dry and warm until it is better.

## BLOODY URINE (RED WATER) IN CATTLE.

**Causes.**—It is caused by the animal eating some weeds that act on and irritate the kidneys, or it may be caused from stones in the bladder or kidneys, or from a severe strain to the back.

**Symptoms.**—The urine or water is of a smoky, red color, and the animal will pass water often, and strain after making it, and, in some cases, makes a great lot of it.

**Treatment.**—Give as a drench :

Glauber Salts.................................1 to 1½ pounds.

Mix in a quart of luke warm water and give as a drench, and follow up with the following powders :

Ground Gentian Root................................½ pound.
Sulphate of Iron......................................½ "

Mix well together and give a large tablespoontul in a mash twice a day, night and morning.

---

## CHAPTER IV.

# TROUBLES IN CALVING AND DISEASES FOLLOWING.

### HOW TO TELL WHEN A COW IS WITH CALF.

During the hot months of spring and summer a cow will come bulling every third week, and occasionally a well-fed cow kept in a warm stable will come bulling during the winter. After she takes the bull and is with calf she ceases to come bulling or running, and will thrive and feed better, and is of a quieter disposition than before. After a few weeks she commences to get larger at the flanks, which is more noticeable on the right side on account of the calf lying mostly on that side. The calt gradually grows, and at the fifth or sixth month becomes alive, and can be seen moving at the side after the cow has taken a cold drink of water. A few months after the cow has been to the bull you can, by pressing your hand in quickly at the right flank, feel the calf, which is easily told by the shape and hardness of the object you feel. Springing commences in young cows about four months previous to calving, and the bag gradually keeps getting larger until calving time, while older cows generally commence to make a bag from four to six weeks before calving. During the time she is making a bag the vulva gradually keeps getting larger. Near the end of the ninth month, when calving time approaches, the ligaments at the sides of the tail and hips relax and leaves a hollow at each side of the tail, this hollow is well marked a day or so before calving. A few days before calving the cow has a wild expression in her eye, and is cross to other animals, and will try to get away by herself, after this the labor pains come on and she is noticed to be straining, then the neck of the womb opens out, the water bag appears and breaks, and if the calf is coming natural and everything all right, the front legs and head appears. The cow generally lies down, and after a few minutes of severe straining the calf is delivered, and the cleaning, placenta or afterbirth generally comes away at the time of calving, or very soon after.

### DROPSY OF THE WOMB BEFORE CALVING.

This is due to some derangement of the afterbirth, and there is an over abundant amount of fluid secreted around the calf, in some cases several pailfuls collect in the womb around the calf.

**Symptoms.**—The belly keeps getting larger and larger until the cow seems almost as broad as she is long, and on account of so much fluid forming she becomes weak and has difficulty in getting up and walking around.

**Treatment.**—There cannot be much done in this disease only keep the strength up, give a teaspoonful of nitrate of potash or saltpetre in a mash every third day until she calves, this is to act on her kidneys, which helps to get the water out of the womb. Feed plenty of good, strong, nourishing food to keep her strength up and she will be all right after calving.

### PARALYSIS OF HIND QUARTERS BEFORE CALVING.

This disease is generally noticed in poorly-fed, unthrifty cows, especially if they are exposed to cold or wet, and is caused by the calf in the womb pressing on the nerves that go to supply the hind quarters with power.

**Symptoms.**—The cow appears healthy—is eating and chewing her cud, but is not able to rise on her hind parts.

**Treatment.**—Give half-pound doses of Epsom salts once or twice a week, according to how it acts on the bowels, and give the following powder:

Ground Gentian Root .............................. ½ pound.
Nitrate of Potash or Saltpetre .............................. ¼ "
Powdered Nux Vomica .............................. ¼ "

Mix and give a tablespoonful in a slop twice a day. Give plenty of good food; keep her warm; have good bedding under her, and turn her from side to side twice a day until she calves, then she generally comes all right. Never attempt to put her in slings, just let her lie until she is able to get up.

### TROUBLES MET WITH IN CALVING.

Troubles met with in cows when calving, such as deformities, or calves coming in unnatural shapes, is fully explained in connection with "Difficulties met with in a mare foaling." The deformities and positions are about the same, and the principle laid down to take foals away is used in taking calves away. Always, if the calf is coming front end first, have the front feet and head coming together, and if the hind end is coming first, do not attempt to turn the calf, but bring it out with hind feet first. Never be too eager to use hooks, because small ropes are better and there is less danger of tearing the womb; and in cases where there has to be any cutting done, it is best to get an experienced hand, for the parts of the calf have to be skinned inside, commencing at the legs and skinning to the shoulder blade, and then taking it off with the leg; then take out the ribs and insides, and so on, with the other parts until you have enough of the calf cut away so that you can get it out all right.

### CLOSURE OF THE NECK OF THE WOMB AT CALVING TIME.

This is where calving time has come, and labor pains are on the womb, but the neck of the womb keeps contracted or closed, and will not allow the calf to come out of the womb.

**Treatment.**—Give the following drench:

| | |
|---|---|
| Epsom Salts ........................ | 1 pound. |
| Sweet Spirits of Nitre................ | 1 ounce, or 4 tablespoonfuls. |
| Fluid Extract of Belladonna.......... | 1 dram, or 1 teaspoonful. |

Mix in a quart of luke warm water and give as a drench. Keep the body warm with blankets and half a pail of hot salt, in a bag, over the back. On examining the neck of the womb with your hand you will find that you can only get one or two fingers worked into it. Take a small piece of sponge or cloth and saturate it with fluid extract of belladonna, then shove it well into the neck of the womb as if it were for a plug; change this two or three times a day to put more of the fluid extract of belladonna on it. Do not use any rough treatment, for the belladonna in a day or so will dilate or open the neck of the womb enough so that she will calve herself. In case this treatment should fail, take a penknife and nick around the inside of the neck of the womb in several places on the upper side, then use the belladonna as described above—and this will open it.

### RUPTURE OF THE WOMB OR THE PASSAGE OUT FROM THE WOMB.

This occurs at the time of calving in the same way it does in the mare when foaling, and for symptoms and treatment look at rupture of the womb, or the passage out from the womb, in the mare.

### TURNING OUT OF THE VAGINA OR PASSAGE LEADING FROM THE WOMB.

**Causes.**—Both in cows and mares is from standing in the stall with their hind feet too low before calving or foaling, and while lying down, on account of being so full, the womb presses back against the passage and turns it out ; it is liable to come out in cows a few days after calving from straining, or it may be caused from constipation, in either cows or mares where they strain in making manure.

**Symptoms.**—There is a bulging out of the passage about the size of a man's head.

**Treatment.**—Bathe the parts well with luke warm water until it is nice and clean, and shove it back to its place and stitch the edges of the vulva together by putting in a couple of stitches, just leaving space enough at the bottom for the mare or cow to make water. Put the stitches deep in the vulva and allow them to come out of their own accord, which generally takes a week or so. If it is a cow, give a pound of Epsom salts and a tablespoonful of ginger in a quart of luke warm water, and raise her stall two or three inches higher at the back than in the front; feed on rich food, as it will not make such a bulk in her stomach. If it is a mare, give her a pint of raw linseed oil, and raise her stall two or three inches higher at the back than in the front; and in either case be careful until after they foal or calve. In either case, if they are about to have their young, watch them close, and if the labor pains come on, cut the stitches out and give her assistance, and after she is delivered of her young shove the parts back and stitch up the vulva again for a few days, then she well be all right.

### WOMB, CALF OR FOAL BED TURNED OUT.

This is where the calf or foal bed is turned inside out, and hangs down from the vulva. This is not often seen in mares, but is a common occurrence in cows.

**Causes.**—From the animal lying with her hind end too low, and while the womb is in its dilated or enlarged state, after calving or foaling, the body being low behind, the bowels and stomach presses the womb back up into the pelvic, or hip cavity, and as soon as it gets up in this part it causes the animal to have pain and strain, which soon turns the womb inside out. To prevent this from taking place, it is always well to keep the animal standing for a few minutes after having her young, so as to allow the womb to go back into its place, and after this, if she lies down, see that her hind end is not too low.

**Symptoms.**—The animal seems very weak and has a large, red mass hanging out behind, sometimes larger than a large wooden pail.

**Treatment.**—In all cases, as soon as it is noticed, return it, for the sooner it is done the easier it is put back and the less danger there is of losing the animal, for the longer it is out the more it swells. If the cleaning is still attached to the womb—as it is in some cases—remove the cleaning, which is easily done, before returning the womb, by separating it from one button at a time. After this bathe well with warm water, and when it is nice and clean, place a clean sheet or blanket under it and have it held up by two men, one on each side, while you are returning it; after everything is ready for returning it make the cow, or mare, get on her feet, and have her stand so that her hind end is a few inches higher than her front end, then have the men who are holding the sheet raise the womb a little higher than the vulva, this makes it easier to shove in. After this begin turning the womb in, commencing at the edge of the vulva, returning it gradually until all of it is in the passage, then, with your hand closed, press it against the end of the womb and shove it right back to its place and hold it there for a few minutes with your hand and arm. During the time you are returning it be careful not to run your fingers through it. After you draw out your arm place three or four good, solid stitches across the vulva, leaving a little space at the bottom for the water to come out. Cover the animal so that she will be warm and keep a half pail of hot salt in a bag to her back. If it is a cow give the following:

| | |
|---|---|
| Epsom Salts.......................... | 1 pound. |
| Sweet Spirits of Nitre ............... | 1 ounce or 4 tablespoonfuls. |
| Tincture of Laudanum................ | 1 " 4 " |

Mix in a quart of water and give as a drench, after this give one ounce or four tablespoonfuls of tincture of laudanum and ten drops of aconite in a pint of water every three hours until she stops straining, also keep her standing on her feet for a few hours, with her hind end raised three or four inches higher than her front end. In two or three days after the pains are gone take the stitches out and allow her to stand on the level floor again. If it is a mare give one pint of raw linseed oil instead of the Epsom salts, but the rest of the treatment is the same. In both cases remove the stitches in two or three days, when the animal quits straining and seems all right.

### HOW TO TAKE AWAY THE CLEANING FROM A COW.

**Cause.**—Is from a congested and swollen state of the buttons to which the cleaning is attached to the womb inside.

**Treatment.**—As soon as the cow is noticed not to clean give the following drench:

Epsom Salts........................1 pound.
Fluid Extract of Belladonna..........1 dram, or 1 teaspoonful.
Sweet Spirits of Nitre......... ......1 ounce, or 4 tablespoonfuls.

Mix in a quart of luke warm water and give as a drench. Blanket well and keep half a pail of hot salt in a bag to her back; give her a hot bran mash and leave her quiet, and when the medicine operates she will generally clean all right, in thirty hours after you have given her the medicine, if she has not cleaned you will have to take it away with your hand. Roll up your sleeves and oil your right hand and arm, then take hold of the piece of cleaning that is outside with your left hand and pass your right hand into the womb, and by gently pulling the cleaning it will come away quite easy after getting the medicine, it being held in only by the neck of the womb being tightened on it, not allowing it to slip out. In cases where the cleaning has not loosened off of the buttons, keep gradually pulling with the left hand and loosen the cleaning with the right hand off the buttons until it is all worked off. A little practice in removing cleanings soon makes a person perfect at it. When the cleaning is green and too tight on the buttons allow it to remain in another day and give her another drench of the same kind as the first one mentioned, which will make it all right for taking away. Cleaning should never be taken away without first giving the medicine to loosen it from the buttons.

### INFLAMMATION OF THE WOMB (METRITIS).

This disease generally comes on two or three days after calving.

**Causes.**—From getting wet, standing in a draft or anything that will give her a sudden chill, will bring it on.

**Symptoms.**—Slight shivering; the horns, ears and legs are cold, the pulse and breathing quick, she loses her appetite and stops chewing her cud and seems restless on her hind legs, as if in pain; she seems sore on the right side, her vulva is swollen, and she passes bloody looking stuff from it; frequently, after making her water, she seems very thirsty; her bowels are costive, and the urine is of a reddish color; the bag is hot, swollen and and tender, and she does not give as much milk as she should.

**Treatment.**—Give the following:

Epsom Salts........................1 pound.
Tincture of Laudanum.............1½ ounces, or 6 tablespoonfuls.
Fleming's Tincture of Aconite......10 drops.

Mix in a quart of luke warm water and give as a drench; keep her body warm with blankets and half a pail full of hot salt in a bag on her back, and keep it hot by changing it every hour. After the first drench give

Tincture of Laudanum.............1 ounce, or 4 tablespoonfuls.
Sweet Spirits of Nitre..............1 " 4 "
Fleming Tincture of Aconite......10 drops.

Mix in a pint of luke warm water and give as a drench every four hours until she is better. Feed on soft food with boiled flax-seed in it to keep the bowels loose. Give her cold water to drink in small quantities, but often. After this disease has passed off there is sometimes a nasty discharge from the womb of a whitish color, which has a bad smell. This is called whites (Leucorrhœa).

## WHITES (LEUCORRHŒA).

This disease frequently follows inflammation of the womb, but may be caused in other ways, such as handling the womb rough in taking the calf, or cleaning away, or returning a calf bed; anything that will irritate the womb will set up this disease; or it may be caused from the cow being put to the bull too often.

**Symptoms.**—There is a nasty, whitish discharge, which has a bad smell, passes from the vulva, which is often noticed after she makes her water, from the effects of this the cow runs down in condition, gets poor, weak and hide-bound.

**Treatment.**—Give her half-pound doses of Epsom salts dissolved in a pint of luke warm water, twice a week to keep the bowels free. Oil your hand and pass it through the passage to the neck of the womb, then gradually open it up with your fingers until you get it large enough to pass your hand into the womb, then with a pail of luke warm water, soap and a sponge wash the womb and passage out until you get it nice and clean, then with a teaspoonful of sulphate of zinc dissolved in a pint of water bathe the womb well. This will heal the womb and dry up the discharge. In the course of a week, if the discharge is not stopped, dress the womb again the same as above mentioned. Keep giving her the salts once or twice a week, according to how much it acts on the bowels, and, if in the spring of the year, let her out to grass.

### PARALYSIS AFTER CALVING.

This is when the cow apparently seems healthy, only she cannot rise up on her feet, and is generally caused from an injury to the muscles or nerves of the back when she is straining while calving. This disease need not alarm you for, as a general thing, she gets all right in a few days.

**Treatment.**—Give the following:

Epsom Salts.........................1 pound.
Sweet Spirits of Nitre................1 ounce or 4 tablespoonfuls.

Mix in a quart of luke warm water and give as a drench. Keep her body warm with blankets, and apply a quarter of a pound of mustard, mixed in vinegar, over the back every second day. Feed on soft food, with boiled flax seed in it. Milk her out twice a day, and also turn her over from side to side twice a day, but never, under any circumstances, put her in slings.

### MILK FEVER (PARTURIENT APOPLEXY).

This is one of the most fatal diseases cows are subject to, and mostly affects well-fed, fat cows that calve during the hot months of spring and summer, but may affect poor cows. It is also noticed occasionally to affect cows at almost any time of the year; even in the winter, in rare cases, it is noticed.

**Causes.**—The exact causes of this disease are not clearly understood; but it is supposed that on account of the hot weather, and the cow being fat and full of blood, it sets up a fever which affects the nerves, and when the nerves are affected, the milk glands also become affected, and do not secrete the milk, and the milk not being secreted as it should be, leaves the blood charged full of material which should go to form milk, and when the blood becomes full of this material, it affects the brain and nerves, soon causing paralysis. This disease is usually noticed to come on in from one to eight days after calving. The sooner it comes on after calving, the more fatal the disease is. Cows taking it in one or two days after calving seldom get better, but after that time there is more chance of recovery.

**Symptoms.**—At first there is a wild, glary appearance of the eyes, and when you go to milk her there will be very little milk in the bag, which, in most cases, seems soft and flabby. In trying to walk she has a staggering gait. These symptoms will gradually get worse; saliva will run from her mouth, and she will seem greatly excited; keeps staggering, and acts like a drunken man. Finally she gets down, and is unable to rise; her head is turned around to her side; her ears are lopped over, and her eyes now have a peculiar, dull, glassy appearance; the pupils of the eyes are enlarged; she breathes a little heavy; her nose is dry, and she does not take any notice of things around her. If you go to milk her only a little will come out at a time. There is very little or no passage from the bowels, and if you prick her with a pin she cannot feel it. A few hours after this, if she does not get relief, she becomes delirious, moans heavy, lies stretched out, and keeps tossing her head about. The bag keeps getting softer, and after a while, when you try to milk her, you cannot get any. She gradually gets worse and soon dies. We will relate a case where a cow was stricken with milk fever, and owing to the distance we had to go, when called to treat the case, it took four hours to reach her from the time she took sick, and on arriving she was breathing her last. This shows the necessity of every stockowner understanding this disease thoroughly, so he can treat them as soon as they are noticed sick; for this is the only way to treat milk fever successfully.

**Treatment.**—The treatment must be quick. If you notice her sick before she gets off her feet, bleed her; take away from half a pail to a pailful of blood; put her in some shady, cool place and give the following:

| | |
|---|---|
| Epsom Salts | 1 pound. |
| Bitter Aloes | 2 ounces. |
| Nitrate of Potash or Saltpetre. | 1 large teaspoonful. |

Mix in a quart of luke warm water and give as a drench. If you do not bleed her, add to the above drench ten to fifteen drops of Fleming's tincture of aconite according to how fat the cow is; this has a similar action to bleeding. As soon as you give the drench, blanket her, and apply five or six pounds of ice broken up, in a bag, to her head between the horns; this will give her great relief. After this, bathe the bag with luke warm water, and if you can get marshmallow, steep some and add about two cupfuls of it to the luke warm water you bathe the bag with. Bathe the bag three or four times a day and keep trying to milk her as often as you can to get the milk back to the bag again. Apply a half-pound of mustard, mixed with vinegar, over her back every twelve hours. Keep her in a cool place, well blanketed, and keep ice to her head; turn her over from side to side every six or seven hours and follow up with the following drench:

Whisky .......................................... ½ pint.
Nitrate of Potash or Saltpetre ..................... 1 teaspoonful.

Mix in a pint of water and give as a drench every five hours until she gets relief. In milk fever, be very careful in drenching, for, as the cow is unconscious, she is easily choked, simply raise her head enough to allow the medicine to run back into the mouth. If she becomes so paralyzed that she loses the power of swallowing, do not attempt to drench her, but keep on with the other treatment until she can swallow. On account of the bowels being paralyzed, it is well to clean out the rectum, or back bowel, twice a day, by oiling your hand and passing it up, and if there is anything there, clean it out. During the time she is sick allow her to drink cold water, in small quantities, but often. If the cow is going to get better, after a time she looks brighter, raises her head and rests up on her breast bone; milks a little; has a passage from the bowels, and in a few days will get on her feet. Never attempt to raise a cow that has milk fever.

### HOW TO PREVENT MILK FEVER FROM COMING ON.

If the weather is hot and the cow is in good condition, and you are afraid of milk fever, turn the cow into a shady place every day for a week or so before she calves; feed very light and allow her to run out at night, and give her the following medicine:

Epsom Salts .......................... 1 pound.
Sweet Spirits of Nitre ................ 1 ounce or 4 tablespoonfuls.

If she has a very large bag before she calves, milk her every day; after she calves, keep her in during the day and let her out at night for a week, and keep her well milked out. Repeat the above dose and she will generally be all right.

### INFLAMMATION OF THE MILK BAG (GARGET).

**Causes.**—From getting cold in the bag; from an injury; from too great a flow of milk at calving time; from a lump in the teat; or from anything that will stop the milk from being milked out of the bag.

**Symptoms.**—The cow seems feverish; the bag is swollen, hot and tender; she is very thirsty, but does not care to eat; the bowels are a little costive; when you go to milk her it causes her pain; there is very little milk in the bag; and, in severe cases, nothing but a little water will come out. This disease may affect one quarter, half the bag, or all of it. It is most often seen in cows, just after calving, that are kept in high condition. If it is allowed to run on for some time the bag may fester and break, while in other cases, where there is a great deal of inflammation in the bag, mortification may set in, and the part mortified will drop off. The mortification may extend up into the body and cause her death.

**Treatment.**—Give the following:

Epsom Salts. . . . . . . . . . . . . . . . . . . . . . . $\frac{3}{4}$ pound.
Sweet Spirits of Nitre. . . . . . . . . . . . . . . 1 ounce or 4 tablespoonfuls.
Nitrate of Potash or Saltpetre . . . . . . . . 1 teaspoonful.

Mix in a quart of water and give as a drench, and repeat this every second or third day until she is better. This will carry off the fever out of her system. Bathe her bag well with warm water and vinegar three times a day; after bathing, apply white liniment, and as soon as the liniment is on, oil the bag with lard or goose oil to keep the liniment from blistering and also to soften the bag. Milk her three or four times a day and feed light until all the soreness is out of the bag. In cases where the milk stops coming entirely, and the bag is festering, watch for a soft place in the swelling, and, as soon as it forms, lance it and let the matter out. After you lance the bag, if it smells bad, put a few drops of carbolic acid in the water that you bathe it with; this will kill the smell and clean the wound or hole. The rest of the treatment is the same as given above. In case the bag mortifies, give it lots of bathing with the hot carbolic water, as above mentioned, three times a day, then apply the white lotion, and give the drench mentioned above once a week instead of every second day; the mortified part of the bag will gradually rot away and heal up.

### SMALL ROUND LUMPS IN THE PASSAGE OF THE TEAT.

**Causes.**—From a bruise or injury to the passage of the teat in some way, and when it is healing the thickening or lump forms.

**Symptoms.**—There is a small lump in the teat which can be felt between your finger and thumb when you are handling the teat. These lumps may be anywhere along the milk passage of the teat. The first summer the cow is affected with these lumps in the teat they interfere greatly with your milking her, but if she is bred again when she calves these lumps will entirely block the teat and give you a great deal of trouble for you cannot get the milk down, and the bag becomes swollen and inflamed, and in a great many cases she loses the affected quarter.

**Treatment.**—Generally, the first season they are affected with it, you can get the milk out all right, but it is advisable not to breed the cow again but let her go dry and fatten her, for she will be worse next summer. When you cannot get the milk down with your fingers pass a teat syphon or milk tube up the passage of the teat through the lump far enough to reach the milk, and then the milk will run till the quarter is milked out. Use the tube each time you are milking the other teats. These teat syphons can be got at almost any drug store or veterinary instrument store for about 10 or 15 cents. The way to use the tube is to first tie a colored string in the small ring at the side so you won't lose it if it drops out in the straw, then oil it; take hold of the teat with your left hand and with your right hand pass the teat syphon up through the passage to the lump, and when you come to it gradually force it through, which is easily done; continue passing it gently up until the milk runs out, and leave it in until all the milk is out. Bathe the bag twice a day with warm water and vinegar, after bathing apply white liniment and then oil the bag with lard to keep it soft. It is advisable in very bad cases to let that quarter of the bag go dry as soon as you can.

### BLOODY MILK.

**Causes.**—From any injury to the bag, getting cold in it, or from eating irritating weeds.

**Treatment.**—Give the following:

| | |
|---|---|
| Epsom Salts | 1 pound. |
| Nitrate of Potash or Saltpetre | 1 teaspoonful. |

Mix in a quart of luke warm water and give as a drench; give a teaspoonful of saltpetre in a mash every night. After milking bathe the bag with warm water, wipe dry and apply white liniment, then oil the bag with lard or goose grease, and the milk will soon get all right. If you think it is caused from eating irritating plants put the cow in another pasture.

### BLUE MILK.

This is where the milk is watery looking and very blue.

**Causes.**—From little germs, called bacillus cyanogenus, getting up into the teat. The only way to be sure it is caused by these germs is to examine the milk with a microscope.

**Treatment.**—With a small glass syringe inserted into the passage of the teat inject some of the following each time after milking:

Hyposulphite of Soda........................................1 dram.
Water.......................................................1 pint.

Shake well before injecting, and after a few injections the milk will be all right.

### STRINGY MILK.

**Causes.**—From swallowing small germs while drinking out of stagnant pools of water.

**Symptoms.**—A few days after the germs are swallowed the cow's milk will be curdy and stringy looking, mixed with water, and will come out in jerks when milking. It will be like this for a few days, then get all right for a week or so when it will come on again. Generally two or three cows out of a large herd will be affected in the same way.

**Treatment.**—If it is caused from drinking out of low springs or pools keep the cows away from the water by fencing it off, and give two drams, or one teaspoonful of bisulphite of soda in a mash every night, which will soon make the milk all right, and the disease will not come back on her again.

### CHAPPED OR SORE TEATS.

**Causes.**—This is caused from milking with rough hands; or from the cow running through long grass and wetting and irritating the teat; or it may be caused from flies.

**Treatment.**—Each time before milking wash the teats off with luke warm water and a little castile soap, then after you have milked her rub the teats with the following salve:

Vaseline.........................................2 ounces
Oxide of Zinc....................................1 dram.
Carbolic Acid....................................10 drops.

Mix well together and put in a box large enough to hold it. This is a cheap and a grand healing salve for any kind of sores around the bag.

### SMALL WARTS ON THE TEATS.

These are very troublesome when you are milking, but are very easily got rid of if you take the right plan.

**Treatment.**—After the cow is put dry is the best time to treat. Tie the cow up and hobble her two hind legs together above the hocks with a strap so she cannot kick you, then with a pair of large, sharp scissors clip all the warts off as close as you can to the teat. By cutting them off with scissors they will not bleed. After they are taken off dress them once a day with the same salve used for chapped teats and they will not come on again, but if they should come on the next year use the same treatment again.

### CUTS AND FISTULA OF THE TEAT.

This is when the teat has been cut deep enough to cut the milk passage, which allows the milk to keep dripping out through the hole.

**Treatment.**—If the cut is big sew it up with a needle used for sewing wounds; bathe with warm water and apply white lotion every time after milking. The best way to milk a cow while the teats are sore is to insert a teat syphon, or milk tube, up into the teat, and this will let the milk run out without irritating the teat; sometimes after it is healed up there will be a small hole in the side of the teat, which will allow the milk to leak out while you are milking. The best way to fix this is after she has gone dry burn the hole with a pointed stick of caustic potash, which destroys the fistula, then while it is healing up the hole will disappear and be all right the next time she calves.

### COW POX.

This is often seen in cow, and affects herds in all parts of the world. It is somewhat similar to smallpox in people only it is not nearly so fatal. This is an infectious disease, that is, it can be carried from one cow to another. For instance, one man milking eight or ten cows and only one has the disease at first, he will carry the disease to all the others by milking them.

**Symptoms.**—The cow seems feverish and does not give quite so much milk. In a few days, little red, pimple-like spots appear around the teats. In a day or so more, these red spots will rise up in the form of a blister, which contains a watery fluid. If these are not broken during milking, they dry up themselves and form scabs, which, in a few days, drop off, leaving the teat natural. It generally takes this disease from eight to ten days to run its course; but sometimes, when the blisters on the teats are broken by the milker's hand and kept irritated by milking, or from flies, it takes a long time to heal them up.

**Treatment.**—Keep the cow separate from the others, and allow only one person to milk her, and no other, so as to keep the disease from spreading, if in the spring, only let her have grass; if in any other time of the year, feed on soft food with boiled flax seed in it, and give the following powders for her blood and kidneys:

| | |
|---|---|
| Nitrate of Potash or Saltpetre | ¼ pound. |
| Sulphur | ¼ " |
| Ground Gentian Root | ¼ " |

Mix thoroughly and give a teaspoonful night and morning in a mash. Each time before milking her, bathe the teats with luke warm water and soap, then milk her carefully and use the following preparation:

| | |
|---|---|
| Sweet Oil | 4 ounces. |
| Carbolic Acid | 10 drops. |

Mix and apply to the sore parts of the teats each time after milking.

### COWS LOSING THEIR CALVES (ABORTION).

**Causes.**—This generally occurs from slipping on ice; being chased by a dog; or from the hook of another animal.

**Symptoms.**—Labor pains come on; she will get up and down; the water bag appears and breaks; and if the calf is coming straight, it soon appears and comes away all right.

**Treatment.**—If the cow keeps on straining, and the calf does not come, oil your hand, pass it up into the womb and straighten the calf, and it will come away all right. After the calf is taken away, cover her up warm, and if she does not seem very well give her the following:

| | |
|---|---|
| Sweet Spirits of Nitre | 1 ounce or 4 tablespoonfuls. |
| Epsom Salts | 1 pound. |

Mix in a quart of luke warm water and give as a drench. If the cleaning does not come away, use the same treatment as is given in "How to take away the cleanings from a cow." After this, feed on soft food, keep her warm and milk her twice a day; this will bring her back to her milk.

### BARRENNESS IN COWS AND BULLS.

This is a common thing in well-bred cows, especially in Jerseys.

**Causes.**—From their being kept in too high condition; from a diseased state of the ovaries; contracted or diseased state of the neck of the womb; the womb being deformed, such as the neck being twisted to one side; or where there is twin heifers one or the other will be barren. Bulls or cows that are too closely inbred in the same line of breeding for several generations may become barren, or what is known as run out; it is also caused in bulls sometimes from fatty degeneration of the testicles—mostly seen in old bulls—and, also, rig bulls (that is, where only one or neither of the testicles are down in the scrotom) are sometimes barren. This rule also holds good in horses.

**Treatment.**—If it is caused from being in high condition bleed her, take a half pail of blood away the day before taking her to the bull, or give her a physic of one and a half pounds of Epsom salts in a quart of luke warm water as a drench. The idea of this is to cool her blood. Examine her, and if it is from contraction of the neck of the womb, pass your hand up gently and open it by working your fingers in it; if it is from the neck of the womb being to one side, straighten it. In doing this have your hand and arm oiled. In either of these cases put the cow to the bull immediately after fixing it. If it is from inbreeding try and start her to breed by putting her to a mongrel bred bull. There cannot be much done for a barren bull.

---

CHAPTER V.

# DISEASES OF THE GENITAL ORGANS OF THE BULL.

### INFLAMMATION OF THE TESTICLES (ORCHITIS).

**Causes.**—It is generally from an injury, or from serving too many cows.

**Symptoms.**—The bull moves stiff and has a straddling gait, the testicles are swollen and very tender.

**Treatment.**—Give the following:

| | |
|---|---|
| Epsom Salts | 1½ pounds. |
| Nitrate of Potash or Saltpetre | 1 teaspoonful. |
| Ginger | 1 tablespoonful. |

Mix in a quart of luke warm water and give as a drench. Bathe the testicles well with hot water and vinegar and apply a poultice of hot linseed meal and bran, about half and half. Have the poultice held up to the bag by means of strings tied up over the back and it will give steady heat to the bag and draw the inflammation out; it will also support the testicles and ease the pain; keep the poultices hot by changing them twice a day; keep this treatment up until the bull is better, and do not let him get cold afterward. During the time you are treating him feed on soft food and he will soon be all right. If, after he is well, you find that the inflammation has destroyed the seed part of the testicles—which can be told by putting him to cows and if they do not get with calf—castrate him, for he will be of no further use for breeding purposes.

### INJURIES TO THE SHEATH AND PENIS.

**Causes.**—From jumping over a fence and being caught on it, or being caught in any way that will injure the sheath and penis.

**Symptoms.**—There is swelling and soreness of the sheath and penis. If it is a bull he will be unfit for service until he is better.

**Treatment.**—Examine to see that there is nothing in the end of the sheath, and if there is anything remove it at once. Bathe well three times a day with luke warm water, wipe dry and then apply the white lotion. Keep this treatment up until the swelling is all out. In very severe cases give him a pound of Epsom salts.

### CLAPP IN BULLS (GONORRHŒA).

This is inflammation of the lining of the passage of the penis.

**Causes.**—From too frequent service, or from serving a cow that is affected with whites (leucorrhœa).

**Symptoms.**—There is a whitish fluid discharge from the end of the penis and sheath, and they are also very sore to handle, and in making his water it scalds him and causes him pain.

**Treatment.**—Give him one pound of Epsom salts in a quart of luke warm water as a drench to cool the blood; bathe the sheath well with luke warm water twice a day, wipe dry, and inject into the sheath a little of the following mixture:

Sulphate of Zinc........................2 drams or 1 teaspoonful.
Water................................1 pint.

Mix and shake well before using. This is a cheap and effective cure for this disease. Keep the bull away from cows until he is better, as he would give the disease to the cows.

### CASTRATION OF BULLS AND CALVES.

Secure him, either by standing him in a firm stall or throwing him down, which can be easily done by taking a rope about thirty feet long, make a loop in the centre large enough to slip over his head and neck and then tie a knot in it; after the rope is put over his head and fitted on the neck, pass it back between the front legs and bring the ends of it back one on each side around the outside of the hind legs, and back around the inside just above the hock, then bring them forward on the outside of the front legs through the loop; have a man on each side to pull on the ropes, which will soon throw him down, then tie him up solid. This is the best method known for throwing cattle. As soon as you have secured him, take a sharp knife and make a cut along the side of the bag large enough to let the testicle out; be sure the cut extends to the bottom of the bag so it will not form a pocket.. As soon as you have let the testicle out, draw it well up and you will notice a white covering attached to the back part of it, cut this off close to the testicle with your knife, then you can pull up the testicle and cord free; after this, pull the testicle and cord well up, and if the bull is over a year old, tie the cord with a strong, fine piece of string about four inches above the testicle, leaving the ends of the string six or eight inches long so they will hang out of the bag and not heal up in it; leave this string on until it drops off itself; cut the testicles off below the strings; fill the holes full of salty butter and let him go. The reason the string is tied on the cord when the animal is one year old and upwards is because there is danger of him bleeding to death from the cords. The operation of castrating a bull standing up is done by securing him in a solid, narrow stall and operating in the same manner as you would if he was lying down.

The way to castrate a calf is to tie him or have some person hold him; make the cuts in the bag the same as for castrating bulls; when the testicle is out, separate the covering attached to the back part of the testicle with your knife; then draw the cord and testicle well up, and with your knife scrape up and down on the cord until it is scraped off, this will stop the bleeding; fill the holes full of salty butter and let him go. The main thing after castrating bulls, bull calves, boars and dogs is to keep them away from dampness, and if they swell, bathe with luke warm water and soap and open up the cuts with salty butter on your finger. If it swells very much, bathe with luke warm water and salt three times a day, and after bathing apply white lotion. Sometimes, a few weeks after the cuts are healed up, the bag swells and becomes very sore and hot; in this case you may know there is matter forming in the bag. Bathe well three times a day with luke warm water; after bathing apply white lotion and put on a hot poultice of half linseed meal and half bran; fasten the poultice on by means of strings over the back; this is to bring the festering to a head. Change the poultice every time you bathe the bag. As soon as you find a soft spot in the bag, lance it to let the matter out; make a good sized hole in it, large enough to run your finger up into it to clean it out. After this, treat by bathing with luke warm water and soap and applying the white lotion twice a day; keep the cuts open by putting butter on your finger and running it up into the hole once a day until it commences to heal.

### RIG OR ORIGINAL BULLS.

This is when one or both the testicles never come down into the scrotom, or bag. These kind of bulls cannot be castrated like horses, and after they get a little age on them they become a perfect nuisance.

**Advice.**—When you go to castrate a calf and find only one or neither testicles are down, fatten and get rid of it, for it very rarely comes down afterwards. It will save you a lot of trouble if you get rid of it while young.

### HOW TO RING A BULL.

Secure the animal by throwing him, or having him in a solid, narrow stall; take a piece of sharp-pointed, clean, hard wood, or a sharp piece of bright steel large enough to make a hole for the ring; put the hole through in the soft part of the nose, just in front of the hard cartilage that separates the nostrils, which is easily felt. After the hole is through, open and oil the ring, slip it through, close it and put in the screw. After the ring is in turn it every day until the wound is healed. These rings can be got at any hardware store.

### SWELLING OF THE POINT OF SHEATH IN STEERS.

**Causes.**—It is mostly seen where they are grazing on a pasture field where there is a lot of limestone, or when there is much lime in the water they drink, and on account of the steer not putting out his penis while making water, just letting it dribble out of the sheath, the lime in his water collects and forms a small limestone which soon gets large and irritates the sheath, causing it to swell.

**Treatment.**—If he is a quiet steer, let him stand, and have someone to hold him by the horn and nose, while you, with your fingers oiled, pass one of them up into the sheath, and by working the stone around you can soon remove it, then oil the sheath inside and outside with lard and it will be all right. In case you cannot do this with the steer standing up, throw him down and secure him as for castration and remove it in the same way.

## CHAPTER VI.

# DISEASES OF THE EAR AND EYE.

Diseases of the ear of the ox are very rarely met with, and are similar to those of the horse, and for any information concerning them refer to diseases of the ear in the horse.

### CANCER IN THE EYE.

The eye of the ox seems to be a favorite place for cancers and is very often met with.

**Causes.**—The causes of it are the same as other cancers, that is, the cancer germs get into the blood, for the disease first begins in the blood, but afterward locates and shows itself in the eye, although some say it will come on from an injury.

**Symptoms.**—The first symptom is dullness of the eye, with tears flowing from the corners; there will be a bulging out of the eye, and if you look close you will see in the back part of the eye a small growth; the animal may thrive fairly well for a while, but will fall off in condition as the growth comes on account of the pain in the eye; soon the growth gets so large that it will destroy the whole eye and hang down on the cheek; it gets very angry and red looking, and will bleed freely if the least thing touches it; the cancer keeps on growing, and in a short time the bones around the eye become diseased, and when they become diseased, they also become enlarged, and have a very bad smell.

**Treatment.**—By removing the eye in the early stages of the disease you can effect a cure. This is done by throwing and securing the animal; have the head held solid, and with a knife cut around the eye and loosen it from the eyelids, then stick a small hook into the eye and pull it out as far as you can, then take a piece of carriage trimmers' twine and slip it around the back part of the eye and tie it tight—this will stop the bleeding—then cut the eye off in front of where you tied the string; saturate a piece of cotton batting with Monsell's solution of iron and insert it in the hole where the eye came out of; take the batting out the next day and bathe the eye twice a day with luke warm water and soap, after bathing apply white lotion; if the parts are raw and angry looking touch the spots with caustic potash every day. This may effect a cure, but in a case where it grows again, or where the growth is very bad before operating, or the bones diseased, have the animal destroyed immediately.

### FOREIGN SUBSTANCES IN THE EYE.

Sometimes chaff, barley-beards or small pieces of stick get into the eye and become lodged there.

**Symptoms.**—The animal suffers very much; tears run down over the cheek; the eye becomes very much inflamed and dim, and if allowed to run on the sight will soon become covered with a white scum. If you catch the animal and examine the eye closely you will find out what is in it.

**Treatment.**—In all cases catch the animal and examine the eye closely, and when you find out the cause of the trouble remove it, then bathe the eye every day with new milk or luke warm water; wipe dry and apply, in and around the eye every time after bathing, the eye wash mentioned in the receipts at the back of this book. Keep this treatment up till the eye is better.

### A GROWTH ON THE HAW OF THE EYE.

This is a red growth in the inner corner of the eye, caused by some irritation of the haw of the eye, which is a piece of cartilage or tough membrane that fits across the inner corner of the eye.

**Symptoms.**—At first the eye looks sore and angry in the inner corner and runs water freely, afterwards followed by the red, angry looking growth growing out of the corner of the eye. The growth may vary from the size of a marble to a small hen's egg.

**Treatment.**—Secure the animal by throwing it the same as mentioned for castrating bulls. Have the head held firmly on the ground, take hold of the growth with a small hook, or anything that you can hold it with; pull the growth out of the eye far enough to get under it with a pair of scissors and clip it off; there is usually very little bleeding or trouble with it afterward; bathe the eye with new milk once a day, wipe dry and apply the eye wash.

### SORE OR INFLAMED EYES IN CATTLE.

**Causes.**—From getting cold in the eyes or from an injury.

**Symptoms.**—Tears run freely, and the eyes are very weak and red looking, and if not relieved a scum soon forms over the sight of the eye.

**Treatment.**—Bathe well twice a day with new milk or luke warm water, and each time after bathing wipe dry and apply the eye wash.

### CATARACT OF THE EYE IN CATTLE.

This is very rarely met with in cattle, and for information regarding it look up cataract of the eye in horses, for the causes, symptoms and treatment are the same.

### INJURIES TO THE EYELIDS.

It may occur in a good many ways, as a kick from a horse, a hook from a cow, or from catching on something.

**Treatment.**—If they are torn much stitch them up with a sewing needle, used to sew up wounds, and carriage trimmers' twine, which is the best twine for this work, and treat it afterwards by bathing with new milk or luke warm water and applying the eye wash. Do this twice day and it will soon heal up and the stitches will work out themselves.

---

## CHAPTER VII.

# FRACTURED BONES, WOUNDS, SPRAINS OF JOINTS AND TENDONS.

### FRACTURED BONES.

Fractures occur in various ways, from the kick of a horse, getting caught on a fence while jumping, falling, being chased by dogs, or being struck with anything hard enough to fracture a bone. As a rule, when the fracture is so bad that the bone is shattered, or a piece of the bone is stuck through the skin, it is best to destroy the animal, and if it is fat enough it makes good beef, provided it is killed in time.

### FRACTURE OF THE LOWER JAW.

This is a fracture that generally occurs from a kick or a blow of some kind, and is first noticed by the animal not being able to eat, and the mouth will seem crooked. It is recommended in cases of this kind to set the jaw in place, and have it held there with wire fastened around the teeth in the jaw. This can be done better if it is the front of the jaw that is fractured. When you have set the jaw with wires feed on soft food, such as gruels, that the animal can drink down. If the animal is fit to kill it is best to butcher it.

### BROKEN NECK OR FRACTURE OF THE NECK BONES.

To show how simple this may occur we will relate a case that came under our own personal observation. A cow that had been kept in the stable during the winter and fed well was let out one day by the owner, and she, feeling good, was playing about when the dog was put after her, and while running away from the dog she kicked up her hind feet and lowered her head, in doing this she caught her nose on the ground, which threw her over onto her head and neck. We heard the bone snap, and by the time we got to her she was dead. On examining her we found that one of the bones of the neck was broken in the fall, which caused her death instantly. In any case where the bones of the neck are fractured enough to press on the spinal cord it will cause death instantly.

### FRACTURES OF THE BONES OF THE BACK.

This may occur from something falling on the animal, or by slipping and falling, or from another animal jumping on it while standing crooked.

**Symptoms.**—There is paralysis of the hind quarters, attended with pain ; the animal will moan and refuses to eat anything. In severe cases the back will be swollen, and the mark of what caused the fracture can be seen.

**Treatment.**—It is best to kill the animal, but if you wish to trv to treat it, keep it quiet, feed on soft food and keep the bowels regulated by giving small doses of salts ; turn it from side to side twice a day, and be careful while turning it not to hurt its back.

### FRACTURE OF THE BONES OF THE HIP.

In some cases we have a hip knocked down from a blow, or from running through a narrow doorway and striking the hip. This is not dangerous, only it spoils the look of the animal when its hip is knocked down. If it is sore after being knocked down bathe twice a day with luke warm water and apply white liniment until the soreness is out ; if the bone heals all right do nothing more to it. Sometimes we have a case where the broken piece of bone does not heal to the other ; it soon begins to fester around it, and the parts become swollen and sore. You must then open it with a sharp knife and remove the broken piece of bone. Fracture of the under part of the hip bones generally occurs from the animal slipping on ice when the legs straddle out. As soon as it gets up it walks off very stiff, and the legs are straddled out behind while walking or standing. The treatment for this is to keep the animal very quiet by tying it in a stall until the bones unite, which generally takes four or five weeks.

### FRACTURES OF THE RIBS.

This is always the result of a kick or a blow of some kind. In a severe case the animal cannot raise to its feet; there will be a dinge in the side, and on shoving it in and out you can hear the bones grating on each other. In slight cases the animal will be able to get up all right, but will be stiff and sore. In most of these cases the animal will cough a little, and breathe short and quick.

**Treatment.**—All that is needed is quietness, good care and food. If the animal is not able to raise turn it over from side to side twice a day. If the rib is broken so bad that it penetrates the lung and sets up inflammation there is no hope of recovery.

### FRACTURE OF THE SHOULDER BLADE OR SHOULDER BONE.

Fracture of these bones is indicated by the extreme lameness and pain it causes the animal, and on moving the leg you can hear the bones grating on each other. In a case of this kind it is best to destroy the animal, but if the fracture is not so severe, and it is a young animal, keep it very quiet and feed well, and it will come all right in the couse of time. The less you bother with it the better.

### FRACTURES OF THE BONES BELOW THE KNEE.

There is crookedness of the leg, lameness and extreme pain, and when you move the leg you can hear the bones grating on each other.

**Treatment**—Get the animal in a quiet place, set the leg in shape, and have some one to hold it while you bandage it with a starched bandage, which is a long strip of cotton dipped in starch used for starching clothes. On drawing the bandage out of the starch draw it between your fingers to clean out as much of the starch as you can, then wrap it moderately tight around the leg, put lots of the bandage on, and have some one to hold the leg and bandage straight for an hour or so until the starch hardens the bandage. After that the bandage will hold the leg to its place. Leave it on four or five weeks until the bones are healed. Keep the animal quiet until the bones are well knit together. If the leg should swell with this bandage take it off and put it on looser.

Fractures above the knee are sometimes treated by this method, but not nearly so successfully.

### FRACTURES OF THE BONES ABOVE AND BELOW THE STIFFLE JOINT.

The animal may not able to stand, but if it is the leg will be hanging loose. By moving the leg you can hear the broken bones grate on each other.

**Treatment.**—In very severe cases it is best to destroy the animal, or, if is a fat animal, kill it for beef. If you attempt to treat it you will not be able to do anything for it only keep the animal quiet, and leave it lying down or standing up, whichever it prefers.

### STIFFLE OUT IN CATTLE.

For this disease we refer you to dislocation of the patella (stiffle out) in horses, for the causes, symptoms and treatment are the same in both. This does not occur so often in cattle as it does in horses.

### FRACTURES OF THE BONES BELOW THE HOCK.

For this we refer you to fracture of the bones below the knee, for the causes, symptoms and treatment are the same in both cases.

### SPAVIN.

This is generally seen in working oxen, or cows. There is lameness and an enlargement on the inner side of the lower part of the hock joint, similar to spavin in horses. Blister with the following :

| | |
|---|---|
| Biniodide of Mercury or Red Precipitate | 2 drams. |
| Powdered Cantharides or Spanish Fly | 3 " |
| Vaseline or Lard | 1½ ounces. |

Mix thoroughly and clip off the hair on the inside of the hock over the enlargement, and rub on half of this blister, rubbing it in well, and tie the animal short so it cannot lick it ; grease the blistered part the third day after blistering, then let it go for a month, and then wash it off with warm water and soap and keep repeating the blister until the animal is over the lameness, which generally takes three or four months.

### SPRAINS IN ANY PART OF THE ANIMAL.

**Causes.**—Generally from the animal stepping crooked, or a dog worrying it, or from fighting.

**Symptoms.**—There is heat, swelling, pain and stiffness, or lameness according to the part of the body it is in.

**Treatment.**—Bathe the parts well with luke warm water and vinegar three times a day ; after bathing wipe dry and apply the white liniment. If the sprain is in a joint of the legs, by bandaging it each time after bathing will help to relieve the pain and support the joint.

### WOUNDS OF ALL KINDS.

For wounds we refer you to the explanation of wounds given in horses, for they happen in a similar manner and are treated the same; but, in sewing the skin in cattle you will find it tougher and a little harder to sew than in the horse.

### AIR UNDER THE SKIN CAUSED FROM A WOUND.

Sometimes from a very small wound in the ox, air gets under the skin into the tissue which connects the skin to the body. The amount of air which gets in varies greatly, sometimes only a small amount works in just around the wound, while again we have seen cases where so much air would work in that it spread all over the body under the skin and make the animal look double its natural size. The way to be sure it is air, is to rub your hand over the skin and it will make a peculiar crackling noise.

**Treatment**—The main thing to be done in this, is to tap the skin in several places around where the air is with a penknife and let the air escape; rub the skin to get all the air out, and after this give the skin a good rubbing once a day with white liniment, then give the wound the treatment given for wounds, and in a few days the air will all disappear. This disease is sometimes met with in the horse and is treated the same.

### MAGGOTS IN NEGLECTED WOUNDS.

**Symptoms.**—The wound smells bad, is dirty, and if you stir it, the maggots can be seen moving around in it.

**Treatment.**—Give the wound a thorough cleaning, by washing it out with luke warm water and soap, after this apply creolin lotion, this will destroy them. In very bad cases, if this does not effect a cure, give the wound a good dressing with spirits of turpentine, and afterwards bathe twice a day and apply the creolin lotion.

### RHEUMATISM.

This is a kind of inflammation that affects the joints and tendons.

**Causes.**—From bad blood where there is too much acid in it; from getting wet or from lying on the damp ground in the spring of the year, will cause it.

**Symptoms.**—There is swelling and soreness in the joint affected; it may affect one joint for a while, then move to another, and so on.

**Treatment.**—Rub the affected joint well with acid liniment once a day and give the following:

Nitrate of Potash or Saltpetre.......................... ¼ pound.
Common Soda.......................................... ¼ "
Salicylic Acid........................................... ¼ "

Mix and give a tablespoonful twice a day in its feed.

---

## CHAPTER VIII.

# DISEASES OF THE SKIN.

### WARTS AND SMALL GROWTHS ON THE SKIN.

Some cattle are more subject to warts than others. If the warts have a neck the best way to get rid of them is by tying a small, strong string tightly around the wart, as close to the skin as you can, and by leaving the string tied tight on it it will stop the blood circulating in the wart, which will cause it to die and drop off. If the wart is flat and has no neck that you can tie the string on cut it off with a sharp knife and burn it with a stick of caustic potash. Small lumps, or tumors, in the skin are very common in horses and cattle, but are easily got rid of. If it is a horse put a twitch on his nose and have one of his front feet held up; if it is a cow tie her up solid, then cut a hole in the skin over the lump and skin around it, then lift it up and cut it off at the bottom. There is not much danger from bleeding so long as you do not cut into a large vein, which you can see right in the skin. The after treatment is to bathe with luke warm water twice a day and apply the white lotion after bathing until it heals up

### LICE ON CATTLE.

They may be cattle lice or hen lice.

**Symptoms.**—The animal keeps rubbing itself until it rubs the hair off in places, and if you examine closely you will see them in the hair. Cattle affected with lice do not thrive well.

**Treatment.**—The treatment is very simple and cheap. Here is the most effective remedy known if it is properly used:

Creolin.............................. ½ ounce, or 2 tablespoonfuls
Water................................ 1 pint.

Mix and shake well and there will be enough in this to go over a cow twice. Before applying it give the animal a good brushing with a stiff brush; pour the wash into a flat dish where you can get at it, and with a brush or sponge, dipped in the wash, rub it all over the animal. Do this twice a week until the lice are all killed. Twenty cents will buy enough creolin to kill the lice on twenty head of cattle.

### MANGE IN CATTLE.

For this disease refer to mange in horses, for the causes, symptoms and treatment are the same.

### RINGWORM.

This is a common disease in young cattle and calves.

**Causes.**—From a parasite, or germ, getting into the skin and working around the bottom of the hair, causing it to fall out in round patches. This disease affects young cattle more than older ones, but may affect them at any age.

**Treatment.**—The cheapest and best remedy for this is crude petroleum oil painted over the spot and a little over the edges of the ringworm, in the sound skin, to keep it from spreading. Paint this on with a feather every day, or every second day until the ringworm disappears. Be careful in handling ringworms on cattle, as you are liable to get them yourself from the cattle.

### HORN FLY.

These are also called Austrian flies and are a small, black, hard fly. They first started on this continent in the Eastern states in the year 1887, and since then have spread all over the country. They start to bother the cattle during the warm days of May and continue bothering them until the frost comes in the fall. These flies become very numerous on the cattle, and they pierce holes in the skin and suck the blood. While they are resting they light on the horns, and sometimes the base of the horn is literally covered with with them.

**Treatment.**—Apply wagon grease, tar, or some oily substance around the base of the horn every few days to keep them from resting on the horn. Wash the animal's body twice a week with creolin wash, which is very cheap and effectual.

Creolin.................. ..........2 ounces or 8 tablespoonfuls.
Water..............................1 gallon.

Mix, shake well and rub them over twice a week with a cloth or sponge dipped in this wash and it will keep the flies from bothering them.

## WARBLES.

This trouble is only found to affect cattle, and is caused by a large fly, called the gadfly, stinging the animal around the back; this fly lays an egg down in the skin each time it stings, which develops into what is known as the warble.

**Symptoms.**—Small lumps appear in the skin, principally over the animal's back, and coming on spring, these lumps, which contain the grub, or warble, becomes pierced, and the grub gradually works out and falls on the ground and in a few days matures into another gadfly which flies off to sting cattle again during the summer.

**Treatment.**—As soon as you notice the lump, cut the skin and squeeze the grub out. By killing the grubs in this way you will soon get rid of the pest.

## SNAKE BITES.

The bite of some snakes is very poisonous and causes the animal to have great depression; it becomes very weak and feeble, the eyes are dull and the pupils enlarged, the ears and legs become cold, and in severe cases death soon comes on if not treated. If you see the animal as soon as it is bitten cut the piece out and burn the wound with a hot iron to kill the poison; give half-pint doses of whisky or brandy mixed in a pint of water every three or four hours to stimulate the animal and counteract the depression and weakness caused by the poison in the system.

## BITES OF INSECTS, SMALL SNAKES, HORNETS, ETC.

After the bite, or sting, there is noticed a soft swelling, which is sore. Rub the parts with white liniment three or four times a day; which will soon draw the poison out and take down the swelling.

## FROST BITES.

Frost bites in cattle are treated the same as frost bites in horses.

## BURNS AND SCALDS.

Burns and scalds on any animal are treated by applying carbolic oil to the burnt or scalded part, take four ounces of sweet oil with ten drops of carbolic acid in it. Put this on twice a day and it will stop the pain and heal the parts.

CHAPTER IX.

# DISEASES OF THE FEET—HOW TO DEHORN, BLEED AND DRENCH.

### LAMINITIS (FOUNDER).

This is inflammation of the sensitive structures of the foot, or what is commonly called the quick of the foot.

**Causes.**—Are from overfeeding, overheating, or from driving a long distance on a hard, stony road.

**Symptoms.**—The animal persists in lying down ; the feet are hot and sometimes swollen around the top of the hoof and sore to press on ; the animal is greedy to drink on account of being feverish, but does not care to eat much, and if you force the animal to move it just slides its feet along, seems very stiff and its belly is all drawn up from trying to favor its feet.

**Treatment.**—Keep the animal as quiet as possible, and poultice the feet with hot linseed meal and bran—about half-and-half. The way to do this is to take an old grain bag, cut about a foot off the bottom of it and pack the hot poultice in the bottom ; place the foot in it, and then tie it up around the fetlock and foot so it cannot fall off. Do this to all the feet and change the poultice twice a day; keep this up until the animal gets all right. As well as this, give a pound and a half of Epsom salts in a quart of luke warm water; also give a teaspoonful of saltpetre, or nitrate of potash, in a mash night and morning. Sometimes, if the weather is warm, by standing the animal in a stream of water with a mucky bottom, for a few hours every day, will soon bring them all right without anything else.

### SORENESS OF THE FEET FROM ANY CAUSE.

Treat just the same as you would for founder, by poultices and keeping the animal quiet. If the toes are too long, cut them off with a chisel and mallet.

### FOUL IN THE FOOT (FOOT-ROT).

This is an inflammation of the skin and parts between the trotters or toes, and after this there are ulcers or small boils form and break out all around the top of the foot and between the trotters. The foot becomes very much swollen in some cases and causes the trotters to spread wide apart. The animal suffers great pain and can scarcely put the foot to the ground, and if allowed to run on without being treated at once, it becomes very tedious and hard to treat. The hind feet are more often affected than the front ones.

**Causes.**—Are from something becoming wedged in between the trotters or toes, such as hard clay, manure or a piece of stick, bone or any such like substance, it is more often seen where cattle stand in a filthy place or have to walk through a dirty, soft place.

**Treatment.**—As soon as noticed examine the foot and remove any substance found between the trotter or toes, wash the foot thoroughly with luke warm water and soap, after this apply a good warm poultice of linseed meal, poultice every night and keep the animal in a nice dry place and after you take the poultice off in the morning and before you put it on at night give the foot a good dressing with the following:

Carbolic Acid.............. ...........1 dram, or 1 teaspoonful.
Water........ .......................1 pint.

Shake well together each time before using and apply as mentioned, when applying get it worked in between the trotters or toes as much as you can. Keep this treatment up until it is better, in very bad cases it takes a long time to get better, but keep at it. Another very good wash to use in place of the carbolic water is

Creolin........ .....................½ ounce or 2 tablespoonfuls.
Water......................... ......1 pint

Mix and use the same as the carbolic water, if one remedy should fail try the other; during treatment be sure and keep the animal quiet and in a dry place, and feed well to keep its strength up.

### FISTULA OF THE FOOT.

No matter what part of the foot is affected it is just the same, it is caused by a bruise or from something running into the foot and dirt getting up in the hole or from diseased bone.

**Symptoms.**—There is lameness and a discharge from a small hole which has no tendency to heal, and if it is from a small piece of diseased bone the discharge smells very bad.

**Treatment.**—In all cases pare out the hoof or horn around the sore spot, so as to allow whatever is in it to have a chance to run out, and then poultice until you draw out whatever is in the hole causing the trouble. The best poultice for this is hot linseed meal. After you get it out, the hole will soon heal up of its own accord.

## NAIL RUN IN THE FOOT.

This generally occurs where cattle are running around old buildings where boards with nails in them are lying about.

**Symptoms.**—There is severe lameness which comes on all of a sudden; the animal appears to be in great pain and can scarcely touch its foot to the ground.

**Treatment.**—Pull the nail out, pare out around the hole made by the nail and poultice with hot linseed meal; keep the animal quiet until the soreness is all out; change the poultices twice a day. If it should fester then pare down around the hole until the matter comes out and then poultice well to draw it all out; keep the animal quiet until the hole heals up. After you quit poulticing stuff the hole with tar and cotton batting to keep the dirt from working up into it.

## DEHORNING.

This is an operation which is carried on to a great extent in Canada and other countries, and is gaining the favor of stock owners rapidly. It is a very simple, although a painful operation while it lasts, and is, as a general thing, attended with very good results. While this is a painful operation so are all other operations, such as castration and docking, but as long as it is done with a view to benefitting the lives of the cattle themselves, and also their owner, it is not considered inhuman. It is best not to perform this operation on cattle under one year old for the horns will often grow again, and before that time they never do much harm. The best time to dehorn is in the spring, during the months of March and April, so that the horns will be well healed up before the flies come to bother them, or in the fall of the year, just after there has been frost enough to kill the flies. The operation is a simple one, and is performed in this way: Build a stanchion, similar to the old way of tying cattle, in a solid doorway, or any other such place where you can run the cattle into it, one at a time. Have the stanchion built good and strong, also have the sticks in it good and tight together, just large enough for the animal's neck to fit in when it is closed; have a narrow stall, built out of good strong plank, at the side of the stanchion where you can run the cattle in, this will keep them from swinging the body around while you are dehorning them. When you have the animal fast in the stanchion put a rope halter over its head, and have the head and neck well pulled forward by means of a double pulley so as to get good purchase to hold the animal in its place while operating; have the pulleys attached to something about eight or ten feet straight in front of the animal, and as near the ground as possible, this will hold the head in better position; have a man take hold of the nose and ear at one side while you saw the horn off with a stiff-backed, fine-tooth carpenter saw, taking about one-eighth of an inch of skin off with the horn, then take off the other horn in like manner. By taking the horn off in this place it is easier sawed, bleeds less, heals nicer and there is no danger of it growing again. As soon as you are through with this animal put in another. In cases where you only have one to dehorn throw and secure it, and take the horns off in the same manner. Another way by which they may be taken off is to use large dehorning clippers. These clippers can be got at a hardware store, and the directions how to use are along with them. We recommend sawing the horns off, except in cases of very young cattle. After two or three years of age the horns become brittle, and in pinching them off with the clippers there is danger of fracturing the bones of the head. Always keep the animal from being chased before and after the operation, for there is more danger of bleeding when they are excited. If they bleed much apply a little of Monsell's solution of iron with a feather, which will stop the bleeding. Keep the animal quiet and do not allow it to be out in any cold storms, and be careful when feeding it not to throw dust or chaff on the head so it will get in the holes, which would be apt to cause festering. If you want to kill the horns on calves it must be done when they are about a week old. When the little horn first appears take a stick of caustic potash, dip it in water and rub it well into the skin around where the little horn is coming through. One burning generally kills the horns; if not, repeat it heavier in a few days.

### HOW TO BLEED A COW.

Tie a small rope around the neck, just in front of the shoulders, so it will raise the jugular vein, then take the largest blade of an ordinary fleames, hold it lengthwise, fair in the centre over the vein, then hit the fleames a sharp tap with a piece of hardwood; hit hard enough to cut the vein, and catch the blood in a pail. Take from half a pail to a pailful of blood away. When you have enough blood away, let the rope slack, run a pin through the two edges of the cut and wind a string around the pin in the form of a figure eight and tie it there. Keep the animal in the stable, and feed out of a high manger, for twenty-four hours, then remove the pin and allow the animal to go. In this operation, as in all others, have everything clean for fear of blood-poisoning.

### HOW TO DRENCH CATTLE.

In cattle always mix your drenches in a large quantity of water, because it will wash out of the paunch quicker, and have a quicker and better action when given this way. Have an assistant to hold the horns while you take hold of the nose with your left hand, and hold the head a little above a level; with your right hand put the bottle well back into the mouth and allow it all to run down without taking the bottle out of the mouth, unless the animal should cough; if it does, let go of its head until through coughing, then continue the drenching. Be careful in drenching, especially in lung troubles, for they are easily choked.

---

## CHAPTER X.

# DISEASES OF THE NERVOUS SYSTEM.

### CORN STALK DISEASE.

**Causes.**—From eating corn stalks which have minute germs underneath the leaves; these germs are so small that you cannot see them without the aid of a microscope. Corn stalks that are affected with these germs do not grow so well, and ripens long before the other corn.

**Symptoms.**—There is first, symptoms of impaction of the third part of stomach or manyplies, after that the brain becomes affected and the animal becomes delirious, this is followed by stupidness; the animal will shove its head forward against the stall and pay no attention to anything, and after a few days, dies.

**Treatment.**—Give the following:

| | |
|---|---|
| Epsom Salts | 1 pound. |
| Bitter Aloes | 1 ounce. |
| Sweet Spirits of Nitre | 1 ounce or 4 tablespoonfuls. |
| Fleming's Tincture of Aconite | 10 to 15 drops. |

Mix in a quart of luke warm water and give as a drench, and afterwards follow up with:

| | |
|---|---|
| Fleming's Tincture of Aconite | 10 to 15 drops. |
| Sweet Spirits of Nitre | 1 ounce, or 4 tablespoonfuls. |
| Ginger | 1 tablespoonful. |
| Common Soda | 1 " |

Mix in a pint of water and give as a drench every four hours until it is better. Give luke warm water to drink and feed on soft food, keep the body warm, and if the animal's head is affected, keep ice to its head in a bag. Smut on corn is very bad feed, as it is apt to derange the stomach and cause diarrhœa, and if the animal gets too much smut it will set up a disease similar to ergotism.

### INFLAMMATION OF THE BRAIN (ENCEPHALITIS).

This disease is not so often met with in cattle as it is in horses.

**Causes.**—From a severe blow on the head, or from falling and striking the head; irritation of small tumors growing around the brain and pressing on it; certain kinds of food containing ergot or narcotic principles will cause it, or from eating grains from a distillery.

**Symptoms.**—The first symptoms are the animal will be dull and drowsy and stands with its head pressed up against a wall or fence; its legs keep moving as if it was going to walk right through whatever its head is pressed against; when walking it has a staggering gait; its bowels are costive and its urine is of a dark-red color. After these symptoms pass off the animal gets delirious and acts as if it were mad, it bellows, stamps its feet, grates its teeth, froths at the mouth, runs about wildly, and, if in the stable, rears up into the manger.

**Treatment.**—In the first stages, when the animal is dull, bleed it, taking a half pailful of blood away, and give

| | |
|---|---|
| Epsom Salts | 1 pound. |
| Bitter Aloes | 1 ounce. |
| Ginger | 1 tablespoonful. |
| Common Soda | 1 " |

Mix in a quart of luke warm water and give as a drench. Keep the animal in a quiet, shady place; apply a mustard plaster to the back; keep the body warm and apply ice to its head in a bag. Give

| | |
|---|---|
| Sweet Spirits of Nitre | 1 ounce or 4 tablespoonfuls. |
| Fleming's Tincture of Aconite | 10 to 15 drops. |

Mix in a pint of cold water and give every five hours, until the animal is better. Give plenty of cold water to drink in small quantities and feed on soft food. Give an injection into the anus of half a pail of luke warm water and soap twice a day to help to start the bowels.

### SUNSTROKE.

This very rarely occurs in cattle. The causes, symptoms and treatment are the same as those of the horse, only in giving a physic use one pound of Epsom salts along with the bitter aloes.

### LOCKJAW (TETANUS).

This disease is rarely met with in cattle.

**Causes.**—Following operations, or wounds of any kind, and sometimes it comes on from causes unknown.

**Symptoms.**—If it is from a wound just when it is healing up the muscles all over the body is contracted and hard; the animal has a stiff way of walking; the tail will be stiff, and keeps working like a snake; the jaws become partially set; the animal cannot reach down to the ground; the eyes have a peculiar look and seem to be turned back in the head and set. In some cases the animal can eat, while in other cases it cannot, according to how much the muscles of the jaws are affected.

**Treatment.**—This is one of the diseases that does not need much medicine because in giving the medicine it excites the animal and does more harm than good, and, anyway, medicine does not seem to do any good in this disease. Give the following:

| | |
|---|---|
| Epsom Salts | 1 pound. |
| Common Soda | 1 tablespoonful. |
| Ginger | 1 " |
| Fluid Extract of Belladonna | 1 dram, or 1 teaspoonful. |

Mix in a quart of luke warm water and give as a drench once a week. Keep the animal perfectly quiet and free from noise. Feed on food made into gruel, so that it can drink it down. If it is caused from a wound, bathe the wound twice a day and fill the place full of green salve. If the wound is where you can poultice it, poultice it every night with linseed meal. This disease generally takes three or four weeks to run its course, and if you can keep the animal's strength up till then it will pass away, while in severe cases they die in a week or so.

## CHAPTER XI.

# CONTAGIOUS DISEASES IN CATTLE

### CONTAGIOUS PLEURO-PNEUMONIA.

This disease is contagious, or catching, and is inflammation of the lungs and covering of the lungs; this is how it gets its name. This disease was more common at one time than it is now. It was first noticed in Prussia in 1802, Russia in 1824, England in 1841, and America in 1843. It is a very contagious disease in cattle, but never affects other animals. If an animal once gets over this disease it will never get it again. When there is an outbreak of this disease it spreads very rapidly, by the germs of the disease being carried about in different ways.

**Symptoms.**—The first symptom is, the animal gets very feverish; the temperature goes as high as 105 degrees. The animal will remain feverish for a week or so, and also have a cough, as if from a slight cold. After this the lungs become inflamed and sets up inflammation of the lungs and their covering. By listening at the sides you will hear the peculiar grating sound that is heard in inflammation of these parts. The animal breathes heavy and quick, falls off rapidly in condition, refuses to eat, becomes hide bound, and there is a discharge from the nostrils of a whitish color, which has a very bad smell. The pulse runs up higher and becomes weaker; the nose is dry, and the animal lies on its breast bone most of the time. Sometimes there is diarrhœa, then costiveness; the eyes become dull and the animal soon dies. When one animal in a herd becomes affected with this disease the whole herd will soon become affected and die, and if it is not checked in this herd it soon spreads all over the country.

**Treatment.**—If there should be several animals die in the same district, and you suspect this disease, send for one of the government veterinary inspectors, who looks after all contagious diseases that break out in America. The treatment he will follow, after he is sure it is this disease they have, is: He will have the affected herd immediately destroyed, and, the people who have been attending the cattle and those who are on the farm, will not be allowed to leave the farm for ninety days, and all the stables where the cattle have been will be disinfected and no other cattle will be allowed on the farm for ninety days after the slaughter. Medical treatment is of no avail in this disease, and you are not allowed to treat them.

### CONSUMPTION IN CATTLE (TUBERCULOSIS).

This disease is a contagious one, caused by germs called the bacillus tuberculosis. This disease has been known for centuries back, and there has been laws passed calling for the destruction of affected animals, and also forbidding the meat to be used as food. This disease is known in all the civilized world. It may affect the lungs, bowels, liver, kidneys, bladder, brain or spinal cord, or any part of the body. The germs in the affected cattle come away from the lungs by coughing, or flows away in the saliva from the affected animal's mouth, they fall on the grass, in mangers, pails and such like, and other animals following them up may breathe the germs into the lungs by eating or drinking out of the same pail or manger, or off the grass where the diseased cattle have been, and this is how the disease is communicated from one to the other. These germs may also pass out of the system into the milk, and animals or people that drink this milk are liable to take the disease, so you see the danger of having a diseased cow around.

**Symptoms.**—At first the disease comes on very slowly after it is taken into the system. If the disease affects the lungs there is a short, dull cough which may be noticed more in the morning, after exercise, or drinking, later on in the disease the cough becomes more troublesome, the animal runs down in condition, the breath has a bad smell, there is a dribbling of saliva from the mouth, the animal becomes hide bound, the hair stands out and it is a pitiful looking sight, and in a few months pines away and dies. The time it takes the disease to run its course varies from three months to a year. This disease is noticed more in thoroughbred cattle than it is in grade cattle. If it affects the bowels the animal will run down in condition, will have diarrhœa sometimes, then costiveness changing every few days; the other symptoms are the same only when the disease does not affect the lungs the animal has not such a cough. If the disease affects the brain or spinal cord it causes paralysis, and death soon follows. If any of the other parts or organs of the body are affected it causes symptoms peculiar to that organ when affected, and the animal slowly pines away and dies. The way to test cattle to find out whether they are affected with tuberculosis or not is to use the test known as the tuberculine test, which is done by injecting tuberculine into all the herd of cattle that are supposed to be affected with the disease and having them starved for twenty-four hours after the injection, then take their temperature. The temperature of the cattle not affected with the disease will be normal, while the temperature of the affected ones will be raised two or three degrees. In case you suspect this disease in your herd of cattle it is best to send for the government veterinary inspector to come and use the test, for he thoroughly understands how to test them, and the affected ones will be destroyed. In this way you prevent the disease from spreading in your own herd of cattle, and by doing this you also protect the lives of your family, yourself and those around you, for this disease can be communicated from cattle to people by drinking the milk or eating the meat of affected cattle. In opening cattle that are affected with this disease the organ that is affected will be found to be eaten away with the disease, and its place is taken by a lot of small lumps inclosed in a membrane, and if you cut into them they are found to be full of other little cheesy lumps about the size of a pea. In very bad cases tumors will sometimes form and be full of yellow matter.

### HYDROPHOBIA, MADNESS OR RABIES IN CATTLE.

This disease originates spontaneously in dogs and cats and can be communicated to other animals or man by the poison caused from the bites of these animals while they are mad. The saliva of the mouth contains the poison and this is how it is caused from a bite. Every animal that they bite does not go mad, but about one-fourth of the animals bitten do go mad.

**Symptoms.**—A few days after being bitten the animal loses its appetite, is very restless and anxious looking, then there is increased restlessness, loud roaring at times, bunting at things and pawing with its feet; saliva keeps dripping away from the mouth, there is a peculiar wild look in the eyes, and the animal keeps continually straining to pass manure, but very little comes, and in a short time it becomes paralyzed in the hind quarters, falls down and death soon relieves it. If the above symptoms are present, and a mad dog has been through that section of the country, you may as well destroy the animal, for it is mad and is dangerous to have around.

**Treatment.**—If the disease has set in, destroy the animal, and in going around it be very careful not to let it hook or bite you; but if you go to treat the animal just after it has been bitten by a mad dog, take a sharp knife and cut a piece right out of the wound, then take a stick of caustic potash, or nitrate of silver, and burn the wound well with it; but if you have not got these, burn the wound well with a red hot iron, which will also kill the poison.

### LUMPY JAW (ACTINOMYCOSIS).

This disease is contagious and spreads among cattle. It is caused by germs known as "actinomycosis." This disease generally affects the upper or lower jaws, but may affect the tongue and other parts of the body. The way this disease is communicated from one animal to another is from the affected animals slavering on the grass or over feed and other animals take it up when eating or drinking. These germs pass down into the bowels, where they are taken up into the blood and carried around until they locate in the jaw. They may also be taken into the system from the slaver getting into the wound on another animal, and it is thought that this disease may be carried to and affect man by eating the flesh of an animal affected with this disease.

**Symptoms.**—If it affects the jaw there will be a hard, bony lump form opposite the roots of the teeth, either in the upper or lower jaws. The disease gradually works in the jaw, the lump grows, and in the course of time the disease gets so bad that the teeth loosen and fall out, and on account of the disease being so bad that the animal cannot eat, it falls off in condition and dies. If it affects the tongue, it is generally the thick part at the back that is affected. It thickens and hardens the tongue so much that when this is affected, it sometimes receives the name of wooden tongue; in this case the animal will fall off in condition, from not being able to eat, and will soon die.

**Treatment.**—If more than one animal in a herd becomes affected with lumpy jaw, and after examining the teeth you find nothing wrong with them and you cannot see the mark of any injury outside around the jaws, and if they are not sore to handle, then be suspicious of this disease and treat in the following manner: Separate the affected animals from the sound ones; treat the affected ones by throwing and securing them, then skin a piece of the skin off the upper part of the lump, and take a quarter-inch auger or a trephine, which is an instrument used for boring into the bone, but an auger will do just as well; bore two small holes well into the bone, which is easily done when the bone is diseased; pour tincture of iodine into the holes and let it soak well around the diseased bone; let the animal up, and in a few days afterward catch it again and fill the holes with iodine, and keep putting in the iodine every fourth day until the holes heal up. If this treatment fails to kill the lump destroy the animal and burn it, for it is dangerous to have it in your herd, and its meat should not be used for food, as it might set up the disease in man.

## ANTHRAX.

This is an infectious disease caused by germs, called the anthrax bacilli, getting into the blood. This disease affects cattle in all parts of the world, and is only noticed in cattle that are grazing on low, swampy land that have pools of stagnant water on it. When once the disease gets into a pasture field it will remain there for years, and the only way to get rid of it is to break the field up and drain it.

**Symptoms**—It more frequently affects young cattle than older ones; the attack is very sudden, and an animal apparently well the night before will be found dead in the morning. In some cases as soon as the animal is affected it drops down, goes into convulsions and dies. In other cases it will last longer, the pulse will run up to from 80 to 100 beats per minute; the animal will not eat; the whole surface of the body, legs and ears are cold, and it is very dull, stupid and weak. In a short time this dullness gives way to uneasiness, it champs its jaws, kicks and paws the ground and appears to be in terrible agony; it has very much difficulty in breathing, the nostrils are enlarged and the mouth open; the lining of the mouth, nostrils, rectum or back bowel and vagina are of a blue color, the manure is first thin and watery looking, then covered with slime and blood; the symptoms gradually get worse, and in a few hours it dies a miserable death. The germs are taken into the system from eating grass around a swamp or drinking the water from stagnant pools. They find their way from the bowels into the blood, and work around until they locate themselves in the bowels or tissues under the skin. When an animal dies from this disease it bloats up, decomposition setting in very quickly, and there is a blood-stained fluid flows from the mouth, nose and anus. If you have had an animal or two die while pasturing on this kind of land, and showing the above symptoms and appearance after death, send for the government veterinary inspector, he will examine the blood, and if the germs are found in it you will then be sure it is anthrax. Burn the carcasses of the dead cattle and remove the sound ones from the pasture; drain the field and break it up is the best way to get rid of the germs. Sheep are also subject to this disease and have the same symptoms.

**Treatment.**—There is no treatment for this disease.

## BLACK QUARTER.

This is an infectious disease and is very fatal, and affects cattle from six months to three or four years' old. It is caused by germs getting into the system. This disease is noticed mostly in cattle grazing on low-lying, swampy lands.

**Symptoms.**—About three days after the germs get into the system the animal becomes dull and feverish, the temperature rises as high as 107 degrees; there is lameness in some of the legs, generally in one of the front ones, which is caused from a swelling which usually forms just behind the front leg, but may be in any other part of the body. The symptoms gradually get worse; the animal acts as if it had severe spells of colic, but keeps looking around to where the swelling is; breathes heavy; becomes very weak and soon dies. The swelling, after it is first noticed, grows very fast, and when you rub your hand over it, it makes a crackling noise as if it contained air. If you cut into the swelling the animal does not feel it, and nothing comes out only a dark red, frothy looking substance, which has a bad smell. If you skin the animal where the swelling is, there will be dark brown, dark red and yellow colored patches, and the parts around where the swelling is will be soft and easily torn.

**Treatment.**—There is no treatment for this disease, only remove the sound animals, destroy and burn the affected ones, and drain and break up the pasture.

## TEXAS FEVER.

This disease affects cattle all through the southern parts of the states, and is caused from germs getting into the blood. The germs are carried from one animal to another by cattle ticks.

**Symptoms.**—At first there is dullness, loss of appetite, and the animal will leave the herd and stand or lie down alone. The temperature at this stage will be up from 105 to 107 degrees, and stays about that all through the disease. The bowels are constipated, and anything that is passed is covered with bile, and near the end of the disease the urine or water is of a dark red color.

The pulse and breathing are both quickened ; the animal becomes very stupid and lays down most of the time, and in a few days death relieves it In some cases they die in three days, while in others they live for several weeks, and if an animal does recover it takes it a long time to do so.

**Treatment.**—In order to keep the disease from spreading do not allow cattle that have these ticks on them to go into other herds, and cattle affected with this disease should be put on a pasture by themselves, and the ticks should be picked off and killed, then give the animal the following : Fifteen grains of sulphate of quinine three times a day, either in a drench mixed with a pint of water, in a capsule or on its tongue with a spoon.

---

CHAPTER XII.

# DISEASES THAT ARE LIABLE TO AFFECT ANY ANIMAL.

### TUMORS.

Tumors are growths which are not inflamed nor sore to handle, and are of various kinds ; they do not fester and break, and run matter as abscesses do, but simply grow; they may affect any part of the body in any kind of an animal, and in some cases there will be a small amount of matter in the center of the lump. Tumors, as a general thing, grow very slowly, and are very hard when you feel them.

**Treatment.**—At first they may be stopped growing and absorbed by blistering

**For Tumors on Cattle**—Use the following blister :

Powdered Cantharides or Spanish Fly..................2½ drams.
Vaseline or Lard........................................1 ounce.

Mix and rub in well what you can get on the tumor ; tie the animal's head so it cannot bite the blister ; grease the blistered part in three days, and in two weeks blister again, and continue until you have blistered it three or four times, and then if this does not stop the growth of the lump and reduce it down it is best to cut it out. Tincture of iodine is also good to use in some cases. Rub the lump well with it once a day till you get it well blistered, then grease and let it go for a few days, then wash the grease off with luke warm water and soap and commence using the iodine again.

**For Tumors on Horses.**—Use the following blister:

| | |
|---|---|
| Powdered Cantharides or Spanish Fly................ | 1½ drams. |
| Vaseline or Lard.................................... | 1 ounce. |

Mix and apply with the same directions as is given for the blister on cattle, if this does not fix it try the iodine the same as for cattle. In case you have to cut the lump out, throw the animal down and secure it, make a good, long cut in the skin over the tumor and cut around it until you have it cut out, after the tumor is out and it is bleeding, take a large piece of dry cotton batting about the size of the tumor and shove it in where you took the tumor out, then sew up the skin the same as sewing a wound, and leave it sewed up for twenty-four hours, then take a couple of stitches out of the lower part of the wound and take the batting out and dress it by bathing it off with luke warm water and soap twice a day, after each bathing dress the wound by inserting melted green salve with a feather up in the wound and applying white lotion around the outside and inside of the cut. Keep this treatment up until it is healed. If the tumor lies near a large vein such as the jugular vein, and you cannot cut it out, keep blistering it, and if this does not help it, paint the lump over with tincture of iodine every day the same as above mentioned. This treatment will do for tumors either in the horse or in cattle.

## CANCER.

Cancers are angry looking growths generally, affecting cattle, sheep and dogs, and is sometimes noticed in horses, around the head and neck, but may affect the other animals in any part of the body.

**Symptoms.**—At the first appearance of the cancer—before it gets into the blood—it does not hurt the animal's health, but after the cancer once gets settled in the blood the animal falls off in condition, the growth grows very rapidly, and the least irritation causes it to bleed. In some cases the disease gradually eats the surrounding tissue until it kills the animal, and if it is near a bone it gradually works into it, and when the bone becomes affected it has a very bad smell. When the disease has arrived at this stage it is called bone cancer.

**Treatment.**—In the first stages of the disease cut it out good and clean, and by doing this you can sometimes prevent its spreading. It is best to throw the animal and secure it before you attempt to cut the cancer out, and as soon as you have taken it out burn the place good with a red-hot iron, then bathe it twice a day with luke warm water and apply the green salve and white lotion after each bathing. If, at any time afterwards, there should appear to be a red spot growing, tie the animal up and give the spot another good burning with the red-hot iron, and keep this treatment up until it is healed. In some cases when the disease has run on too long before it has been cut out, or after it is cut out, it grows again worse than at first, or where it has been entirely neglected and allowed to run on so long that the bones become affected and smell bad destroy the animal and burn the carcass, as it is not fit for beef and would be very apt to communicate cancer to anyone eating the meat.

### ABSCESSES.

Abscesses are of two kinds, one festers and has matter in it, while the other, which is called a serious abscess, is only a bruise containing water, or serum. These kind of abscesses are generally found around a horse's shoulder, but may affect any part of the body.

**Symptoms of a Serious Abscess.**—About twenty-four hours after the bruise there will be a swelling varying in size from an egg to that of a man's head, and when you press on it you can tell there is fluid in it, and on opening it you will find a watery looking fluid mixed with blood.

**Treatment of a Serious Abscess.**—Cut a whole in the bottom part of the lump large enough to run your finger in and clean all the fluid out, then bathe the part well with luke warm water and salt, and insert melted green salve up into the lump through the hole with a feather twice a day, also run your finger into the hole once a day to keep it open until it heals inside, each time after bathing rub the outside of the swelling with white liniment to help get the thickening out and keep the swelling down.

### ABSCESSES THAT CONTAIN MATTER.

**Causes.**—From distemper; from bruises; or from a cut healing up and there being some substance left in the cut which afterwards festers.

**Symptoms.**—Abscesses differ from tumors, as they fester and form matter; are hot and sore to handle; cause the animal to be feverish and in pain, and after a time they come to a head and either break or have to be lanced.

**Treatment.**—Bathe well with hot water and salt, or vinegar, twice a day, and where it is very sore poultice it once in a while to help to bring it to a head. Each time after bathing apply white liniment around the swelling; this also helps to bring it to a head. When the abscess is ready to open there will be a soft place on it where the hair is falling off and pulls out easily; take a sharp penknife or lance and make a cut in it large enough to put your finger in and clean all the matter out, and treat with green salve, bathing and applying white liniment the same as is given in the treatment of serious abscesses.

## RUNNING SORE (FISTULA).

Running sore, or fistula, may form in any animal or in any part of the body.

**Causes.**—From the animal getting a deep wound and some substance being left in the wound after it is healed which should have been taken out before the wound healed up. It is also caused from diseased bone, or a small piece of broken bone which keeps festering.

**Treatment.**—In all cases where it has been a wound that is healed up and then turns to a running sore, you may come to the conclusion there is something at the bottom of the wound that is causing the trouble, and if it is in a place you can operate on it without throwing the animal down, by twitching it, do so (this can generally be done in the horse), but if not, throw it down and secure it, then take a goose quill and probe into the hole to see which way it goes and the depth of it, then with a sharp knife cut a hole large enough for your finger to go in; cut the hole to the bottom of the wound and clean out whatever there is in there, then treat the wound by bathing with warm water twice a day. After bathing each time put melted green salve up in the hole and apply white lotion in around the cut; keep this treatment up until it is healed. If it is caused from a diseased or fractured bone, cut down same as above and take the piece of bone out. If it is a diseased bone, scrape it out well with a hard steel spoon, and then treat the wound the same as given above, and it will soon be all right.

## STRUCK BY LIGHTNING.

The shock is instantaneous, and affects the brain and nerves. If the shock is heavy enough it causes immediate death, but if it is not severe enough to cause death it leaves the animal prostrate, unconscious and paralyzed.

**Symptoms.**—When it does not kill the animal it falls paralyzed and unconscious, the muscles relax and are soft and flabby, and in some cases there will be twitching of the muscles, the animal will breathe hard and slower than natural, and in most cases there will be spots of hair singed off the body.

**Treatment.**—Dash cold water on the head, rub the body and legs well to get the circulation up, then apply a thin mustard plaster on the back of the head and sides of the neck; blanket the animal well to keep the body warm, and keep the head cold by applying cold cloths, and as soon as it is strong enough to swallow, if it is a cow or horse, give it half a pint of whisky in a pint of warm water; if it is a sheep, dog, or pig give a wine glassful of whisky in half a pint of warm water, pour the drench down slowly so as not to choke the animal; give a drench every hour, and turn it over from side to side every once in a while, and attend to its general comfort until it is better. After the shock has passed off, if the animal seems to be paralyzed, for a cow or horse give a teaspoonful of powdered nux vomica twice a day on its tongue with a spoon, or in its feed, this is for a nerve stimulant. Give a sheep, dog, or pig one-quarter of a teaspoonful of powdered nux vomica twice a day until it is strengthened.

## THE APPEARANCE OF AN ANIMAL KILLED BY LIGHTNING.

If a thunder-storm has just passed over, and the animal, which was previously healthy, is found dead shortly afterwards, the muscles are soft and flabby and you can move its legs any way you like, and in some cases there will be spots of hair scorched off the body, this indicates that the animal was killed by lightning.

## ERGOTISM.

This disease is most often seen in cattle, and shows itself mostly in the winter and spring of the year.

**Causes.**—From eating ergotized grass, which is nothing more than diseased grass. (Ergot also affects rye wheat). The diseased grass is cut along with the hay and other food, dried and brought into the barn and fed during the winter with the good hay and other parts of food. This is how the animals get it. The ergot affects the grass, etc., on the same principle as smut does the corn. Ergot is mostly seen in grasses and rye wheat, but may affect other kinds of grain that is grown on low, rich land in wet, hot seasons.

**Symptoms.**—After cattle have eaten it and it gets into the blood it has such a peculiar contracting action on the heart, arteries and other vessels that it weakens the circulation of the blood so much that in some cases it stops it entirely in the legs, ears and tail, and as soon as circulation in these parts stop they die, rot and drop off, and later on in the disease the brain becomes affected and the animal goes into convulsions and dies.

**Treatment.**—Change the food entirely, bathe the affected parts with luke warm water twice a day, rub dry and apply white lotion to them after bathing. Give the following powder:

| | |
|---|---|
| Nitrate of Potash or Saltpetre.............. .......... | ½ pound |
| Ground Gentian Root................................ | ½ " |
| Sulphate of Iron....................................... | ¼ " |

Mix and give a teaspoonful three times a day to a horse or cow; to smaller animals give one-half of the above dose.

### HOW TO DRY AN ANIMAL THAT IS GIVING MILK.

For a cow, give her a pound of Epsom salts in a quart of luke warm water as a drench, then bathe the bag once a day for a week with luke warm forge water (this is best obtained at a blacksmith shop where they cool the irons). After bathing, milk out a little of the milk once a day, on the ground, for a few days, then milk a little out every second or third day for a few days, then once every week or so for a while, and then stop milking. The last time you milk her, milk the bag out dry. During the time you are putting the cow dry, feed on dry, hard food. For a mare, give eight drams of bitter aloes and a teaspoonful each of ginger and common soda, dissolved in a pint of luke warm water, as a drench. After you give the physic drench, let her stand in the stable for a day or two, bathing the bag and milking her the same as is given for the cow. In drying a sow give a teaspoonful of sulphur in her feed twice a day, and if it is a quiet sow bathe the milk glands with luke warm forge water once a day for a week, and allow her to run out. In drying a ewe give a tablespoonful of Epsom salts dissolved in a half teacupful of luke warm water, once a day for a week, and bathe her bag with forge water, and milk her out in the same way as is given for the cow. Why forge water has such a good effect in drying up the secretion of milk is on account of the iron in it, which gets into it while cooling the irons in the water, and this iron has an astringent action on the milk glands. The reason the physic is given is to clean the milk out of the blood.

### DISEASES OF THE HEART, ARTERIES AND BLOOD.

These diseases in cattle are the same as in the horse, and for causes, symptoms, and treatment we refer you to the pages in Part II. of the horse where they are fully explained.

---

## NOTICE.

All the doses of medicine which are mentioned in this book in the diseases and treatment of cattle, if not specially mentioned for the age and size of the animal, are intended for an average sized cow, steer or bull, so in giving doses to young cattle you must regulate the dose to the age and size of the animal. Calves under one year old would take about one-third of the dose; two-year-olds would stand one-half the dose; three-year-olds and upwards take full dose mentioned, and cattle over the average size, such as large bulls, cows or steers, can take a little more than the dose mentioned.

# PART IV.

# DISEASES AND TREATMENT OF SHEEP, PIGS, DOGS AND POULTRY.

## CHAPTER I.

## DISEASES AND TREATMENT OF SHEEP.

Sheep belong to the class of animals called ruminants, which means animals that chew their cud. In regard to the structure of sheep, they are on the same principle as that of cattle, only smaller. The skin of the sheep is different to that of the ox, as it grows wool instead of hair, and has in the skin numerous small glands which secrete oil, which is used to lubricate and soften the wool.

### COLD IN THE HEAD (SIMPLE CATARRH).

This is a common disease in sheep, and consists of an inflamed state of the lining of the nose and the cavities of the head.

**Causes.**—From being left out in a cold rain and exposure to cold weather, and is mostly seen in the fall and spring, during the seasons of cold rains. It is often seen during washing and shearing time.

**Symptoms.**—There is a discharge from the nose, and the affected sheep keeps snuffling, sneezing and coughing; does not feed well and seems very dull.

**Treatment.**—The treatment is very simple in most cases. Keep the sheep dry, warm and clean, and this often effects a cure in a few days without any medicine; but if the disease does not pass off, give the following mixture:

Nitrate of Potash or Saltpetre........................ ¼ pound.
Sulphur................................................ ¼ "
Ground Gentian Root.................................. ¼ "

Mix thoroughly and give a teaspoonful twice a day on the tongue with a spoon, in its feed or as a drench; feed hot mashes, and take special care to see that it does not get wet, for the disease might settle on its lungs. If the discharge from the head gets clogged around the nose wash it off with warm water and soap. In some cases, where the above treatment does not stop the disease, change the treatment to half a teaspoonful of ground sulphate of iron three times a day in its feed, or on its tongue with a spoon. Iron acts on discharges of any kind in helping to stop them, and also builds up the system, for it is a great tonic.

### SORE THROAT.

**Causes.**—The causes are similar to those of catarrh, only the throat being the weaker part it settles there instead of elsewhere.

**Symptoms.**—There is swelling and soreness around the throat, and the animal holds its head and neck poked out, and when it coughs it seems to try to save itself all it can. The animal is dull, eats very little, and when drinking the water runs out through its nose.

**Treatment.**—Give

Nitrate of Potash or Saltpetre........................ ¼ pound.
Sulphur................................................ ¼ "
Powdered Alum......................................... ⅛ "

Mix thoroughly and give a small teaspoonful on its tongue with a spoon three times a day. As well as giving the powder rub the throat three times a day with white liniment. Feed on soft food raised high so it will not have to bend its head over to eat. Give all the cold water it wants to drink in small quantities at a time; also take good care of it, not allowing it to get wet or cold, and it will soon get all right.

### BRONCHITIS.

This is inflammation of the lining of the bronchial tubes.

**Causes.**—From being chased by a dog; inhaling smoke, or getting wet are the principal causes of bronchitis in sheep.

**Symptoms.**—The affected sheep will be dull, breathe heavy and quick, the ears will lop over, and, like the ox, they will lie down most of the time while suffering with lung trouble. If you listen at the bottom of the windpipe a wheezing noise will be heard.

**Treatment.**—Keep the animal very quiet in an airy place and give the following:

Sweet Spirits of Nitre................½ ounce, or 1 tablespoonful.
Fleming's Tincture of Aconite.........3 drops.

Mix in a teacupful of luke warm water and pour this down as a drench. In drenching do not raise the head too high, and be very careful not to choke the animal. If the sheep wishes to drink give cold water in small quantities; also, small, hot mashes and grass, or hay to eat. Give the sheep the above drench three times a day for a day or so until it gets relief. If the sheep seems very weak then change the drenches to something more stimulating.

Whisky........................................1 wine glassful.
Ginger........................................1 teaspoonful.

Mix in a half pint of oatmeal gruel and pour this down very carefully. Give this drench three times a day until it gets strong; also, be sure to keep it warm and dry, and allow it to have plenty of fresh air without being in a draft.

### FILARIA BRONCHITIS.

**Causes.**—From small germs getting into the system and settling in the lining of the bronchial tubes and inflaming them. This disease is not very common, but sometimes we have an outbreak of it where the sheep are pasturing on a low-lying pasture, and, as a general thing, if one of the flock gets affected with it, nearly all the herd will be showing symptoms of the disease.

**Symptoms.**—They have a dry, husky cough; fall off in condition, and with the slightest exertion the animal will breathe heavy, and on placing your ear to the bottom of the windpipe you will hear a wheezing noise, the sheep will not feed well, scarcely ever chews its cud and lies down most of the time. If you have several sheep showing the above symptoms, and they are pasturing on a low-lying pasture you may come to the conclusion it is this disease they have.

**Treatment.**—Turpentine seems to have a special action in cases of this kind, because as soon as it gets into the blood it spreads through the system and has a tendency to kill all kinds of germs. Give the following:

Spirits of Turpentine................½ ounce or 1 tablespoonful.
Raw Linseed Oil......................½ teacupful.

Mix and give as a drench every third day until the disease disappears. Another good treatment is to burn sulphur in a stable where the sheep are and allow them to breathe the fumes of the sulphur. Do this twice a week and it will kill the germs and stop the disease. The way to tell when they have as much as they can bear is to have a man stand in the stable with them, and when he has all he can stand let the sheep out of the stable. Sheep can stand about the same amount of the fumes of sulphur as a man.

### INFLAMMATION OF THE LUNGS AND ITS COVERING.

**Causes.**—This generally comes after sore throat when the sheep get cold, or from a neglected cold in the head, or from getting chilled from being in a cold rain. Sometimes it is caused from a sheep getting cold after lambing.

**Symptoms.**—The sheep lies down with its ears lopped over, breathing heavy, its legs and ears are cold, and it does not eat nor chew its cud, but seems feverish and greedy to drink. Sometimes it will cough. If you separate the wool on its side and listen with your ear you will hear a peculiar grating sound in the lungs, and the affected sheep will moan once in a while as if in distress. The sheep gets very weak, falls off in condition and will die in a short time if it does not get relief. In some cases the sheep will linger nine or ten days.

**Treatment.**—Put the sheep in a dry, warm, well ventilated box-stall or stable. If the sheep has just been shorn cover the body well with blankets, but if not just keep dry. Give the following medicine :

Sweet Spirits of Nitre................½ ounce or 1 tablespoonful.
Ginger................................1 teaspoonful.
Fleming Tincture of Aconite..........3 drops.

Mix in a half pint of water and give as a drench. Give this three times a day, being careful not to lift its head too high and choke it, and give plenty of time for the animal to swallow the drench. In a day or so, when the animal is better, but seems weak, give the following drench, as a stimulant :

Whisky.........................................1 wine glassful.
Ginger.........................................1 teaspoonful.
Oatmeal Gruel..................................½ pint.

Mix and give as a drench three times a day until the sheep gets strong and commences to feed all right again. During the time the sheep is sick give cold water to drink in small quantities, but often. If it will eat mashes with flaxseed in it give that, but if not give any other kind of grain it will eat.

### TROUBLES OF THE TEETH.

Sometimes, when the sheep is a year or two old, there will be a cap of one of the grinders, while shedding the teeth, stick on the new tooth and not fall off; this causes the animal difficulty in eating.

**Symptoms.**—The sheep does not do well; would like to eat; takes food into its mouth and chews it, but throws it out again and goes off, and although it seems to be failing and does not eat, still it does not seem sick. It will sometimes be noticed to be working its tongue around its mouth.

**Treatment.**—Catch the animal and examine the mouth, and if it is a cap you can tell it, for it will be higher up than the other teeth in the row. Remove the cap with a pincers and the animal will soon be all right.

### PIECES OF WOOD OR ANY SUBSTANCE LODGED IN THE TEETH.

In eating, the sheep gets hold of a piece of stick or some other substance and it gets caught between the teeth or around the lips or gums, and it cannot work it out with its tongue.

**Symptoms.**—The animal is not eating and keeps its tongue working around in its mouth as if trying to work something out.

**Treatment.**—Examine the mouth, and if you find anything caught in it, remove it, either with your fingers or a pincers.

### OLD SHEEP LOSING THEIR TEETH.

Sometimes an old ewe that has lost some of her teeth gets with lamb; you want to keep her over another year, and she only has two or three front teeth, which interfere greatly with her eating, and she runs down and gets so poor that you think she will not pull through.

**Treatment.**—Catch the ewe and pull out the remaining front teeth and let her gum it, and as soon as the gums heal up she will do well and be able to eat, and will be good for a year or so. The reason why the sheep did not do well was because all the food she could get to eat was merely what she could catch with the few stubs of teeth that remained in her mouth, and when they are pulled out she can take the food in all right with her gums and chew it with her back teeth, which are generally all right at this age.

### CHOKING IN SHEEP.

This is not nearly so common in sheep as it is in cattle, but sometimes it occurs when they are fed on sliced roots.

**Symptoms.**—The animal stops feeding, froths at the mouth, coughs and keeps working its tongue, and in some cases is bloated and the eyes bloodshot.

**Treatment.**—Pour down a little raw linseed oil as a drench; in some cases this will start whatever is lodged in the throat causing the trouble. Feel along the bottom of the neck, and if you can find the obstruction try to work it up or down with your hands, but be sure you do not mistake the larynx or Adam's apple for the lump. If this treatment fails, get a probang that is used for choking in cattle and pass it down in the same way as is explained for choking in cattle, being very careful in passing it down, for the sheep is a more tender animal than the cow. Oil the probang well before passing it down.

### BLOATING (TYMPANITIS) IN SHEEP.

This disease in sheep affects the first part of the stomach or paunch.

**Causes.**—From a change of food of some kind; getting a feed of wet clover; getting a feed of grain; by chilling the stomach with a big drink of cold water; or by eating frozen roots or grass.

**Symptoms.**—The sheep is uneasy; keeps getting up and down and seems in pain; breathes heavy and keeps moaning; on examining the sheep it is found bloated on the left side; if you separate the wool on the left side over the paunch and tap on it, it gives a hollow, drum-like sound, showing that there is gas inside.

**Treatment**—The treatment must be quick. Give the following mixture:

| | |
|---|---|
| Epsom Salts | ¼ pound. |
| Sweet Spirits of Nitre | ½ ounce, or 2 tablespoonfuls. |
| Common Soda | 1 teaspoonful. |
| Ginger | 1 " |

Dissolve in a pint of luke warm water and give as a drench. If this does not give relief in one hour, give:

| | |
|---|---|
| Raw Linseed Oil | 1 teacupful. |
| Spirits of Turpentine | ¼ ounce, or 1 tablespoonful. |

Mix and give as a drench. If this does not give relief in two hours, follow up with the following:

| | |
|---|---|
| Sweet Spirits of Nitre | ½ ounce, or 2 tablespoonfuls. |
| Common Soda | 1 teaspoonful. |
| Ginger | 1 " |

Mix in a half pint of luke warm water and give as a drench every two hours until it gets relief. In very urgent cases, where the bloating is so bad that it endangers the sheep's life, tapping is performed by measuring first half way between the point of the hip and the last rib, and about three inches down from the back bone, then clip the wool off that space about the size of your hand and cut a small hole in the skin just large enough to allow the trocar and cannula to go in; the one used for tapping cattle will do for the sheep by oiling it first and pressing it through the hole that you cut in the skin downwards and inwards right into the paunch. Leave the cannula in and pull the trocar out, and the gas will all escape through the hole in the cannula and save the sheep's life. As soon as the gas is all out, draw out the cannula and let the hole heal up itself. If you have not a trocar and cannula, run a penknife in the same place as described; this will let the gas out.

### IMPACTION OF THE FIRST PART OF THE STOMACH WITH FOOD.

**Causes.**—From the sheep getting at some food that it is not used to eating, such as grain, or being turned into a good pasture after feeding in a poor one. Feeding a poor quality of peas, straw or clover hay, especially if it is a little musty or over ripe, will sometimes cause it.

**Symptoms.**—The sheep will not be eating, looks dull, grates its teeth, does not chew its cud, and has a short, quick grunt while breathing, caused from the food in the stomach pressing on the lungs. The animal walks about in a stupid manner, and has very little passage from the bowels, and what does pass is hard and dry. By pressing on the left side over the stomach it is found to be hard, and by tapping on it, it makes a dull, heavy sound, showing that there is food lying in the first part of the stomach or paunch.

**Treatment.**—Give the following:

| | |
|---|---|
| Epsom Salts | ½ pound. |
| Ginger | 1 teaspoonful. |
| Common Soda | 1 " |
| Sweet Spirits of Nitre | ½ ounce or 2 tablespoonfuls. |

Mix in a pint of luke warm water and give as a drench, then wait for twenty-four hours to see if this dose will physic and clean out the stomach; if it does not physic in that time, give:

| | |
|---|---|
| Raw Linseed Oil | ½ pint. |
| Spirits of Turpentine | ½ ounce or 1 tablespoonful. |

Mix and give as a drench. After the physic operates, to strengthen the animal give:

Whisky.......................................1 wineglassful.
Ginger.......................................1 teaspoonful.
Common Soda.....................................1 "
Oatmeal Gruel.....................................½ pint.

Mix and give as a drench three times a day until the sheep gets strong.

### COLIC IN SHEEP.

This is cramps of the bowels.

**Causes.**—From eating frozen grass or roots; from taking a cold drink of water; or from eating anything that disagrees with the bowels.

**Symptoms.**—The sheep will lie down, moan, strike its feet up against its belly, jump up and seem quite easy for a few minutes, and then the pain will come on again. This is the only disease where the sheep acts like this.

**Treatment.**—Relieve the spasms by giving

Tincture of Laudanum................½ ounce or 1 tablespoonful.
Ginger..............................1 teaspoonful.
Common Soda........................1 "

Mix in a half pint of luke warm water and give as a drench every hour until the animal gets relief. In some cases, after you give a few doses and it does not get relief, try this drench :

Raw Linseed Oil.....................½ pint.
Spirits of Turpentine.................½ ounce or 1 tablespoonful.

Mix and give as a drench and this will sometimes give immediate relief.

### INFLAMMATION OF THE BOWELS.

This is a common disease in young sheep.

**Causes.**—From eating a lot of snow; from eating dirty pea straw; or from a severe chill caused from being out in a cold rain and getting the wool very wet.

**Symptoms.**—There is severe pain, the animal gets up and down and keeps pawing first with one foot and then other, the legs and ears are cold, and when you press on its belly it causes it to moan with pain. This disease differs from colic for the sheep does not get easy spells but the pain continues all the time.

**Treatment.**—Give the following as soon as noticed :

Tincture of Laudanum................½ ounce or 1 tablespoonful.
Fleming's Tincture of Aconite..........3 drops.

Mix in a half pint of raw linseed oil and give as a drench. Turn the sheep on its back and rub in one-quarter of a pound of mustard, wet up with vinegar, on its belly where the wool is short. If it is a ram be careful that you do not get any of the mustard around the point of the sheath, for it will cause the sheath to be sore and irritate him, afterwards grease where you blistered; keep it very dry, and in a comfortable, warm place, and give the following drench every hour after the first one until the sheep gets relief:

Tincture of Laudanum................½ ounce, or 1 tablespoonful.
Fleming's Tincture of Aconite..........3 drops.

Mix in a half pint of luke warm water and give as a drench.

### TAPE WORM IN SHEEP.

Tapeworm usually affect young sheep, but may affect old ones. The disease is generally seen in sheep that are pasturing on low, rich pasture fields, where dogs that are affected with tapeworm often run across and leave their droppings, for this is the way the disease is carried, and spreads from one flock to the other. When one sheep becomes affected with tapeworm, the worm grows rapidly until it assumes the length of from ten to fifty feet, and is made up of flat joints about half an inch long. After the tapeworm gets large, frequently these joints become broken off and pass away with the manure, this joint is alive when it comes away and remains alive and crawls about on the grass, and very often another sheep comes along and picks it up, and as soon as it reaches the bowels it begins to grow and forms a new tapeworm, this is the way the disease spreads among the flock of sheep.

**Symptoms.**—The animal falls off in condition, and the eyes have a peculiar, dull look. The animal has a good appetite, but what it eats does not do it much good, for the nourishment all goes to nourish the tapeworm. As soon as these symptoms are noticed in one or more young sheep, look out for tapeworm and watch their droppings very closely, and if you see joints of flat worm about half an inch long coming away, you can be sure that they are affected with tapeworm. This is a dangerous disease, and sometimes before the owner is aware, he will lose a lamb or two, and it is best treated as soon as first noticed.

**Treatment.**—Separate the diseased sheep from the ones that do not show the symptoms and put them in a stable, not allowing them to eat anything for twelve hours; at the end of this time they are ready for treatment. Give the following:

Oil of Male Shield Fern................1 dram or 1 teaspoonful.
Raw Linseed Oil......................½ teacupful.

Mix and give as a drench to each of the affected sheep. After this drench feed them light, keeping them in the stable so you can watch their droppings, and if it works on any of them as it generally does, you cannot miss it, for there will be a bunch of worm as large as your two hands come away. If it should not act on the sheep, starve it again for twelve hours and repeat the dose, and if it is a large sheep, give a teaspoonful and a half of the oil of male shield fern. Keep this treatment up until the worm has passed away. Watch the other sheep to see if any of them become affected. The dose for a small lamb is:

Oil of Male Shield Fern..................½ dram or ½ teaspoonful.
Raw Linseed Oil..........................¼ teacupful.

NOTE.—This disease affects sheep and lambs far more than stockowners have any idea of. Often these pieces of white, flat worm are seen coming away with the manure, without considering the danger the flock of sheep are exposed to; they allow it to run on, not treated, until a few lambs or sheep die, then treatment is given to the balance after there is a heavy loss; so you see the importance of watching things like this. A trifling cost and a little trouble will often save heavy losses in your flock.

### FLUKE DISEASE IN SHEEP.

This is a disease of the liver and is very common in England, but not very common in this country, not so much so as it is in the cattle of this country. This disease is fully described in connection with fluke disease in cattle, for the causes, symptoms and treatment are the same. You will find a thorough explanation given there.

### GRUB IN THE HEAD OF SHEEP.

This is a common disease in some localities, especially if the sheep are pasturing on low-lying swampy lands where there are pools of stagnant water.

**Causes.**—The way this disease spreads is by allowing a sheep that has died of grub in the head to lie and be eaten by dogs of the neighborhood, and when they are going across pasture fields they leave their droppings, which contain the grub, in the pasture. The grub, being still alive, crawls onto the grass, and the sheep, while eating the grass, takes the grub into the stomach, and in this way it gets into the blood along with the nourishment and passes around in it until it comes in contact with the brain, where it settles itself in the upper side of it as near the centre as it can lodge.

**Symptoms.**—The symptoms are very peculiar. The sheep holds its head to one side and a little higher than natural, and for a few days will go around in this manner acting very strange. The symptoms become worse; the animal will take spells of running, and if there is a fence in its road it will run up against it, or if there is a ditch, it will run into it as if it were blind and fall over, then take a fit of jerking for a minute or so, then get up and be apparently all right for half an hour or so, when another fit will come on. In other cases, if the grub settles a little to one side of the brain the animal will keep running around and around in a circle. These symptoms gradually get worse, and the fits come on oftener, and if not relieved will die in a few weeks.

**Treatment.**—As soon as first noticed the grub can be killed and absorbed away by giving

Spirits of Turpentine.................. ½ ounce or 1 tablespoonful.
Raw Linseed Oil...................... ½ teacupful.

Mix and give as a drench every second or third day for a week or so. This will kill the grub, which will gradually absorb away itself, and the sheep will soon be all right. During treatment, keep the sheep in a small field where it cannot hurt itself; catch it every day and feel around the top of the head for a soft spot in the bone over the brain by pressing hard on it, and when you find it take a sharp knife and cut the skin off the soft spot about the size of a twenty-five cent piece, and then cut the diseased bone, being careful not to touch the brain. As soon as you have cut around it, raise the piece of soft bone out and leave it out. This soft piece of bone, being over where the grub is, will at once give the sheep relief, for the grub will bulge up in the hole and take the pressure off the brain. Do not attempt to remove the grub, for nature will remove it in a few days itself. The only thing to be done after the bone is cut out is to keep the sheep in a cool place where flies will not bother it and cause maggots. If they should, wash it out clean with luke warm water and soap, then apply the white lotion with a few drops of carbolic acid in it once or twice a day and it will be all right. The hole in the bone will soon heal over. Providing you should lose a sheep with this disease, bury it deep so the dogs will not get at it and carry it to other sheep. Never, in any case, pour spirits of turpentine in the nostrils, for it will do no good, is cruel, and sometimes kills the sheep itself. Always give the turpentine as above mentioned.

### SCAB IN SHEEP.

This disease somewhat resembles mange in horses. It is a very serious disease in some countries, but is not much seen in Canada or United States.

**Causes.**—It is a contagious disease and is caused from little germs or parasites getting down into the skin. The disease spreads by these germs or parasites getting from one sheep to another.

**Symptoms.**—There is extreme itchiness, and the sheep is continually rubbing itself against something. The wool, on account of the germs or parisites working in the skin, falls off in patches, the sheep falls off in condition, and if you examine the scabs with a microscope you will see the minute germs or parasites. When one sheep is affected in a flock it soon spreads and affects them all. Be careful in handling them, as you are apt to get the disease yourself.

**Treatment.**—Clip the remaining part of the wool off short enough to get the medicine down to the skin. A very cheap and effective remedy is creolin water, made by adding two teaspoonfuls of creolin to a pint of rain water. Shake well and rub in all over the body with a brush. Do this every second day until the disease stops and the wool starts to grow. If this should fail, there are regular sheep dips for this disease which can be bought at drug stores; it is done up in a package and it contains the full directions how to use it. As soon as you notice the disease among your flock, separate the affected sheep from the sound ones, and thoroughly clean the stable out and burn sulphur in it with the doors closed.

### WOOL FALLING OFF SHEEP IN THE SPRING.

**Causes.**—From feeding on hot food, such as pea or wheat meal, and keeping them in a place that is too hot.

**Symptoms.**—In the spring, just about lambing time, the wool will become loose and fall off in patches. The animal does not seem itchy to any great extent, but will rub some.

**Treatment.**—As soon as noticed, change the feed and put them in a cooler place, being careful they do not catch cold, and, instead of the rich food, feed roots of some kind to get the blood cool. In very bad cases give:

| | | |
|---|---|---|
| Nitrate of Potash or Saltpetre ........................ | ¼ | pound. |
| Sulphur ........................................ | ¼ | " |
| Ground Gentian Root ................................ | ½ | " |

Mix well together and give a teaspoonful on the tongues of the affected sheep every night. To prevent sheep from loosing their wool, keep them in a cool place,feed chopped oats and plenty of roots, such as mangels, to keep the blood cool.

### SHEEP TICKS.

Sheep ticks are best got rid of about shearing time, by catching the lambs and sheep and giving them a good rubbing all over with creolin water, using two tablespoonfuls of creolin to a pint of water, by rubbing it in well with a brush one dressing will generally cure the whole flock. If in other times of the year, it is best to use a regular sheep dip which can be bought at almost any drug store, the full directions are given in the package.

### MAGGOTS IN WOUNDS ON SHEEP.

Maggots are a very common thing in sheep when they have a wound that has been neglected, or they will in some cases form around their bag when it has festered or bealed.

**Treatment.**—Wash the parts well with soap and warm water; then apply creolin lotion, containing creolin, two tablespoonfuls to a pint of water. The first time you apply the lotion put a good lot on to kill the maggots, then bathe twice a day and apply the creolin lotion to the parts each time after bathing, until healed.

### DIARRHŒA IN SHEEP.

This disease does not occur so often in sheep as it does in cattle.

**Causes.**—From a very cold drink when the animal is dry; from a sudden change in the feed, or from eating anything that is frozen or very green, will sometimes cause it.

**Symptoms.**—The sheep will be dull; will not eat much, and passes a thin, watery manure often, and the hind legs and tail gets wet and dirty looking.

**Treatment.**—Give the following:

Castor Oil .............................. ½ teacupful.
Tincture of Laudanum ................ ½ ounce or 1 tablespoonful.
Ginger .............................. 1 teaspoonful.
Common Soda........................ 1 "

Mix in a half teacupful of luke warm water and give as a drench, then follow up with:

Tincture of Laudanum................ ½ ounce, or 1 tablespoonful.
Common Soda........................ 1 teaspoonful.
Ginger .............................. 1 "

Mix in a half pint of luke warm water and give as a drench every three hours until the diarrhœa stops. In very bad cases add one-quarter ounce, or one tablespoonful of tincture of catechu in each drench and this will make a sure cure. Give luke warm water to drink with dry flour dusted in it to make a sort of a thin gruel.

### BITES FROM DOGS WORRYING SHEEP.

Bathe the wound off with luke warm water and soap twice a day each time after bathing, then wipe dry and apply the white lotion and this will soon heal it up.

### SORE EYES IN SHEEP.

**Causes.**—From something getting into the eye; from an injury to the eye; from getting cold in the eye, or anything that will cause it to be irritated.

**Symptoms.**—The eye is partly closed and water runs from the corner, and when you open the eye it is found to look red, very sore, and inflamed, and may have a scum over it.

**Treatment.**—Catch the sheep and examine the eye, and if there is anything in it, remove it. To get the soreness and inflammation out of the eye, bathe twice a day with new milk or warm water, and then apply the eye wash mentioned among the receipts in the back of this book in and around the eye. Keep this treatment up until it is all right.

### BROKEN LEGS IN LAMBS OR SHEEP.

**Causes.**—From being run over by anything on the road, or from a kick or an injury of any kind that will cause a fracture of the bones. Providing that the fracture or break is so bad that the bones pierce through the skin, it is best to kill the sheep. If it is fat, butcher it and use it for meat. But if the fracture is not so bad and the sheep is young and you want to save it, set the bones to their place and apply a starch bandage, which is a bandage saturated in starch, and in drawing it out of the starch, draw the bandage between your fingers, so as to scrape as much of the starch out of the bandage as you can, for it will harden quicker and better; then wrap it moderately tight around the leg over the place where it is broken, being sure to have the bones set straight, then have someone to hold the sheep and keep the leg straight until the bandage hardens, after this the bandage will hold the leg to its place; leave the bandage on for a month or six weeks. Keep the sheep as quiet as possible during treatment. Watch the sheep's leg where the bandage is on for fear it would make the leg sore or be on too tight; if it should, change the bandage. Leave the bandage on until the sheep can walk on the leg all right, then cut the bandage off.

### INFLAMMATION OF THE TESTICLES IN RAMS.

This is sometimes seen in rams and is caused from an injury of some kind, such as the bunt of another sheep or a kick of some kind.

**Symptoms.**—This disease is very painful. The ram walks stiff; the bag will be swollen and sore to handle; he will not eat much, and lays down most of the time to relieve the testicles as much as possible. If the inflammation is allowed to run on the bag and testicles will become blackened and mortified; it will pass up the cords of the testicles into the belly and soon kill the ram.

**Treatment.**—As soon as the trouble is noticed, separate him from the rest of the flock; keep him in a quiet, cool place and poultice the bag with a hot poultice of half linseed meal and bran, change the poultice every three hours, and each time while changing the poultice bathe the bag with hot vinegar for a while before applying the next poultice. If the ram is fat give the following:

Epsom Salts ........................ ¼ pound.
Tincture of Laudanum .............. ½ ounce, or 2 tablespoonfuls.

Mix in half a pint of luke warm water and give as a drench. In cases where the bag festers and forms matter, which you can tell by the feel of it, you can tell when it is ready to open by feeling for a soft spot in the bag, and as soon as you find the soft spot lance it to let the matter out, and continue poulticing until all the matter is drawn out and the ram seems better. After the inflammation is pretty well out poultice only about half the time.

### HOW TO CASTRATE RAMS.

Throw the ram down and have him held firmly, take a sharp knife and make a cut lengthwise near the bottom of the bag so it will not form a pocket afterwards, cut the hole large enough for the testicle to slip out, as soon as the testicle is out you will notice where the covering is attached at the back part of the testicle, separate the covering from the testicle with your knife, pull the testicle up three or four inches and tie a strong string around the cord, tie it tight enough to stop the blood, leaving the ends of the string four or five inches long so they will hang out of the cut and not heal up in the bag, in a few days they will rot off the end of the cord and drop out themselves;

as soon as this is done cut the cord off half an inch below the string, then operate on the other testicle in a similar manner. As soon as both testicles are cut off fill the holes up with salty butter and let the sheep up. If he swells afterwards bathe the bag and cuts with luke warm water and open the cuts by putting salty butter on your fingers and run them up into the cuts. The main thing after castration is not to let the sheep get wet. In castrating young ram lambs, perform the operation in the same manner as in rams only it is not necessary to tie the cord, but cut it off with a sharp scissors and this will stop it from bleeding and be better than tieing. As soon as the cords are both cut off fill the hole up with butter and let the lamb go, and if it swells afterwards give it the same treatment that is given for swelling in rams after castration.

### CUTTING LAMBS' TAILS.

This operation is generally performed on ewe lambs and on ram lambs that are being kept for breeding purposes; the operation is best done in the latter end of May when the lambs are from two to six weeks old. The best way to do this is, catch the lambs and have someone to hold them while you take hold of the tail to find the second or third joint; after deciding at which joint you wish to cut, place the knife at that joint and cut it right off with one stroke, which is very easily done. They usually do not bleed much, but if they do put some of Monsell's solution of iron on, with a feather, and if you have not this, burn with a red hot iron by touching it to the end of the tail.

### LAMBING.

Sheep are usually put in with the ram in the month of November, and a good way to tell when a ram has served a ewe is to shake dry red paint on the ram's breast and when he has served a ewe the paint will be on her back. The length of time the ewe carries the lamb from the time the ram served her till she is delivered is five months.

### SIGNS OF LAMBING.

In the young ewe about the third month she commences to to make a bag; in the old ewe it is about the fourth month when she begins to make a bag; after this time the bag keeps getting larger until a day or so before she lambs; her bag and teats get hard and full for a few hours before lambing; she is very cross to dogs and other animals; her eyes have a wild look and she tries to get away by herself; finally the labor pains come on and she seems in great pain, the water bag soon appears and breaks, then if the lamb is coming all right the front legs and head will come out and she will soon be delivered of her lamb.

### TROUBLES MET WITH IN LAMBING.

Sometimes the lamb will be coming with the front legs out and the head turned back. In a case of this kind, raise the ewe's hind end up high, then oil your hands and shove the front legs back, and pass your hand inside and turn the head straight, then draw on the head and legs and it will come all right. Sometimes a lamb will be coming with only its head and neck out, the front legs being turned back inside, in this case raise the ewe's hind end up high, oil your hands and shove the head and neck back into the womb and straighten the legs, then draw gently on them and the head and it will come all right. Sometimes one front leg and the head will be out; the lamb cannot come in this position; raise the ewe's hind end up high and shove the leg and head back into the womb and straighten the leg that is bent back and bring it up with the other leg and the head, then it will come all right. Sometimes in a case where there are twin lambs, a leg of each will be out, in this case watch when you shove them back and be sure you get hold of the two legs that belong to the one lamb, because sometimes a mistake is made and they draw on a leg of each lamb and injure both the lambs and ewe. Sometimes the ewe will try to lamb but nothing will come—the only thing you can feel in the passage is the tail and rump of the lamb; the trouble in this case is, that the lamb is coming backwards with the hind legs turned forward under it. In this case raise the ewe's hind end up high, oil your hand and shove the lamb upwards and forwards in the womb, then slip your hand down along the hind leg of the lamb until you can hook your finger around it, then draw it upwards and then backwards until it sticks straight out in the passage, as soon as you have this leg up do the same to the other, and when you have both of them up, draw the lamb out backwards. When you find a lamb coming backwards never attempt to turn it, for this is impossible in the ewe, but take it away backwards. As soon as a ewe has lambed get her on her feet to allow the lamb-bed or womb to go back to its place, also try to get her to lick the lambs. Sometimes in a young ewe where the passage is so small you cannot get your hand in, you will need a boy with a small hand to turn the lamb, while you instruct him. Sometimes the ewe will try to lamb and the neck of the womb will remain closed, not allowing the lamb to come out; in this case medicine will have to be used. Give the following:

Epsom Salts ........................................ ¼ pound.
Fluid Extract of Belladonna ........................ 15 drops.

Mix in a half pint of luke warm water and give as a drench, then follow up with the following:

Fluid Extract of Belladonna ........................ 15 drops.

Mix in a half pint of luke warm water. Give this drench every two hours until it acts on the neck of the womb and opens it so she can lamb without difficulty.

### THE CLEANING NOT COMING AWAY AFTER LAMBING.

In a case of this kind give the ewe a hot drink, and then leave her alone, allowing her to lie down, and if it does not come away then, give:

Epsom Salts ........................ ¼ pound.
Sweet Spirits of Nitre ............. ¼ ounce, or 1 tablespoonful.
Fluid Extract of Belladonna ........ 15 drops.

Mix in half a pint of luke warm water and give as a drench, and as soon as the medicine operates it will loosen the cleaning and it will come away all right.

### LAMB BED TURNED OUT.

**Causes.**—From a ewe, after lambing, lying with her hind parts too low, and the womb gets shoved up in the pelvic or hip cavity, which causes her pain, and on account of this she commences straining and does not stop until she turns the lamb bed inside out.

**Symptoms.**—The ewe seems very weak and will generally be lying down with the bed out behind. The bed is about as large as a man's head and is covered all over with little processes like buttons; it is to these little buttons the cleaning is attached.

**Treatment.**—As soon as noticed, if there is any cleaning attached to the button-like processes, take it off, then bathe the bed well with luke warm water and place a sheet or clean bag under the bed to keep it up and also keep it clean. Have a man on each side to hold the hind end of the sheep up, and, having your hands well oiled, start to turn it in, commencing at the vulva and keep turning it in until it is all in the passage, then shove the womb back to its natural place with your hand before you leave it. With a needle and twine used for sewing wounds, put two or three good stitches in, leaving only enough space for her to make her water through, and in a day or two, when she is all over the straining, take the stitches out. After the womb has been returned she will likely strain some; to relieve this, give the following:

| | |
|---|---|
| Epsom Salts ........................ | ¼ pound. |
| Tincture of Laudanum................ | ½ ounce, or 1 tablespoonful. |
| Fleming's Tincture of Aconite.... .... | 3 drops. |

Mix thoroughly and give as a drench. After this, give a warm bran mash, and give the following drench every hour until she is through straining:

| | |
|---|---|
| Tincture of Laudanum........... .... | ½ ounce or 1 tablespoonful. |
| Sweet Spirits of Nitre .......... .... | ¼ " 1 " |
| Fleming's Tincture of Aconite.......... | 2 drops. |

Mix in half a pint of water and give as a drench.

### INFLAMMATION OF THE MILK BAG (GARGET).

**Causes.**—This is generally noticed after lambing by getting cold, or from an overflow of milk; from an injury; or sometimes, when a ewe loses a lamb, from not being milked enough will cause it.

**Symptoms.**—The bag becomes largely swollen with milk—is hard, hot, tender and inflamed. The ewe seems feverish and is in pain; if you try to milk her at this stage, only a watery, curdy milk comes out. If it is not checked now, the inflamed part of the bag will commence to fester and form matter, and will either break or have to be lanced. Sometimes, instead of the bag festering, it becomes mortified, and if not checked, the mortification will go up into the belly and kill the ewe.

**Treatment.**—It is always best, if a ewe loses a lamb, or when the lambs are being weaned, to watch the bag and milk her out once in a while until she goes dry. When the bag is swollen and inflamed, milk her out once or twice and bathe her bag three times a day with warm water and vinegar, and in a day or so she will come all right. When it has been neglected until it festers, watch the bag and keep bathing it with warm water and vinegar. As soon as a soft spot comes in the bag, lance it to let the matter out, then continue bathing and applying the white lotion twice a day until all the swelling is gone and the bag is healed up. In cases where the bag becomes mortified, give it lots of bathing with warm water and vinegar, and keep applying white lotion three times a day, each time after bathing, until the mortification is checked, and then the mortified part will separate from the healthy part in the course of time, and drop off, then bathe well with warm water and soap twice a day and apply white lotion each time after bathing. As soon as it is healed up it is best to fatten the ewe and butcher her. During treatment, feed well and take extra care of her to keep her strength up until she gets better.

---

# DISEASES OF YOUNG LAMBS.

## WEAKNESS.

**Causes.**—Sometimes, where the ewes have been poorly fed, not getting any grain, and has a couple of lambs, the lambs will naturally be weak and unable to stand, and seems very dumpish; or this weakness may be caused from a lamb coming on a cold night and getting a chill; or where a ewe has bother lambing and the lamb has to be pulled away.

**Treatment.**—If it is a cold night, take the lamb where it is nice and warm, and give it a good rubbing until it is perfectly dry, and pour down with a spoon the following:

Whisky ........................................ 1 teaspoonful.

Mix in a tablespoonful of its mother's milk. Give this amount every hour or two until it is strong enough to suck.

## CONSTIPATION.

This is where the young lamb's bowels get stopped.

**Causes.**—Sometimes from a lamb getting cow's milk in its full strength. Cow's milk should be weakened down about one-third by adding water and sugar before giving it. It is noticed in lambs sometimes, when no causes can be given.

**Symptoms.**—The lamb seems to be a little fuller at the sides than natural, is dull in appearance and keeps straining occasionally as if trying to pass something from the bowels—but nothing comes. Soon it gets in pain and refuses to suck.

**Treatment.**—Give the following:

Raw Linseed Oil .............................. 1 tablespoonful.
Whisky .......................................... 1 teaspoonful.

Mix this in a tablespoonful of the ewe's milk and pour it down with a spoon once a day until the bowels seem all right. Give an injection of half a teacup of luke warm water and a little soap into the back bowel with a small syringe three times a day until the bowels move, also keep the lamb good and warm.

### DIARRHŒA.

**Causes.**—Sometimes from a chill, or from the milk not agreeing with it.

**Symptoms.**—The lamb will often pass a watery manure from the anus, and the tail and hind legs soon gets wet and sticky. The lamb will not suck and seems quite dull, and soon becomes very weak if the diarrhœa is not checked.

**Treatment.**—Give a teaspoonful of whisky and one of castor oil in a tablespoonful of its mother's milk, mix and shake well and pour it down with a small spoon. If not relieved in four hours, give five drops of tincture of laudanum and a teaspoonful of whisky mixed in a tablespoonful of the ewe's milk. Give this with a spoon, being careful not to choke the lamb; repeat this dose every four hours until the diarrhœa is checked, also be sure and keep the lamb in a warm, comfortable place until it is all right. As well as attending to the lamb in those cases, see that the mother's bag is kept milked out so it will not become inflamed.

### FOOT ROT IN SHEEP.

In some parts of the country this is a very common disease and resembles foul in the foot in cattle

**Causes.**—This disease is sometimes brought on from the feet growing too long and splitting up and setting up inflammation in the feet. Sometimes when they are pasturing on a low, damp pasture from continually getting the feet wet, it irritates the foot and sets up the disease. It may be caused from the sheep walking through mud or dirty places, and the mud or dirt getting up between the trotters and getting hard, which irritates the foot and sets up the disease. In some localities this disease is of an infectious nature, that is to say, where the matter from the feet of an affected sheep gets on another sheep's feet it will set up the disease.

**Symptoms.**—The sheep are lame and stiff when walking, and if the disease runs on the feet will swell and little boils will form around the top of the hoof, which break and run matter, and the sheep will become exceedingly lame. If the disease is allowed to run on for a time, the hoofs become loose and fall off, and the sheep will then die from exhaustion. When one sheep becomes affected, the same cause will usually affect more; so you see it is advisable to try and find out the cause and remove it, for the same cause that brings it on in one sheep is likely to bring it on in all of them.

**Treatment.**—Separate the affected sheep from the sound ones and put the affected ones in a quiet, dry place, and if it is caused from the sheep being in a wet place, remove the sound ones to a dry field. Catch the affected sheep, lay them on their side, and bathe the feet well with luke warm water and soap, cleaning all the dirt from between the trotters. As soon as you have bathed the feet, poultice them with a hot poultice of half linseed meal and bran; leave the poultice on all night, and poultice every night until the sheep is better. Each time before putting on the poultice, and after taking it off, dress the foot with white lotion with a few drops of carbolic acid in it. If this does not effect a cure, then try the following mixture:

| | |
|---|---|
| Sweet Oil | 4 ounces. |
| Carbolic Acid | 20 drops. |

Apply this the same as the lotion before putting on the poultice and after taking it off.

---

CHAPTER II.

# DISEASES AND TREATMENT OF PIGS.

Pigs are not subject to as many diseases as cattle or sheep.

### CHOKING.

This is not a very common thing in pigs, but is sometimes met with when they try to swallow some hard substance which is too large for the throat.

**Symptoms.**—The pig keeps coughing, and saliva runs from the mouth; when it tries to eat or drink, the food or water will run back out of its mouth, and if it does not get relief will soon die.

**Treatment.**—Pour down a little raw linseed oil, and then if you can feel the obstruction in the tube along the neck, try and work it around with your hand to get it to slip down. If the obstruction is caught in the back part of the mouth, remove it by prying the mouth open with a stick and working it out with another stick or a long pincers. If the obstruction is so solid that you cannot get it out or rub it down after giving the oil, take a probang, which is used for cattle when choking, and pass it back through the throat into the œsophagus, and push the obstruction down into the stomach. Before attempting to pass the probang, tie a rope around the upper part of the mouth and have the head held up, then place the gag across the mouth (the same as is done when cattle are choking) and run the oiled probang down.

### SORE THROAT.

This is more often met with in young pigs from three to six months old, but may occur at any age.

**Causes.**—Generally from getting cold; by changing from a warm to a cold pen; or from getting wet in cold weather.

**Symptoms.**—The pigs will sneeze and cough; in drinking, the water will run out through the nose; the throat will be swollen and sore to press on. When one pig becomes affected others generally get it from the same cause.

**Treatment.**—Make the pigs as comfortable as possible, by having the pen dry and lots of bedding in it. Feed on soft, warm food, with a large tablespoonful of sulphur in it, which will be enough for six small pigs. Give the sulphur twice a day. When the pigs get so bad that they do not even attempt to eat, take a quarter of a pound each of sulphur and nitrate of potash or saltpetre, mix together and throw half a teaspoonful back on the tongue three times a day; this will gargle the throat. Rub white liniment around their throats each time you give the medicine, and if it is a valuable pig, and a bad case, poultice the throat with hot poultices of half linseed meal and half bran. Change the poultice every couple of hours until the pig is better.

### ACUTE INDIGESTION.

This is where the stomach is distended with food and gas. It is mostly seen in pigs six to twelve months old.

**Causes.**—Generally from a pig that is poorly fed getting into a field of peas or grain and getting a big feed, or sometimes when the pigs are not used to grain and on being brought in to fatten they are given a large feed of peas or wheat which will sometimes set up this disease.

**Symptoms.**—The pig refuses to eat, seems bloated, very uneasy, and in pain. If the sickness is after any of the above causes, you may then be sure it is acute indigestion.

**Treatment.**—The dose for a pig from six months to a year old is as follows :

| | |
|---|---|
| Epsom Salts | 3 tablespoonfuls. |
| Common Soda | 1 teaspoonful. |
| Ginger | 1 " |
| Sweet Spirits of Nitre | 1 tablespoonful. |

Mix in a half pint of luke warm water and pour down the pig after it has been caught and turned on its back. Give only luke warm water to drink and no feed until the medicine operates. If it is not better the next day repeat the dose. In giving the drench do not pour it down too fast for fear of choking the pig. If the cause is from getting a large feed of wheat or peas it sometimes swells so much that it ruptures the stomach, in this case nothing can be done. As soon as the rupture takes place the pig soon dies; if you want to be sure of it being a rupture, open the pig and you will find a hole in the stomach.

### STUNTED OR CHRONIC INDIGESTION.

This is a common thing in young pigs where they are being weaned and started to feed and before they get used to it they are fed on strong food such as peas, wheat and corn ; this is too much for the young pig's stomach and sets up indigestion. It is also caused from worms.

**Symptoms.**—The pig seems to eat plenty but does not grow or thrive.

**Treatment.**—To prevent this disease from coming on pigs, when they are being weaned they should never be fed on hard, hot feed, but should get milk and swill until about three months old, then bring them gradually to hard feed by giving a little at a time. In young pigs, they are so much troubled with worms that they should get a handful of hardwood ashes and charcoal put in the feed twice a week, one handful being enough for six young pigs, also give them plenty of salt, for pigs getting ashes, charcoal and salt in this way seem to thrive half as well again. When the pigs are stunted change the feed from hard to soft feed, and for six pigs give ot :

| | |
|---|---|
| Sulphur | ½ pound. |
| Common Soda | ½ " |
| Nitrate of Potash or Saltpetre | ¼ " |

Mix thoroughly and give a tablespoonful in their feed night and morning until they begin to thrive, also give them plenty of hardwood ashes, charcoal and salt in their feed, as is explained above.

### CONSTIPATION.

This is liable to affect pigs at any age, but is more often noticed to affect stunted pigs.

**Causes.**—In young pigs it is generally caused from chronic indigestion, or worms in the stomach, while in older pigs it is caused from feeding on dry food without getting exercise, that is, being kept in a small pen.

**Symptoms.**—The pigs are very dull; refuse their food; lie down most of the time, and seem fuller than natural; they will strain to pass something but nothing comes. In young pigs, they will not thrive well; the rectum, or back bowel, will be bulged out, and in some cases will be turned out entirely; they seem full and do not eat as they should, and if not attended to will become stunted and pine away until they die.

**Treatment.**—In young pigs, catch them and pour down a tablespoonful of Epsom salts and a half teaspoonful each of ginger and common soda dissolved in half a teacupful of luke warm water; give this every day until the bowels get nice and loose; after the physic, continue the treatment by giving the hardwood ashes, charcoal, sulphur and salt, as recommended in chronic indigestion. In older pigs, give:

| | |
|---|---|
| Epsom Salts.................................... | 2 tablespoonfuls. |
| Ginger ............... .... .................. | 1 teaspoonful. |
| Common Soda..... ......... .................. | 1 " |

Dissolve in a teacupful of luke warm water and pour down as a drench, after catching the pig and turning it on its back; also give the hardwood ashes, charcoal, sulphur and salt as mentioned in chronic indigestion. In aged pigs (that is where they are two or three years old) use the same treatment as is given for medium sized pigs, only give a larger dose. They can stand about a quarter of a pound of Epsom salts with a teaspoonful of common soda and ginger dissolved in a pint of luke warm water; repeat this dose every second day until it works the bowels. As well as giving the medicine, give them exercise every day which will help the physic to operate.

### DIARRHŒA.

This is the very opposite of constipation.

**Causes.**—It is generally caused from a sudden change in the food; from eating something that is frozen; or from excitement by being chased.

**Symptoms.**--The manure runs away like water; the pig is dull and refuses to eat its food, but is very thirsty and greedy to drink.

**Treatment.**—Change the food, and give a gruel drink of

| | |
|---|---|
| Dry Flour | 1 teacupful. |
| Common Soda | 1 tablespoonful. |
| Ginger | 1 " |

Mix in a half pail of luke warm water and give as a drink three times a day. The above is the proportion to give six small pigs. In a case where it is in young pigs sucking a sow, give the sow

| | |
|---|---|
| Dry Flour | ½ teacupful. |
| Common Soda | 1 teaspoonful. |
| Ginger | 1 " |

Mix in a half pail of luke warm water and give three times a day. As well as the above treatment, keep the pigs warm, dry and clean, and they will soon get all right.

### WORMS.

This is mostly seen in young pigs.

**Symptoms.**—The pig eats plenty but does not seem to thrive well, and sometimes you will see small worms pass away in the manure.

**Treatment.**—Give a handful of charcoal and hardwood ashes in the food twice a week; also give half a teaspoonful of sulphur for each pig twice a day in their food and this will generally kil the worms.

### TURNING OUT OF THE RECTUM OR BACK BOWEL.

**Causes.**—This is generally caused in young pigs by getting dry food to eat, and the bowels become costive, and while straining to pass manure, it turns the back bowel out; or where pigs rise on their hind legs with their front feet upon the front of the pen every time you go to feed them; or from going to jump a low fence and getting caught, and while lying with the belly over the fence, the back bowel becomes turned out.

**Symptoms.**—There is a bulging out of the back bowel from half the size of a hen's egg to even larger than a hen's egg, which will be red and angry looking, and after a time become blackened.

**Treatment.**—As soon as noticed give the back bowel a syringing out with luke warm water and soap so as to clean any hard manure out of it, then wash it off clean and oil the bulged out part of the bowel and shove it back to its natural place. As soon as this is done, if it is a small pig, give a tablespoonful of epsom salts mixed in half a teacupful of luke warm water and pour down as a drench, and give one-half teaspoonful of sulphur mixed in sloppy feed twice a day, this will keep the bowels free and the pig will generally be all right. Have the pig fastened in its pen so it cannot jump up on its hind legs; in a case where the bowel is blackened or comes out two or three times after putting it in, do not return it but leave it alone and the piece of dead bowel will drop off of its own accord, and the pig will be all right in a week or so, but give the physic and sulphur as mentioned to keep the bowels loose.

### BLIND STAGGERS.

This is a disease that affects the brain and nerves.

**Causes.**—From pigs being kept in a dirty, ill-ventilated, poorly drained, small pen. The blood gets in such a bad state that it becomes stagnant. This disease is mostly seen in pigs under a year old.

**Symptoms.**—The pig will be dull, stands in a corner with its ears lopped over, will not eat, and when it goes to walk will run against anything in its way. In a short time, the ears, nose, and around the head turns to a blue color which is caused by the stagnant blood. The bowels are costive, and the pig becomes duller and duller, until convulsions come on—and it dies. When one pig becomes affected in a pen where there are thirty or forty pigs kept, what caused it in that one will cause it in the other pigs, and we have seen cases where farmers would have lost half a dozen pigs in a few hours, so when you find the disease is affecting your pigs, let them out of the pen where they are kept into open air, and clean out the pen thoroughly.

**Treatment.**—As soon as one pig becomes affected let them all out of the pen into the open air for a few hours, and afterward put them in a clean pen. In treating the affected one, drag it out into the open air and rub it well with cloths to get the circulation started, and give the following:

For a pig 3 months' old, give 1 tablespoonful of Epsom Salts.
" 3 to 6 " " 2 " " "
" 6 to 12 " " 3 to 4 " " "

As well as this put from a teaspoonful to a tablespoonful or two (according to the age of the pig) of good brandy, also a teaspoonful of ginger and common soda in the drench. Dissolve in half a teacupful of luke warm water, and pour down the pig as a drench after turning it on its back. As soon as the drench operates it will relieve the pig. In some cases persons have been known to cut off the ears and tail with a view to getting the circulation of the blood started, but the hand rubbing is better and is not so cruel and does not disfigure the pig.

### FOUNDER OR SORE FEET.

**Causes.**—This is a very common thing in pigs, especially when fattening them and feeding them on hard, dry feed, when they are in a small pen with very little straw on the floor. Driving them on a hard road or from getting a feed of wheat when not used to it will cause it.

**Symptoms.**—The pig will be dull and lie most of the time and when it moves around it is so sore on its front feet that it walks with its hind feet well under it with the front feet stuck out ahead of it, in some cases the feet are so sore that the pig squeals when it is forced to put its weight on the front feet. From the severe pain of the feet and not being able to get around to eat its food it soon falls off in condition and becomes very gaunt.

**Treatment.**—If it is in the summer time, turn the affected pigs out where they can root in the ground, and pour water in the hole where they lie every day to keep the ground wet, as this will help cure the sore feet, as well as this mix half a pound each of sulphur and nitrate of potash or saltpetre, and give a teaspoonful of the mixture in a slop twice a day to each of the affected pigs. If it is in the winter time, keep the pig in a warm, dry place. Give it a physic of Epsom salts (use the proportion given in blind staggers), feed light and give the mixture above mentioned, also poultice the front feet every night and the pig will soon get all right.

### WOUNDS.

**Causes.**—From a bite of a dog or other pigs; from going through a barb wire fence and getting caught on a barb, or from running against a nail and tearing itself.

**Treatment.**—Where it is a very bad rip or tear, it is best to catch the pig and sew it up with the same kind of a needle and thread as is used to sew up wounds in horses. Put the stitches three-quarters of an inch apart, then fill the wound up with green salve; let the pig go and it will soon heal up, for pigs' flesh heals very quickly; but watch the wound that maggots do no get in it, if they do, wash the wound out well with luke warm water and soap, then apply the white liniment to kill the maggots, and fill the wound up with green salve and it will soon heal up.

## CASTRATING.

Old boars can be castrated at almost any time of the year, but it is said that the colder the weather the better, as long as the weather is dry and they are kept in a dry place. Catch the boar, turn him on his back and tie him securely with a rope so he cannot get up or hurt you, then take hold of the testicle with the left hand, and with a sharp knife in the right hand make a cut in the bag large enough to allow the testicle to come out; make the cut in the underside of the bag so it will not form a pocket afterwards. As soon as the testicle is out, separate the covering from the testicle, where it is attached to the underside, by cutting it off. Pull the testicle and cord out three or four inches and tie a strong string tightly around the cord to prevent it from bleeding; leave the ends of the string four or five inches long so they will hang out of the cut. As soon as the cord is tied, cut the testicle off half an inch below where it is tied, then operate on the other testicle in the same manner. If it is a young boar you are castrating, instead of tying the cord with a string, cut it off with a sharp scissors and this will stop the bleeding. In very young pigs, after the testicle is out scrape the cord with a knife until it is worn off, or cut it off with a sharp scissors. In all cases after you are through castrating the pig, and before you let him up, fill the cuts where the testicle comes out of with salty butter. If the cuts swell much and the pig seems stiff and sore in a few days after he is castrated, catch him and open up the cuts with some butter on your finger and allow the matter that has formed in the bag to run out; this is one thing that should never be neglected after castrating any animal.

### HOW TO CASTRATE PIGS THAT ARE RUPTURED IN THE BAG.

**Causes of Rupture.**—Generally from a young pig standing on its hind feet with its front feet up on the front of the pen when you go to feed them; in other cases it is noticed as soon as it is pigged.

**Symptoms of Rupture.**—The bag is enlarged by the bowels coming out into it, and when you catch the pig and press on the enlargement, you can press the bowel back into its place, but as soon as you let go it will fly out again.

**How to Operate.**—It is best to castrate ruptured pigs when they are young—about five or six weeks old. If it is a large boar, starve him for twenty-four hours before operating, so his bowels will be empty, for they will slip back to their place easier while you are castrating him. Have help enough to hold the hind end of the pig well up while you press the bowels back, which is easily done when they are empty, and his hind end being higher than his front they will stay back better while you are operating on him. Take the testicles out in the same manner as you would a pig that was not ruptured, only make the cuts as small as you can. After each testicle is out, sew up the cut with a needle and thread (the same as used for sewing up wounds); put the stitches in about one-quarter of an inch apart, so the bowels cannot slip out. If it is a small pig, let him run with the others; but if a large pig, keep him in a quiet place and do not give him much to eat for a few days until the cuts swell, which will keep the rupture back. Leave the stitches in until they rot out of their own accord.

### BLACK TEETH IN YOUNG PIGS.

These are very small black teeth which are found in the sides of the mouth when young pigs are pigged, and continue growing for some time, but after the pig gets larger they drop out of their own accord. These teeth sometimes grow in such a manner that they cut and poison the tongue and make it so sore that they cannot eat, and in a little while, if not attended to, the tongue becomes so badly swollen that often the pigs will die from starvation and blood-poisoning. We have seen cases where three or four pigs in one litter died from black teeth.

**Treatment.**—Catch the young pigs and examine each one of them, and if they have black teeth, pull them out with a pincers and they will soon be all right. It is a good practice to catch and examine a litter of young pigs, for if the black teeth are taken out in time it will often prevent some of them from dying before you notice them sick.

### LICE ON PIGS.

These lice are large and resemble ticks on sheep, only they can run very fast. We have seen cases where the pigs were literally covered with them. Pigs that have lice on them do not thrive well, and they are a great preventive to fattening pigs, because they irritate and cause them to be continually rubbing and scratching themselves.

**Treatment.**—Wash the pig well with creolin water, which is a sure, cheap and simple remedy. The strength of creolin to be used is two tablespoonfuls to a pint of water. Rub the creolin water in all over the pig with a stiff brush. It generally takes two washings to make a complete cure; the second application is to be put on about a week after the first one. As well as this, clean the pen thoroughly, shake lime around the floor, and put fresh, clean straw in for them to lie on.

### FRACTURE OF A PIG'S LEG.

**Causes.**—This may be caused in various ways, such as being hit by a stone, being kicked, or from getting the foot through a hole in the floor and giving the leg a wrench.

**Symptoms.**—The pig cannot use its leg in walking; it will hang loose. If you take hold of the leg and twist it you can hear the ends of the broken bones grating on each other.

**Treatment.**—If it is a fat pig, and about ready to kill, it is best to butcher it. If it is a pig you want to save, and the ends of the broken bone are not out through the skin, try and treat it. Take a long bandage, soak it in starch (same as is used for starching clothes), when you are drawing the bandage out of the starch, draw it between your fingers and scrape it with a knife to get as much of the starch out as you can, so it will harden quicker; roll the bandage up so it will be handy to put on the leg, then set the broken bone to its place, and put the bandage on the leg moderately tight. After it is on, hold the leg and bandage straight until the bandage hardens, after that it will hold the bone to its place. Keep the pig very quiet, and feed it so it will not have to stir around. Leave the bandage on for three or four weeks, until the pig can use the leg all right, then remove the bandage by cutting it off. If the break is in the hip, or some place where you cannot bandage it, leave the pig in a very quiet place, and sometimes the broken bone will knit together itself.

### BROKEN BACK IN PIGS.

This is very often seen in sows when they are very thin and weak after suckling pigs. It will happen very easily sometimes; a very little tap on her nose will sometimes break a sow's back; getting hit over the back, slipping, or from something falling and hitting her over the back will sometimes cause it. It may occur in other pigs in the same manner, but they are not so liable to be hurt as weak sows just after you wean the little pigs.

**Symptoms.**—All at once she will loose power of her hind quarters and drag them after her. If you prick the hind parts with a pin she cannot feel it, and will lie there quite helpless.

**Treatment.**—Put the sow in a small, clean pen with a good bed, and feed her so she will not be hungry and try to move around. Give one or two teaspoonfuls of sulphur a day in her feed to keep the bowels loose. Keep her as quiet as possible and she will probably get well in the course of time.

### HOG CHOLERA.

This is an infectious blood disease, and is sometimes called anthrax; it is noticed to break out in herds of hogs all over the country.

**Symptoms** -The first symptoms are dullness, drooping of the head and ears, loss of appetite, and the pig will go away by itself to lie down; the pig seems very feverish, hot, and in some cases will lie quiet and die very suddenly, while in other cases, as the symptoms advance, the pig has pains in the bowels, will lie on its side, moan with pain, then jump up, run around, squeal and grunt in a very painful manner. The pig at times gets easy spells and becomes quiet; if you catch and examine it you will find that the skin of the belly, thighs, front legs, throat, and around the nose will be of a purple color, and in some cases on account of the high fever, the skin breaks out in a rash. In the last stages of the disease there is diarrhœa, and the manure which is very thin, is of a black color and has a very bad smell. The pig rapidly loses strength, gets a cough, begins breathing very heavy, and in a few hours is so weak it can hardly stand. In some cases the pig dies in from six to ten hours after being smitten with the disease, while in other cases it lives for a few days. The rash that comes out on the skin soon causes it to slough, and the skin will drop off in places, giving the animal a bad appearance—and will soon die. Although this is a very fatal disease and most of the pigs die that are affected with it, occasionally we have a case get better when the treatment is taken in time.

**Treatment.**—As soon as any of the pigs are noticed sick, separate the sick ones from the others that are not sick, and put them in a dry, clean pen, and give the following medicine to each pig, using your own judgment as to the proportion to give the different sized pigs:

| | |
|---|---|
| Epsom Salts | 2 to 4 tablespoonfuls. |
| Sweet Spirits of Nitre | 1 teaspoonful to 1 tablespoonful. |
| Sulphur | 1 " 1 " |

Mix in half a teacupful of luke warm water and pour down as a drench after turning the pig on its back. If this drench operates on the bowels before the disease gets too bad, it will often save the life of the pig. If this disease breaks out in your herd of pigs, send for the government veterinary inspector; he will come and examine the pigs to make sure about the disease and help you to prevent the disease from spreading.

## DIFFICULTIES MET WITH IN SOWS PIGGING.

The average time it takes a sow, after being put to the boar, before she has pigs is three months, three weeks and three days. In a month and a-half after being put to the boar, she begins to get larger and continues getting larger until pigging time. Just before she pigs she is noticed to be uneasy, and starts to make her bed: if anything disturbs her she makes a fuss and seems excited. After her bed is made she seems sick and lies down for a short time, then the labor pains come on—she will strain and the water bag will appear and break, if the pigs are coming all right, she will soon be delivered of one; the front feet and head should come first but sometimes they come with the hind feet first; in a few minutes more she will strain again and another pig will be delivered, and so on until they are all delivered. After this there is usually no trouble unless the sow has too much bedding and she smothers the little pigs. Sometimes, when a sow is pigging, a pig will come crooked and get lodged in the passage and she cannot pig without a little assistance. The pig may be coming head first with the front feet turned back, or coming with its hind end first and its legs turned in under it, and all you can feel while examining is the tail and rump. If the sow is large enough for you to pass your hand into the passage, oil your hand and pass it up; if the front legs of the little pig are turned back shove it back into the womb; catch the legs with your finger and bring them up even with the head, then pull on the legs and it will come all right. If it is coming backwards, and the hind legs are turned under it shove the pig back into the womb and straighten out the legs so as to have the hind feet coming first. In case the sow is too small for a man's hand to go into the passage, get a boy that has a small hand and have him oil it and pass it up into the passage, and by you telling him what to do he can bring it away all right. In working with sows always keep your hand well oiled, and try and not bruise the passage, for it will swell and make it worse for you to work at.

### MILK FEVER IN SOWS.

This disease is occasionally met with in sows, but the causes of it is not clearly understood.

**Symptoms.**—They are generally noticed as soon as the sow is through pigging, when the little pigs go to suck they cannot get any milk; if you try to milk her no milk will come; the teats are soft instead of being full and hard, showing that there is no milk being secreted in the teats, the sow seems very sick, is feverish and does not take notice of her little ones, nor eat, but is very thirsty and will drink a great deal if she can get it to drink.

**Treatment.**—Give the following for a large sow:

| | |
|---|---|
| Epsom Salts | ¼ pound. |
| Sweet Spirits of Nitre | ½ ounce, or 1 tablespoonful. |
| Ginger | 1 teaspoonful. |

Mix in half pint of luke warm water and pour down as a drench, first turning the sow on her back and prying her mouth open with a stick to get the drench down. After giving the drench follow up with the following powder:

| | |
|---|---|
| Nitrate of Potash or Saltpetre | ¼ pound. |
| Sulphur | ¼ " |

Mix thoroughly, and for a large sow, give a teaspoonful on her tongue with a spoon twice a day, or in soft feed. Give her plenty of luke warm slops to eat, bathe her milk glands three times a day with luke warm water, rub the parts dry each time after bathing and keep the little pigs sucking to try to bring her back to her milk. By attending to her in this way for a few days she will get all right. While the sow is sick keep the little pigs alive by feeding them milk from a newly calved cow; weaken the milk down one-third with luke warm water and sweeten it with sugar, pour this down with a spoon, or let them suck it out of a bottle, also keep them sucking the sow as much as you can.

### INFLAMMATION OF THE MILK GLANDS IN A SOW (GARGET).

This is generally noticed right after pigging.

**Causes.**—From catching cold, or from losing some of the pigs and not having enough left to keep the bag sucked out, and on account of this the glands get so full of milk that they swell become sore and inflamed.

**Symptoms.** -The sow seems dull, feverish and does not care to eat, and when the pigs go to suck her it causes pain and they do not get much milk. On examing her the milk glands will be found swollen, hard, hot and tender.

**Treatment.** Give the same medicine inwardly with the same directions as is given for the treatment of milk fever in sows, besides this bathe the milk glands well three times a day with luke warm water and vinegar, each time after bathing rub the glands dry, and in severe cases where the young pigs are not sucking her, rub the glands with white liniment, and after putting on the the liniment oil the parts with oil or lard to soften the glands and keep the liniment from blistering. In cases where the disease runs on, and the glands fester and form matter, they will have to be lanced to let the matter out; then treat by bathing with luke warm water and soap twice a day, and apply the white lotion each time after bathing until the glands heal up. In bad cases it is best to wean the young pigs.

### PIG BED OR WOMB TURNED OUT.

This is generally noticed right after pigging, where the sow is left lying with her hind end lower than her front, which causes the pig bed to work back into the passage, and when it does, it causes her pain, and she starts to force and forces the pig bed inside out.

**Symptoms.**—The womb, or pig bed, is turned out behind, which is larger than a man's head, and in a very short time becomes very red and swollen, and is a miserable looking sight; it should be put back as soon as noticed before it swells much.

**Treatment**—As soon as noticed, bathe the womb well with luke warm water to take the swelling down and make it clean and warm; then have a couple of men to raise the sow's hind end straight up while you start turning the womb in at the passage, and keep turning it in until it is all in the passage; then oil your hand and arm, if the sow is large enough for your hand to pass into the passage, but if not, get a boy with a small hand, and press the pig bed right back into its place. As soon as you have done this, sew up the vulva or the opening of the passage to the womb; put two or three stitches across it; put them in good and deep, just leaving space enough at the bottom for her to make water. Leave the stitches in for a day or two until she stops straining, then cut the stitches, pull them out and she will be all right. After you have put the bed back, if she keeps straining, give her the following:

| | |
|---|---|
| Tincture of Laudanum................ | ½ ounce, or 1 tablespoonful. |
| Sweet Spirits of Nitre........... ... | ½ " 1 " |
| Fleming's Tincture of Aconite. ....... | 2 to 3 drops. |

Mix in a teacupful of luke warm water and pour down as a drench, after turning her on her back and prying her mouth open with a stick. If she does not get relief in two hours, repeat the drench, and keep repeating it every two or three hours until she is all over the straining.

### HOW TO RING A PIG.

The object of ringing pigs is to keep them from rooting. The best way to ring them is to first make a noose on the end of a small rope, slip the noose into the pig's mouth, draw it tight on the upper jaw, and have the rope over a beam or something to draw the pig's head up tight; have a pincers and ring, which can be obtained at any hardware store, place the ring in the pincers, and while the pig is pulling back, close the ring on its nose so as to catch about one-quarter of an inch of gristle; this is done by pressing on and closing the handles of the pincers. Be careful not to put the ring back farther than one-quarter of an inch, also see that there is no rust on the ring before putting it in. Put in from one to three rings, according to the size and age of the pig. If the nose should fester after ringing, it is best to take the ring out.

### FITS IN YOUNG PIGS FROM WORMS.

**Symptoms.**—The pig is first noticed not to be thriving well, and after a time is noticed to take fits, its legs begin jerking, its head and neck bends back and it keeps champing its teeth, and after working in this manner for a while, falls over as if it were dead, but in a short time gets up and seems all right for a little while, then another fit comes on. The pig keeps on taking those fits every little while for a few days, but finally gets so bad it dies.

**Treatment.**—For a pig two months old give

Raw Linseed Oil........................ ½ teacupful.
Spirits of Turpentine.................... 1 dram or 1 teaspoonful.

Mix well and pour down as a drench. Repeat this dose once a day until the bowels move and the pig seems relieved, afterwards feed lots of charcoal, hardwood ashes, salt and sulphur, the same as is given to stunted pigs to get rid of the worms.

---

## CHAPTER III.

# DISEASES AND TREATMENT OF DOGS.

### MANGE.

This is a very common skin disease among dogs.

**Causes.**—From parasites, or germs, getting down into the skin, and as soon as they do they commence to breed and spread all over the body, mostly affecting the skin on the back, head and neck. This disease will soon spread among dogs and to other animals, even to the human being.

**Symptoms.**—The dog becomes extremely itchy, keeps rubbing and biting himself until the hair falls off, and in some cases the skin gets so sore and irritated that it will bleed; the hair is dry and scruffy-looking, and the dog soon falls off in condition. To make sure of this disease examine the scales under a microscope and you can see the parasites or germs.

**Treatment.**—If the weather is warm clip the hair off the dog good and short, and give him a thorough good washing with luke warm water and soap, after this rub him dry and apply the creolin wash by rubbing it well into the skin all over the body; use two tablespoonfuls of creolin to the pint of water and apply it once a day until the dog stops rubbing himself, the skin heals and the hair starts to grow out; as well as this, give a teaspoonful of sulphur in his milk twice a day if it is a large dog; give the sulphur in proportion to the size of the dog; if it is in the winter time just clip the hair off and apply the creolin wash, but do not wash him with water.

## FLEAS.

Fleas are a very common thing in dogs, especially if they are kept tied up and not properly taken care of.

**Treatment.**—Clean out the place where the dog is kept and whitewash it to get it thoroughly clean and put in fresh bedding, then, if it is in the summer time, wash the dog with luke warm water and soap, then rub him dry, after which rub him thoroughly with creolin wash every third or fourth day until the fleas are killed. One washing is usually enough to kill the fleas. Use two tablespoonfuls of creolin to the pint of water to wash him with.

## CANKER IN THE EAR.

**Causes.**—This is a common disease in dogs that run through long, wet grass.

**Symptoms.**—There is a discharge of matter from the ear which has a bad smell; the dog holds his head to one side, and if you go to catch him to examine the ear he will be very much afraid, showing that it is very sore.

**Treatment.**—Bathe the ear well with luke warm water and castile soap once a day; wipe dry each time after bathing and inject a little white lotion into the ear with a syringe and he will soon get all right.

## DEAFNESS.

This is very common in old dogs.

**Symptoms.**—The dog seems stupid, and when you call him he does not hear you.

**Treatment.**—If it is an old dog, and the cause is from age, there cannot be anything done; but if it is a young dog, and the deafness is caused from wax in the ear or anything like that, wash the ear out with luke warm water and soap, after which drop a few drops of sweet oil into the ear; do this every second day and in a short time it will effect a complete cure.

### SORE EYES.

This is an inflammation of the eye and its covering.

**Causes.**—From something getting into the eye; from a bite or scratch, or anything that will irritate the eye.

**Symptoms.**—The eye is very red, sore and inflamed, and after a time a little scum will form over the sight. The eyelids, in some cases, are swollen and tears keep running from the corners.

**Treatment.**—Catch the dog, examine the eye, and if there is anything in it, remove it at once; after this, bathe the eye with new milk twice a day, and each time after bathing apply the eye wash mentioned in the back of this book until he gets all right. While treating him keep him in a cool, dark place.

### ENLARGEMENTS OR GROWTHS IN THE EYE.

In some cases the eye itself will become so enlarged that it will bulge out of its socket, which gives the dog a very unsightly appearance. In other cases cancerous growths affect the eye, and the cancer grows until it hangs out of the eye onto the cheek.

**Treatment.**—In either of the above cases treat by removing the eye. First, put a muzzle on him or tie his mouth with a strong, small rope so he cannot bite you, then tie his legs firmly. When you have the dog securely tied, stick a hook into the affected eye and pull outwards and upwards, then with a sharp knife cut around the eye and separate it from the eyelid and draw the eye up as far as you can. You will find the eye attached at the back by the muscles, nerve and artery of the eye. Tie a string tight around the back part of the eye, so it will prevent bleeding, then cut the eye off in front of the string, leaving the string on. The after treatment is to bathe the parts with lukewarm water and castile soap once a day; each time after bathing apply the white lotion and it will soon heal up. Keep the dog in a dark, cool, quiet place during treatment.

### WOUNDS OR CUTS OF ANY KIND.

If the wound is torn much, tie the dog's mouth with a rope or muzzle him so he cannot bite you, also tie his legs to hold them firmly, then stitch the wound up with a needle and twine (the same as is used for sewing wounds on horses). There is no necessity for using medicine on it afterwards, but just leave the dog in a dark, quiet place, and allow him to lick the wound, which is the best medicine you can use. If the wound is where the dog cannot get at it to lick it, bathe it twice a day with luke warm water and castile soap, and each time after bathing wipe dry and apply the white lotion.

### TUMORS OR GROWTHS ON THE SKIN.

As soon as they are noticed it is best to cut them out clean. First secure the dog in the same manner as is given in the treatment for "enlargements and growths in the eye," then take a sharp knife and skin the lump right out; let the dog go and do nothing more to it, only let the dog lick it, for it is dangerous to put medicine on cuts or wounds where the dog can lick it, for the medicine might poison him.

### CONSTIPATION IN DOGS.

This is where the bowels have stopped working.

**Symptoms.**—The dog will not eat, seems very dull and dumpish; the belly seems fuller than natural; he is often noticed to strain, but does not pass much, and anything that does come away will be hard and dry.

**Treatment.**—For a small or young dog give

Syrup of Buckthorn.............. .....¼ ounce or 1 tablespoonful.

Pour this down twice a day, night and morning, until the bowels are moved. For a large dog give one-half ounce or two tablespoonfuls of syrup of buckthorn twice a day until it acts on him. Castor oil is also recommended to be given in the same proportions as the syrup of buckthorn. As well as giving the medicine mentioned give an injection of a teacupful of luke warm water with a little soap in it twice a day. Give the dog gentle exercise once a day and see that he has nothing but new milk to drink during the time he is sick.

### DIARRHŒA IN DOGS.

This is not a very common disease in dogs, but is sometimes seen, and is generally caused from a change in his food.

**Treatment.**—Keep the dog very quiet, and in some cases by giving him boiled new milk with a little dry flour in it, it will check the diarrhœa without any medicine. If this fails, give to a small dog half a dram or half a teaspoonful of tincture of laudanum and a teaspoonful of whisky in a tablespoonful of new milk as a drench. For a large dog give one dram or a teaspoonful of tincture of laudanum and a tablespoonful of whisky mixed in half a teacupful of boiled milk and flour and give as a drench. Repeat the drench three times a day until he gets all right. During the treatment keep the dog very quiet and give very little to eat except boiled new milk with a little dry flour sifted in it.

### WORMS.

**Symptoms.**—The dog has a good appetite, but does not thrive well, his hair is dry looking and dusty. Occasionally there will be a worm noticed to pass away in his manure. Sometimes dogs affected very badly with worms will take fits.

**Treatment.**—For a medium-sized dog give one dram of oil of male shield fern mixed with four tablespoonfuls of new milk. Pour it down by putting it well back in the mouth with a spoon or bottle and holding his head until he swallows it. Give this dose every third day until the bowels are moving free and the dog is relieved of worms.

### DISTEMPER IN DOGS.

This is a very common disease in dogs from three to six months old, but may affect them at any age, and is mostly seen during the fall, winter and spring. This disease somewhat resembles distemper in horses.

**Causes.**—It is caused from parasites, or germs, getting into the blood and setting up the disease in the same manner as distemper in horses. By allowing the affected dog to be with other dogs they will catch it from him.

**Symptoms.**—The dog is very dull and sleepy looking, tries to get in a quiet, dark place and refuses to eat, his eyes keep running water, and after a time they become very sore. In a few days his hair becomes dry, there is a discharge from the nose, eyes, and sometimes from the penis, the bowels are costive, the dog falls off in condition, the symptoms gradually get worse, and if he does not get relief he will go into convulsions or fits and soon die. Like other diseases of this kind it must run its course, which should last about nine days.

**Treatment.**—The treatment is very simple. Give the affected dog nothing to eat but new milk and the following medicine:

| | | |
|---|---|---|
| Nitrate of Potash or Saltpetre | 1/8 | pound. |
| Sulphur | 1/8 | " |
| Ground Gentian Root | 1/8 | " |

Mix thoroughly together, and for a large dog give half a teaspoonful of the mixture three times a day on his tongue with a spoon or in his milk. Regulate the dose in proportion to the size of the dog. Keep the dog in a quiet, clean place; wash off the discharge from his eyes and nose with luke warm water, and afterwards wipe dry with a soft cloth. If his bowels are costive, give from a teaspoonful to a tablespoonful, according to the size of the dog, of syrup of buckthorn once a day until the bowels move freely; after that give a dose once or twice a week to keep the bowels free, and in a week or so the dog will begin to get better. Be careful not to let him get cold until he gets all right.

### MAD DOG (HYDROPHOBIA OR RABIES).

This disease originates spontaneously in dogs and cats without being bitten, and is mostly seen in very hot weather and in hot climates. It may also be caused by dogs or cats being bitten by another mad dog or cat.

**Symptoms.**—The affected dog is first noticed to become excited; runs about and bites at everything that comes in his way; saliva keeps running from his mouth and his eyes are bloodshot. After this, the dog will take a spell of quietness, and will go and lie down in an out-of-the-way place. He seems to have a depraved appetite; will eat clay and all sorts of dirt. Suddenly, he will again become excited and run about biting at everything that comes in his way, the same as at first. A mad dog rarely, if ever, goes out of his way to bite anything. In running about he keeps snapping his teeth, and once in a while gives a peculiar howl. He has great difficulty in swallowing, and in some cases cannot swallow at all. These symptoms gradually get worse until the dog dies. The time the disease affects a dog after being bitten varies from ten to fifteen days.

**Treatment.**—As soon as the dog is noticed to be mad, have him destroyed, for he is very dangerous to have around, and, besides, there is no cure for it. In cases where a dog has been bitten by a mad dog or cat, cut the piece out where he was bitten and burn the hole with caustic potash, nitrate of silver or a red hot iron, which will destroy the poison before it gets into the system.

### CHOREA.

This is a disease that affects the nerves, causing the affected part to keep jerking.

**Causes.**—It is very often noticed after a dog has had a very severe attack of distemper, or it may be caused from an injury.

**Symptoms.**—There is a continual jerking of the muscles around the affected part when the dog is resting.

**Treatment.**—There cannot be much done for it after the disease has once set in, but by giving from one-quarter to one half a dram of bromide of potassium twice a day on his tongue with a spoon or in a little new milk, according to the size of the dog, as soon as the disease is noticed, it will nearly always effect a cure.

## DOGS POISONED WITH RAT POISON, STRYCHNINE OR ARSENIC.

**Symptoms.**—All at once the dog seems in great pain; takes fits; slobbers at the mouth; the eyes are bulged out and blood-shot, and he keeps trembling all over. If the dog shows the above symptoms, and you have had poison around the place, or he has been anywhere that he could possibly get it, you may conclude he has been poisoned.

**Treatment.**—The treatment must be quick. Give the dog an emetic to make him vomit; this is the only way you can save his life. The idea is to get him to throw up the poison out of his stomach before it gets absorbed into the blood. Give him one tablespoonful of salt dissolved in a half teacupful of luke warm water as a drench, or one teaspoonful of mustard dissolved in a half teacupful of luke warm water as a drench; either of the above will cause vomiting. But the best of all, if you have it, to make him vomit, is to give him from half a teaspoonful to a teaspoonful of sulphate of zinc dissolved in a half teacupful of luke warm water and give it as a drench. If you can get him to vomit soon enough it will generally save his life.

## FRACTURES OR BROKEN BONES IN DOGS.

**Causes.**—Fractures of bones in the dog may occur in various ways, such as a kick or being struck with anything.

**Symptoms.**—The symptoms vary according to the part affected. If it is a fracture of the neck bones the dog dies instantly; if in the back it causes paralysis of his hind quarters; if it is in the hip or hind leg it is easily told, for the dog cannot use the leg and hops along on the other three while the broken one will be dangling; if it is a fracture in the front leg, he acts similar to when the fracture is in the hind leg. To make sure it is a fracture, catch the dog and move the affected parts and you can hear the ends of the broken bone grinding on each other.

**Treatment.**—If the fracture is in the back, there can be nothing done but to destroy the dog, to put him out of pain. If it is the hip or shoulder bone that is fractured and it is a young dog, keep him perfectly quiet and the bones will soon knit together, but if it is a very old dog there is not much chance of the bones knitting together, although they might. If the fracture is down on the leg set the broken bone to its place and have a long, starch bandage (which is a bandage soaked in starch and then wrung out as dry as possible), roll the bandage around the fractured leg letting it go two or three inches above and below the fracture, wrap it moderately tight and hold the leg and bandage straight until the bandage hardens. Keep the dog quiet and leave the bandage on for three or four weeks until the bones are healed and he can use his leg all right.

### BREEDING PUPS AND TROUBLES MET WITH AT PUPPING TIME.

Bitches usually come in heat about twice a year, but some come oftener than that. She is nine days coming in heat and nine days going off, during that time she will take the dog about nine or ten days. It is best to keep the bitch shut up during the eighteen days, so that no other dog but the one you want to breed to can get at her. If you want to get bitch pups, let the dog to her as soon as she will take him; while if you want to get dog pups, let her go for seven or eight days after she would have taken the dog, before you let the dog to her and you are almost sure to get dog pups. The bitch carries her pups nine weeks. At the end of that time, and a few hours before pupping, she will try to get away to some quiet place and make a nest. As a general thing after she once starts to pup, it does not take her long to have them all. Sometimes we have cases where she cannot pup alone, that is, where one of the pups is coming wrong—sometimes head first with the front legs turned back, and it gets caught at the shoulders. In this case shove the pup back a little, slip your finger in under the front legs and bring them forward, and it will then come all right. If the front legs are coming out and the head and neck are turned back, shove the pup back a little, and straighten the head and neck up along with the front legs, and the pup will soon come all right. Sometimes the pup comes backwards with only his rump and tail up in the passage, in this case shove the pup back and straighten up the two hind feet, and bring the pup away backwards. Before starting to work at a job of this kind, have your fingers and hands oiled so as not to irritate the bitch, sometimes a small hook will be of some use, but if you use one be very careful not to tear her. Bitches are not very liable to disease after pupping and the only thing to be done is to see that she is dry and warm, and that the pups are sucking her all right.

### HOW TO CUT PUPS' TAILS.

This is usually done when the pup is about a month old. The length of the tail to be left on varies with the breed of the pup; find out the length you want to leave it and then find the joint, and cut it through with one stroke of a sharp knife. They do not usually bleed, but if it should, touch the spots with Monsell's solution of iron, or if you have not got that just touch it with a red hot iron which will stop it, but in most cases it will be all right without anything.

### HOW TO CASTRATE A DOG.

This operation is done at all ages—but like other animals it is best done when he is young; it is very simple. Catch the dog and roll him up in a blanket, turning his front legs back and his hind legs forward so as to catch them in the blanket, and have him held on his back with his hind end stuck out of the blanket enough to get at the bag. Take hold of the testicle with your left hand, and with a sharp knife in your right hand, let out the testicle by making a cut in the bag large enough to let it out, separate the covering which is attached to the back part of the testicle with your knife, as soon as this is cut draw the testicle well up, and if he is a young dog cut the cord off with a scissors which will stop the bleeding, but if he is an old dog, tie the cord tight with a piece of strong string, leaving the ends long enough so they will hang out of the bag, then cut the testicle off about one-half an inch below where it is tied, operate on the other testicle in the same way, and fill the holes with salty butter and let him go.

### HOW TO SPAY A BITCH.

This operation is taking the ovaries away from the bitch, to prevent her from coming in heat and getting with pup. It is best to perform this operation when the bitch is between six and nine months old and before she comes in heat for the first time, but it can be done at any time, and we have even done it when she was with pup, and be all right, but, if it is done when she is with pup she will generally lose her pups. It is best to have the bitch thin in condition, not giving her anything to eat or drink but milk the day before the operation. What is needed for this operation is a sharp knife, a sewing needle and string (the same as is used for sewing up wounds), and a probe, which can be made out of a piece of wire about eight inches long having one end rounded so it will not hurt her while it is being passed up into the womb, and a scissors. As well as these you need a large sponge and about eight ounces of sulphuric ether which is used to put her asleep.

Catch the bitch and place her on a table or box where you are going to operate; saturate the sponge with sulphuric ether and hold it tight to her nose, making her breathe right through the sponge. Watch her carefully, and as soon as she is under the influence of it so much that she has no power of herself then take the sponge and ether away from her nose and commence to operate. Have some one to watch her head all the time, and when she commences to stir a little place the sponge saturated with ether to her nose again until she is quiet, and keep doing this all through the operation, but be careful not to give her too much, just enough to keep her asleep. Have the bitch on her back; commence operating by oiling the probe and passing it carefully up into the passage until it is in the womb; make a cut about an inch long in the center of the belly, between the four last teats; oil your fingers and insert the first finger of the left hand into the cut, while with your right hand you take hold of the probe and press it down at the end so the end inside of her will be raised toward the hole in the skin of the belly; with your finger in the cut feel the end of the probe; the part you feel is the womb; now slip your finger forward under the womb and bring the two horns that branch out from the womb up and out through the hole with your finger, then draw one of the horns well up until the ovary, which is about the size of a pea, comes outside, then with the scissors cut it off; do the same with the other horn, cutting off the other ovary; wash the parts off clean with carbolic water—five drops of carbolic acid to a pint of luke warm water—and shove the parts carefully back to their place and sew up the skin by putting in a few stitches; after this take the sponge away from her nose; pour a little cold water into her mouth and on her head and let her come to, then keep her very quiet for a week or so, feeding her mostly on new milk and she will come all right. The operation is very simple when once you have performed it and know how to go about it. A person wishing to become an expert at spaying had better thoroughly examine the womb, horns of the womb and ovaries in a dead bitch before trying the operation on a live one, as it will give a better idea how to go about it.

Another thing to bear in mind in this operation, as well as in other operations, always have everything perfectly clean. When this precaution is taken the operation is generally successtul.

---

CHAPTER IV.

# DISEASES AND TREATMENT OF POULTRY.

### ROUP.

Roup is considered by all poultry men the worst and most dreaded disease that can afflict poultry.

**Cause.**—From germs settling in the throat, or it may be caused from neglected colds.

There are many remedies for roup, but we think the following as is given in "Success With Poultry," is one of the very best:

**Treatment.**—Pen the fowl in warm, dry quarters, and keep out all drafts of cold and damp air. Feed hot bran, mashed potatoes and meat, and medicate the throat, mouth and nostrils with cloride of sodium or common salt, as follows: Take a bucketful of warm water; put a teacupful of salt in this amount of water; then catch the fowl, examine the throat and nostrils, removing all mucous matter out of the nostrils, and then filling a pint cup for each afflicted fowl, hold it by the feet—head down, choke it until the mouth is wide open and then insert the head into the solution. Comb down, so the medicated water may enter the clift in the palate and go out at each nostril and into the throat. Each should be separately treated. Kerosene injected into the nostrils is good; also camphorated sweet oil. Ten drops of coal oil or kerosene added to two drops of water for a flock of twenty fowls will often effect a cure. But when this remedy is applied, do not attempt to prepare one of the flock for table use for three or four weeks thereafter, as the entire carcass will be tainted with coal oil.

IMPORTANT.—In treating roup, be careful to remove any discharge from the nostrils that may collect on the feathers under the wings or on the breast. Be sure and protect the sick fowls from all drafts and feed easily digested food. When the fowls look stupid and droopy, feathers rough and no appetite, reduce their food to even fasting.

### HEN LICE IN A HEN HOUSE AND ON HENS.

When once hen lice get started they increase very rapidly and become a perfect nuisance to the hens and the owner, and in some cases they will spread to other buildings and get on the stock. Hen lice are very small and have a reddish appearance.

**Treatment.**—The first step to be taken to get rid of the hen lice is to drive all the hens out of the hen house, close up the windows and doors, put a pound of sulphur in a pot, put some coals in it to start the sulphur burning, and keep it burning for about half a day. The fumes from the sulphur will go into all the cracks and thoroughly fumigate the place and kill the lice; after this open the doors and windows and clean the place out thoroughly and put plenty of hardwood ashes and coal dust in the hen house for the hens to roll about in; this kills the lice on the hens. In the course of a week drive the hens out of the hen house again and burn sulphur as you did before, this will rid you entirely of the pest, and by watching the hens and giving them plenty of ashes and coal dust to roll in, will keep the lice off them after this. It is also a good plan to feed the hens lots of sulphur and new milk, give every night, two tablespoonfuls of sulphur to a quart of new milk, this would be about the proper dose for twenty-five hens. The sulphur passing off through the pores in the skin helps greatly to kill the lice. Whitewash the walls, ceiling, roosts and nests of the hen house, and put clean straw in the nests; this will effect a complete cure. If the lice are in other buildings treat just the same way as mentioned for a hen house. This treatment for killing the hen lice will also kill the germs of other diseases such as roup, etc., and is a process that a hen house should have every fall and spring to keep diseases from the fowl.

### CHOLERA.

Cholera is very contagious. The cause of it is overcrowding, bad sanitary management, unwholesome or irregular food, etc. Chicken cholera is not very well understood by poultry raisers, and we believe it is a fact that a great many diseases are called cholera simply because it is not understood. Everyone who keeps fowls should be able to tell cholera from other diseases, for without such knowledge it is impossible to treat the disease properly.

**Symptoms.**—The fowl has a very sleepy, droopy appearance; it is very thirsty and has a slow gait and gapes often. Sometimes the fowl staggers and falls down from great weakness. The comb and wattles lose their natural color, generally turning pale and sometimes they are dark. There is diarrhœa with a greenish discharge, or like sulphur and water. The crop fills with mucus and wind, and at last the food is not digested. Breathing is heavy and fast, the eyes close, and in a few hours the fowl dies.

**Treatment.**—The first thing to be done when cholera makes its appearance is to give the coops and yards a thorough renovating; disinfect with carbolic acid, 40 drops to the pint of water. Remove all affected fowls and burn the carcass of dead ones. The best and most effective remedy for cholera that we have ever used is coal oil or kerosene. The coal oil should be given about four times a week, as follows: Take a feeding of corn and wheat and let it soak in the oil a few hours, and then feed it to the fowls, or mix it in soft food, one teaspoonful of oil to every two quarts of corn meal. While treating do not give much water to drink.

### CANKER.

Canker is caused by filthy quarters and musty or unwholesome food. Canker may be noticed by running sores on the head, in the mouth or throat accompanied with a watery discharge from the eyes.

**Treatment.**—Take warm water and a little salt and wash the head and eyes, also swab out the throat and mouth. Remove ulcers with a quill and apply powdered borax to the places left bare. Repeat this treatment twice a day and a cure will be effected.

### CROP BOUND.

The most usual cause is that the fowl has swallowed something that it cannot digest. The foreign substance may be a piece of bone, or anything that obstructs the natural passage.

**Treatment.**—The best treatment for crop bound is castor oil. Give the fowl two teaspoonfuls in one dose. If this does not clear the crop in 24 hours it cannot be cured, and the best remedy is to apply the hatchet just back of the comb.

### GAPES.

Gapes in fowls is caused from drinking dirty water, exposure to wet, damp places, and want of nourishing food.

**Symptoms.**—The symptoms of gapes, as the name implies, consists of constant gaping, coughing and sneezing.

**Treatment.**—Give the bird, daily, unless it recovers, a small piece of camphor, about the size of a grain of wheat, and a few drops of turpentine in the drinking water, or mix in the food about 10 drops to the pint.

### LEG WEAKNESS.

The principal cause of leg weakness, in most strain of fowls, arises from inbreeding or breeding the same strain of fowls for too long a period. It is also caused from overfeeding, which increases the weight of the body out of proportion to the muscular strength of the limbs and usually occurs in the larger breeds.

**Treatment.**—In the first stages, give:

| | |
|---|---|
| Sulphate of Iron.................................... | ¾ pound. |
| Sulphate of Lime... ......... ....................... | ¼ " |

Mix and give the fowl about the size of two peas of the mixture on its tongue twice a day. If, however, the bird is unable to walk it is incurable.

### SCALY LEGS.

This disease is contagious, and one scaly legged fowl will affect the whole flock. The disease is caused by a small insect which burrows under the scales. The best and most effective remedy is to dip the fowl's feet and legs in kerosene or coal oil clear up to the feathers. Some people say that this is severe and think kerosene will burn and make the flesh sore. This is a great mistake, and to satisfy yourself, just take some kerosene and put it on your own flesh, and you will find it will not hurt you at all.

### BIG HEAD IN TURKEYS.

This is a disease mostly seen in turkeys, but occasionally is seen in other fowl. It may affect turkeys at any age, but is seen mostly when they are young.

**Causes.**—This disease is caused from small germs or parasites getting into the system and affecting the head. When one turkey becomes affected it generally spreads through the whole flock.

**Symptoms.**—The first symptom noticed is swelling around the head, usually in front of the eyes. In some cases the swelling only affects one side of the head, and keeps swelling until it closes the eye right up, and the turkey goes around with its head to one side, while in other cases both sides become swollen at the same time, and closes both the eyes. Although these swellings are mostly seen in front of the eyes, it may affect other parts of the head, and even the feet in some cases. If not treated, the swelling gradually increases around the head until it works into the throat and kills the turkey ; if in the foot, it swells and becomes so bad that it finally dies.

**Treatment.**—When this disease gets into a flock of turkeys, if it is not treated it will kill nearly all the turkeys in the flock before it stops. Separate the diseased turkeys from the sound ones, and give to a medium sized turkey three drops of spirits of tupentine mixed in a teaspoonful of raw linseed oil. The best way to give it is to pour it down with a spoon. Give this dose once a day and rub around the swelling with white liniment once a day; this will gradually cure the turkeys if taken in time. If the swelling is very large before you start to treat it, split the lumps open with a sharp knife and squeeze the cheesey matter out of it, then fill the hole and rub the swelling with white liniment ; keep this treatment up until the turkey is all right. Be careful not to let the liniment into the eye while applying it to the head.

# PART V.

# Medicines and Receipts.

## CHAPTER I.

## MEDICINES.

Medicines are got from three great sources, viz.: The animal, vegetable and mineral kingdoms. This part of the book is very important and should be carefully studied by persons interested, as it gives the names of the medicines, the sources they are got from, their action and use, the dose to each of the animals and how long it takes to operate.

### ACONITE.

Aconite is got from a plant that grows in cool, mountainous countries. The form of aconite we recommend and use is Fleming's Tincture, which is got from the root of the plant.

**Its Action and Use.**—Aconite acts on the heart and blood vessels, causing the heart to beat slower, and in this way lowers blood pressure in the system, thus it is good in most kinds of inflammation where the animal is in good condition, and is used now in nearly every case in place of bleeding.

**Dose.**—Fleming's Tincture of Aconite—For horses, 8 to 10 drops; cattle, 8 to 12 drops; sheep, 2 to 3 drops; pigs and dogs, 1 to 2 drops.

### ALCOHOL.

Alcohol is got from certain kinds of wood, grapes, beets, potatoes, etc. It is not used much in its pure state for doctoring stock, but is used in the form of liquors for medicines. One preparation, known as methylated spirits of alcohol is used in making liniments.

**Its Action and Use.**—Its action is as a stimulant, and it also acts on the kidneys as a diuretic, and is mostly used in the form of liquors, such as whisky, gin, beer and ale. For external use see where methylated spirits of alcohol is used in making up white and acid liniments among the receipts.

**Dose.**—For horses (whisky), 1 to 2 wine glassfuls; cattle, 2 to 4 wine glassfuls; sheep, 1 wine glassful; pigs and dogs, 1 to 2 tablespoonfuls. For horses (gin), 1 wine glassful; cattle, 1 to 2 wine glassfuls; sheep, 1 to 2 teaspoonfuls; pigs and dogs, 1 teaspoonful. For horses (beer and ale), ½ to 1 pint; cattle, 1 pint; sheep, ¼ to ½ pint; pigs and dogs, 1 wine glassful.

### ALOES.

Aloes is got from a plant grown in the West India Islands. There are three kinds of Aloes: Barbadoes, Socotrine and Cape. The Barbadoes is the best, and is the kind mostly used for stock. It is a liver-brown color, bitter to taste and is usually bought either in the form of a powder or lumps.

**Its Action and Use.**—Aloes acts as a physic and blood purifier and is used in a great many diseases, especially in the horse.

**Dose.**—For horses, 8 to 10 drams; cattle, 1 to 2 ounces; sheep, 2 to 4 drams; pigs and dogs, 1 to 2 drams. In preparing aloes for a drench it must be dissolved in water that is just luke warm, for if the water is too warm it spoils the action of the aloes, and if it is too cold it will not dissolve properly. In giving it as a ball, see receipt of physic ball in the back of this book. It takes aloes twenty-four hours to operate on the bowels in horses and cattle, and after giving them a dose they should always be left standing quietly for forty-eight hours after giving them the medicine.

### ALUM.

Alum is got from the alum salts, which is a mineral.

**Its Action and Use.**—It is mostly used externally in the form of washes for healing wounds. It is also dusted onto wounds in the form of a powder, and is very drying and healing. It is also used in gargle powders for sore throat, influenza and aphtha or sore mouth.

**Dose.**—The way it is used internally is best seen in the receipt for gargle powders at the back of this book. For making a wash it takes a tablespoonful of alum to a pint of water.

### AMMONIA.

Ammonia is obtained from plants and animals. There are several forms of ammonia, but the one mostly used in the practice is strong liquor ammonia, which is used mostly for making liniments. For how to use it see the receipt of white liniment at back of this book.

### ANISEED.

Aniseed grows in the form of berries on bushes that grow in Spain, Germany and Southern Russia. The berries are dried and ground, this being the form we get them in.

**Its Action and Use.**—It stimulates digestion, sweetens the stomach, and in cases where the stomach is deranged it relieves the gasses that form. It is a splendid thing for animals that are recovering from weakening diseases to give them an appetite and build them up.

**Dose.**—For horses, 1 tablespoonful; cattle, 1 tablespoonful; sheep, 1 teaspoonful; pigs and dogs, ½ to 1 teaspoonful.

### ARNICA.

Arnica is got from a plant that grows in mountainous countries in Central Europe, Asia and America. The form we use is the Tincture of Arnica, which is taken from the root of the plant.

**Its Action and Use.**—It is not used much internally, but is used externally in the form of a cooling lotion for sprains, wounds, etc.—one ounce or 4 tablespoonfuls of arnica to a pint of water and applied twice a day. It is also used to rub race horses' legs before and after bandaging to keep them from getting stiff and sore.

### ARSENIC.

Arsenic is got from the mineral kingdom. In its pure form it is not much used in practice because it is too strong and powerful. The form that is used is known as Fowler's Solution, which is prepared from arsenic, and is in the form of a liquid.

**Its Action and Use.**—It is a powerful tonic for the stomach and system in general, and is especially used after weakening diseases, such as distemper, and will often start the animal to thrive when everything else will fail.

**Dose.**—For horses, ½ ounce or 2 tablespoonfuls; cattle, ½ ounce or 2 tablespoonfuls; sheep, 1 dram or 1 teaspoonful; pigs and dogs, ½ dram or ½ teaspoonful. Give this medicine combined with an ounce of whisky and half a pint of gruel twice a day until the animal gets better and is strong.

### BELLADONNA.

Belladonna is got from a plant known as Deadly Nightshade. It grows wild in some parts of Great Britain, and it is also cultivated to a great extent. The form that is used mostly in practice is the Fluid Extract of Belladonna.

**Its Action and Use.**—It soothes, softens and relaxes the parts applied to, and is greatly used where there is inflammation and pain in almost any part of the body to relieve the pain and check the inflammation. The Extract of Belladonna is strong and must be used carefully.

**Dose.**—Fluid Extract of Belladonna—For horses, 1 dram or 1 teaspoonful; cattle, 1 dram or 1 teaspoonful; sheep, 10 drops; pigs and dogs, 3 to 5 drops. Belladonna is greatly used for diseases of the eyes in the form of an eye wash. For further information look at the receipt for eye wash at the back of this book. It is also greatly used for dilating or opening the neck of the womb, in any animal, where it remains closed when they are ready to be delivered of their young. It is used a great deal for this trouble in cows.

### BLACK ANTIMONY.

Black antimony is got from the mineral kingdom, and the form we get it is in a black, heavy, grayish powder that has neither smell nor taste.

**Its Action and Use.**—It is not used much in the way of medicine, only to color condition powders, etc., but should not be used, as it is very irritating and injurious to the system. When it is used as a coloring material for powders it is in this porportion: one teaspoonful of black antimony to a pound of whatever you want to color.

### BUTTER OF ANTIMONY.

Butter of antimony is also got from the mineral kingdom, and is used in the form of a brown liquid.

**Its Action and Use.**—It is never used internally, for it is an irritating poison, but it is used externally for curing thrush in horse's feet, and for burning growths off around the feet that come from the quick of the foot. It is also good in cases where a nail has run in the foot, after the nail is pulled out and the nail-hole pared out, to drop a few drops in the hole to kill the rust and poison of the nail.

### BENZOIN.

Benzoin is got from a tree called styrax benzoin, which grows in the southern part of Asia. It is got by cutting a hole in the bark and catching the sap which afterwards soon becomes hard and brittle.

**Its Action and Use.**—The form we use this in is known as Compound Tincture of Benzoin, or Friar's Balsam, which is prepared from the gum. Its chief action is for healing wounds by painting the wound over with a feather twice a day, and is especially useful for caulks, or where a nail has run in the foot, to kill the poison. It is not used much internally.

### BROMIDE OF POTASSIUM.

Bromide of Potassium is prepared from seawater, and the form we usually get it in is in a white crystal powder, which has a salty taste.

**Its Action and Use.**—Its chief action is to quiet the nerves in nervous diseases, such as lockjaw (tetanus), or in convulsions, chorea and other painful diseases.

**Dose.**—For horses, 1 teaspoonful; cattle, 1 teaspoonful; sheep, ½ teaspoonful; pigs and dogs, ¼ teaspoonful. The dose can be given three or four times a day, according to the severity of the case.

### BINIODID OF MERCURY (RED PRECIPITATE).

Biniodid of Mercury is got from the mineral kingdom. It is bought in the form of a heavy, bright-red powder.

**Its Action and Use.**—It is chiefly used for blistering bony enlargements, such as spavins, ringbones, sidebones, splints, etc. The proportions in which it is used is 1 to 2 drams to the ounce of vaseline, or lard, according to the breed of the horse and thickness of the skin. Have the mercury and vaseline, or lard, thoroughly mixed before using, and three days after applying the blister grease the part with lard. For further information look at mercury blister among the receipts.

### BUCKTHORN.

Buckthorn is got from the ripe berries of a shrubby bush that grows along the North Pacific Coast. The form we use it in is known as Syrup ot Buckthorn.

**Its Action and Use.**—It is principally used as a mild physic for dogs.

**Dose.**—For a dog give 1 to 2 ounces; cats, ½ ounce. A very good plan is to give a tablespoonful of the syrup every night and morning until you get the desired results.

### CAMPHOR.

Camphor is got from a tall, evergreen tree known as the Laurel Camphor, which grows in Japan, China, and different parts of Europe. The branches are cut and boiled in water and the camphor rises to the top in the form of a gum.

**Its Action and Use.**—It is mostly used in mixtures for coughs, sore throat and heaves. How to use it is best seen in the receipt for chronic cough, and heaves or broken wind.

### CANTHARIDES OR SPANISH FLY.

Cantharides are got from flies which receive the name Spanish Fly on account of so many of them coming from Spain. The flies are of a green color, and an inch in length, and are captured in nets, then killed by pouring boiling water or vinegar on them, after this they are dried in the sun or by artificial heat, and are then ground and made ready for use in practice.

**Its Action and Use.**—Its chief action is as a sweat blister and is one of the principal ingredients in sweat blisters which are used for sweating thickenings or lumps on any part of the body, that are not on the bone, such as curbs, tumors, thickenings left after a wound has been healed up, etc. As to how to use Powdered Cantharides see the receipt for sweat blister at the back of this book. Cantharides can also be got in the form of a tincture, but is not much used. The proportion to use Powdered Cantharides for a sweat blister is, 1 to 2 drams of the Canthardies to an ounce of vaseline or lard according to how heavy the skin of the horse is. For cows use it a little stronger than for the horse.

### CARBOLIC ACID.

Carbolic Acid is got from coal tar and petroleum. The pure Carbolic Acid is got in the form of a crystal, but the way it is generally bought is in the form of a liquid.

**Its Action and Use.**—It has a very cleansing effect on a wound, and is greatly used for that purpose and is also used for fumigating stables by sprinkling a little around on the floor. It is very rarely, if ever used internally. The strength Carbolic Acid should be used in bathing a wound is 10 drops to half a pail of water; for a lotion use 20 drops to a pint of water; for making Carbolic Oil use 25 drops of Carbolic Acid to a half pint of olive or sweet oil, that is about two or three drops of acid to the ounce of oil. It is a mistake to put too much acid in a wash or lotion, for instead of it having a cleansing effect it will burn the wound and stop the healing action. A very good healing salve is made out of 5 drops of Carbolic Acid to an ounce of vaseline.

### CASTOR OIL.

Castor Oil is got from the seeds of a shrubby tree that grows in Africa, Southern Europe, and South America.

**Its Action and Use.**—It is a mild physic, similar to raw linseed oil. It is not much used for horses or cattle, but is sometimes given to dogs, pigs, and young animals.

**Dose.**—For horses, 1 pint; cattle, 1 pint; sheep, ¼ pint; pigs and dogs, 1 ounce or 4 tablespoonfuls. In horses and cattle it takes about twenty-four hours to operate on the bowels, while in sheep, pigs and dogs it only takes about twelve hours to operate on them. The best way to give Castor Oil to pigs and dogs is to give 2 tablespoonfuls every night and morning until you get the action required.

### CATECHU.

Catechu is got by boiling the chips from a tree that grows in Africa and Southern Asia. The form that it is mostly used in practice is known as Tincture of Catechu.

**Its Action and Use.**—It acts as an astringent on the bowels for diarrhœa. Thus it is used in cases of diarrhœa, and it checks it in most animals. It is also used for coloring the White Lotion, and a few drops makes it a nice dark color.

**Dose.**—Tincture of Catechu—For horses, 2 to 4 drams or 2 to 4 teaspoonfuls; cattle, 4 to 6 drams or 4 to 6 teaspoonfuls; sheep, 2 drams or 2 teaspoonfuls; pigs and dogs, 1 dram or 1 teaspoonful. These doses may be given in gruel or a pint of luke warm water three or four times a day until the diarrhœa is stopped.

### CALOMEL.

Calomel is got from the mineral kingdom. It is prepared for medicine in the form of a heavy white powder.

**Its Action and Use.**—Its chief action is as a physic, and also clears the bile from the liver. It is given in cases of jaundice and other liver troubles. It is also used for drying up thrush in the feet of horses, where it is explained how to use it.

**Dose.**—The way to give it to a horse is to combine ½ dram of calomel with 4 drams of bitter aloes and give it in the form of a ball. For how to mix a ball look among the receipts at the back of this book. For cattle give 1 dram of calomel with 1 pint of raw linseed oil.

### CROTON OIL.

Croton Oil is got from the seeds of a tree that grows in the southern parts of Asia.

**Its Action and Use.**—It is a very severe physic when given internally. It is sometimes given to cattle and horses when the bowels are stopped and you cannot get a passage, but is never given until all the milder physic medicines fail.

**Dose.**—For horses, 15 to 20 drops; cattle, 30 to 40 drops; sheep, 5 to 10 drops; pigs and dogs, 2 to 3 drops. For horses and cattle, give it in a pint of raw linseed oil; for sheep, give in a half a pint of oil, and for pigs and dogs, give it in two tablespoonfuls of castor oil.

### CHLORATE OF POTASH.

Chlorate of Potash is got from mixing other medicines together, and is bought in the form of crystals or in a white powder.

**Its Action and Use.**—Its principal action is to thicken the blood in diseases where the blood is too thin, such as in button farcy. It is also very soothing in cases of sore throat.

**Dose.**—For horses, 2 to 4 drams or 1 teaspoonful; cattle, 2 to 4 drams or 1 teaspoonful; sheep, 1 dram or ½ teaspoonful; pigs and dogs, ½ dram. The way it is used for sore throat is to put a teaspoonful on the tongue three times a day.

### CAUSTIC POTASH.

Caustic Potash is got from pearl ashes. It is put up in white pencil-like sticks.

**Its Action and Use.**—It is never given internally, but is used to burn warts and growths by wetting the stick and rubbing it over them. It is also used for burning poisonous wounds to kill the poison, such as dog bites. The sticks must be kept in a corked bottle, for the air dissolves them. While using the stick wrap a piece of paper around the end you hold in your hand so it will not burn your fingers.

### CHLORIDE OF ZINC.

Chloride of Zinc is got from the mineral kingdom. This medicine is generally bought in the form of round white pencil-like sticks.

**Its Action and Use.**—Its principal action is as a powerful caustic for burning off growths, warts, etc. It is not used internally.

### CREOLIN.

Creolin is one of the many products of coal tar which is got from the mineral kingdom. It is bought in the form of a thick, dark fluid and smells like tar.

**Its Action and Use.**—It is used in the form of healing lotions for wounds, scratches, grease and such like diseases. The strength to use it in the form of a lotion is ¼ ounce or 1 table-spoonful to the pint of water, shake well before using. It is also a very effectual remedy for killing lice, ticks or fleas on any animal; also used in mange and scab in sheep; the strength to use it in cases of this kind is ½ to 1 ounce or from 2 to 4 tablespoonfuls to the pint of water, shake well before using. Creolin is a very cheap medicine, it is used a great deal now and is still gaining in favor. It is best to buy the Creolin in its pure state and mix it into washes and lotion as you want to use it, for when it is mixed with water for some time it loses its strength, so you see the necessity of buying it in its pure state and mixing it as you want to use it.

### CRUDE PETROLEUM OIL.

Crude Petroleum Oil is got from the mineral kingdom and is the oil as it comes out of the ground.

**Its Action and Use.**—It is principally used for oiling horses' feet in the form of hoof oils; it is also a great remedy for killing ringworm on cattle, horses and other animals; the way to use it in cases of this kind is to paint it around the ringworm; it is a very cheap and effectual remedy. See ringworm on cattle.

### DIGITALIS.

Digitalis is got from the leaves of a plant that grows in many parts of the country. The leaves are dried and ground, and this is the form we buy it in.

**Its Action and Use.**—It is a heart and lung tonic. It is used mostly mixed in powders that are given in weakening diseases, such as influenza, distemper, and lung troubles. It is also sometimes mixed in powders that are given for heaves.

## GAMBOGE.

Gamboge is got from the sap of a tree that grows in Southern Asia. The form gamboge is used in is a yellow-white powder.

**Its Action and Use.**—It is a powerful physic, mostly used for cattle where mild physics, such as salts and linseed oil, fail. In giving it to cattle it is combined with other medicines in this form:

| | |
|---|---|
| Epsom Salts | 1 pound. |
| Common Salt | 2 tablespoonfuls. |
| Gamboge | 1 ounce. |
| Common Soda and Ginger | 1 tablespoonful each. |

Mix in a quart of luke warm water. By giving it this way it makes a powerful physic; it very rarely fails, and is used in bad cases of constipation of the bowels and impactions of the stomach.

## GENTIAN.

Gentian is got from the root of a plant that grows in the mountainous parts of Europe. The root is dried and ground, and this is the form it is used in.

**Its Action and Use.**—It is a bitter tonic for the stomach and system in general. It is used in all kinds of powders that are given to animals that are weak and run down in condition and require a ton c.

**Dose.**—For horses, 1 tablespoonful; cattle, 1 tablespoonful; sheep, 1 teaspoonful; pigs and dogs, ½ teaspoonful. These doses may be given twice a day in food or on their tongue with a spoon.

## GINGER.

Ginger is got from a plant grown in South America and the West India Islands. The plant is dried and ground, and this is the form it is used in.

**Its Action and Use.**—It acts as a stimulant, relieves the gasses and sweetens the stomach, and is used to a great extent with physic drenches to keep it from griping; and is also used in medicines used for colic, indigestion and a great many other diseases.

**Dose.**—For horses, 1 large teaspoonful; cattle, 1 tablespoonful; sheep, 1 small teaspoonful; pigs and dogs, ½ teaspoonful. These doses can be given every two or three hours.

### HYPOSULPHITE OF SODA.

Hyposulphite of Soda is got from the mineral kingdom, and is used in the form of a white powder or crystals.

**Its Action and Use.**—It is a great blood purifier and is combined with equal parts of gentian, and is used to clean the blood and build up the system after weakening diseases. The way to use it is to take hyposulphite of soda one-half pound and gentian one-half pound, mix well together and give of the mixture as follows:

**Dose.**—For horses, 1 tablespoonful; cattle, 1 tablespoonful; sheep, 1 teaspoonful; pigs and dogs, ½ teaspoonful. The above dose can be given two or three times a day according to the case.

### IODINE.

Iodine is got from sea plants, and is used in the form of a dark brown tincture.

**Its Action and Use.**—It is not often used internally, but is used externally for a sweat blister, for blistering thickened glands by painting it on the lump with a feather once a day until it blisters, then grease the parts and let it go for two or three days until it heals up, then wash it off with luke warm water and soap and blister again as before mentioned.

### IODIDE OF POTASSIUM.

Iodide of Potassium is got from the mineral kingdom, and is used in the form of a white crystal powder.

**Its Action and Use.**—Its chief action, when given internally, is an absorbant, *i. e.*, it is given in dropsy of the belly and chest to absorb the fluid; it is also used where there is a thickening around the throat, legs or milk glands; but is not used to any great extent on account of it being very expensive.

**Dose.**—Mix it with equal parts of ground gentian root and give a teaspoonful to a cow or horse twice a day; one-half teaspoonful to other animals.

### LIME.

Lime is got from the mineral kingdom. Internally it is used in the form of lime water, and is used where the stomach is deranged, also in cases of diarrhœa, and is a good thing to sweeten the stomach.

**Doses.**—For horses, 1 ounce or 4 tablespoonfuls; cattle, 1 ounce or 4 tablespoonfuls; sheep, ¼ ounce or 1 tablespoonful; pigs and dogs, 1 teaspoonful. Lime water is sometimes used for heavy horses by throwing 1 ounce or 4 tablespoonfuls into their drinking water twice a day. Unslacked lime is used for disinfecting stables, etc., by dusting it in its dry form around on the floor.

### LINSEED.

Linseed is used mostly in the forms of linseed meal and raw linseed oil, which is got from flax seed.

**Its Action and Use.**—Raw linseed oil is given as a very mild physic, or what is called a laxative. The dose of the oil is one pint poured down as a drench. In all cases, after giving it to a horse or cow, allow them to stand in the stable the next day and feed light for a few days. Linseed meal is used mostly, when given internally, for fattening cattle and for animals recovering from weakening diseases; but the flax seed itself boiled up is better for feeding young and sickly animals than the linseed meal. Linseed meal is also used for drawing poultices, and is one of the best that can be got, and should always be mixed with boiling water.

### LAUDANUM.

Laudanum is used in the form of a tincture, and is a preparation from opium, which is got from a plant that grows in warm parts of Asia.

**Its Action and Use.**—It is sometimes used externally for rubbing on painful swellings. In using it this way use one-third tincture of laudanum and two-thirds white liniment, apply three times a day after bathing. It is used internally in almost every disease where there is pain (which can be seen by looking through the diseases and treatment of this book), it relieves pain and spasms, and in this way helps greatly to check inflammation.

**Dose.**—For horses, 1 to 2 ounces or 4 to 8 tablespoonfuls ; cattle, 1 to 2 ounces or 4 to 8 tablespoonfuls ; sheep and pigs, 2 to 4 drams or 2 to 4 teaspoonfuls; dogs, 20 to 25 drops. It is given in a pint of luke warm water as a drench, and may be given as often as every two hours in severe cases.

### MONSELL'S SOLUTION OF IRON.

Monsell's Solution is a preparation of iron which is got from the mineral kingdom. It is used in the form of a brown, sticky liquid.

**Its Action and Use.**—Its chief action is for stopping blood in a wound of any kind, and also for scabbing the wound over. It is a great remedy for open joint and leaking of the navel in foals. It is applied to the parts with a feather four or five times a day.

### MARSHMALLOWS.

Marshmallows is got from a plant that grows in this country, generally in the neighborhood of rivers.

**Its Action and Use.**—It is chiefly used in poultice, mixed half and half with linseed meal. It makes one of the most effectual drawing and soothing poultices there is. It is also used when boiled, the tea from it is mixed with luke warm water for bathing the milk-bag, for garget, milk-fever, etc.; it also makes a very soothing bath for sore or irritated wounds or swelling. Use one teacupful of the tea to a pint of luke warm water.

### MUSTARD.

Mustard is got from a plant which grows in most parts of Europe. The seeds are dried and ground, and this is the form we use it in.

**Its Action and Use.**—It is chiefly used for mustard plasters which are applied over the bowels in severe cases of colic or inflammation to relieve the pain and check the inflammation; also in lung troubles, applied over the ribs and chest, and also the back in disease of the kidneys, and around the throat for sore throat. To make a mustard plaster of ordinary strength for a thin-skinned horse take a quarter of a pound of mustard, two tablespoonfuls of flour and enough vinegar to make it in the form of a paste. In very severe cases use the mustard and vinegar without the flour on cattle and horses with a very thick skin. Instead of applying it to an animal with a cloth just rub it into the hair over the parts you want blistered.

### NUX VOMICA.

Nux Nomica is got from the seeds of a small tree that grows in India and Australia. These seeds are dried and ground, and it is used in this powdered form.

**Its Action and Use.**—It is a nerve stimulant, and is used in all cases of paralysis where the nerves have lost their power to strengthen them. The way to use it is to take equal parts of gentian and powdered Nux Vomica, mix thoroughly, and as a dose for horses and cattle give one teaspoonful three times a day in the feed or on their tongues with a spoon; for sheep, pigs or dogs give one-half teaspoonful.

### NITRATE OF SILVER.

Nitrate of Silver is got from the mineral kingdom, and is used in the form of white pencil sticks.

**Its Action and Use.**—It is used for burning off warts, proud flesh in cuts and growths in any part of the body by just wetting the stick and rubbing it to the parts. Keep the sticks corked in a bottle for they dissolve when exposed to the air.

### NITRATE OF POTASH OR SALTPETRE.

Nitrate of Potash or Saltpetre is got from the mineral kingdom, and is used in the form of a white crystal powder.

**Its Action and Use.**—Its chief action is on the kidneys and blood, it causes the kidneys to secrete an extra amount of urine. It is used a great deal in practice in almost all lung troubles, also in cases where the blood is bad and where the sheath and legs are swollen.

**Dose.**—For horses, 1 teaspoonful; cattle, 1 teaspoonful; sheep and pigs, ½ teaspoonful. If given for the kidneys, give once a day; if for lung troubles, see diseases of the breathing organs.

### OLIVE OIL.

Olive Oil is got from the seeds of an evergreen tree that grows in Southern Europe.

**Its Action and Use.**—It is not used internally to any extent, but is used externally for soothing and healing irritated wounds. It may be used in its pure state or be mixed with carbolic acid—10 drops of carbolic acid to 4 ounces of olive oil.

### OIL OF TAR.

Oil of Tar is a product of the pine tree, and the form it is used in is of a dark, thick, sticky liquid with a tar-like smell.

**Its Action and Use.**—It is chiefly used in cases of chronic cough and is a very effectual remedy. Give a teaspoonful three times a day in the horse's feed, or on his tongue with a spoon.

### OXIDE OF ZINC.

Oxide of Zinc is got from the mineral kingdom, and the form we get it in is of a white, fine powder.

**Its Action and Use.**—It is mostly used in making up healing salves, and is used in the same proportion as the receipt that is given in chapped or sore teats in cows.

### OIL OF MALE SHIELD FERN.

Oil of Male Shield Fern is got from a shrub that usually grows along the side of the road in most temperate countries. It is got in the form of a dark, thick, oily liquid.

**Its Action and Use.**—This medicine is a most effectual remedy for worms, especially tapeworms, in all animals.

**Dose.**—For horses, 3 to 4 drams; cattle, 3 to 4 drams; sheep and pigs, 1 to 2 drams; dogs, ½ to 1 dram. In giving it to cattle and horses have them starved for twenty-four hours, then give the above mentioned dose mixed in a pint of raw linseed oil or gruel; feed very light for three days, and if the worm has not come away repeat the dose every third day until it does. In giving it to sheep and pigs, give it in the same manner, only in half the quantity mentioned of raw linseed oil or gruel. In giving it to dogs, give it in half a teacupful of new milk, in the same manner as for horses and cattle. For further particulars how to use it look at tapeworm in the different animals.

## PEPPER.

Black pepper, which is the kind mostly used for animals, is got from the berries of a climbing plant that grows in the West Indies.

**Its Action and Use.**—It is used internally as a stomach stimulant to heat the stomach and bowels, and in this way helps to relieve the pain in colic, indigestion, etc.

**Dose.**—Fo horses, 1 tablespoonful, cattle. 1 tablespoonful; sheep and pigs, 1 teaspoonful. dogs. ½ teaspoonful. In mixing up this drench it is often combined with whisky, and makes a good colic drench.

## QUASSIA CHIPS.

Quassia Chips are got from a handsome tree that grows in the West India Islands. It is odorless but bitter to taste.

**Its Action and Use.**—It is used mostly as an injection for pinworms. To prepare it for injection refer to pinworms in horses.

## SULPHURIC ACID.

Sulphuric Acid is a product of the mineral kingdom, and is got in the form of a light brown liquid.

**Its Action and Use.**—At one time it was used a great deal as a caustic for burning warts, etc., but is not so much used now as it is too irritating, its place being taken by better caustics, such as chloride of zinc, nitrate of silver and caustic potash. It is used externally in the form of liniments. How to mix and use it, refer to the receipt for acid liniment in the back of this book.

### SULPHURIC ETHER.

Sulphuric Ether is prepared from sulphuric acid and rectified spirits of alcohol, and is used in the form of a clear liquid with a very strong odor.

**Its Action and Use.**—It is used a great deal for putting animals asleep for operations, especially for dogs. As to how to use it, refer to "how to spay a bitch." It is sometimes given for indigestion or colic, both in horses and cattle, to relieve the pain and gases.

**Dose.**—For horses and cattle, 1 ounce or 4 tablespoonfuls mixed in a pint of luke warm water, and can be given every two hours. Where the pain is severe, it is a very effectual remedy.

### SULPHATE OF COPPFR.

Sulphate of Copper, blue vitriol or blue stone is got from the mineral kingdom, and is got in a blue crystal form.

**Its Action and Use.**—Its chief action when given internally is for checking discharges, such as nasal gleet or chronic catarrh, and whites or leucorrhœa. When used for this purpose, refer to the receipt to be used internally in the above named diseases. When used externally, it is used for wounds that are not healing well and have proud flesh in them by grinding it fine and dusting it on the wound every two or three days according to how much it burns it.

### SULPHATE OF IRON.

Sulphate of Iron, or commonly called green vitriol, is got from the mineral kingdom, and is used in the form of a crystal, similar to sulphate of copper, only of a lighter shade in color.

**Its Action and Use.**—It is one of the best mineral tonics that we have, and is used combined with gentian, in equal parts, for almost every case where the system is run down and needs building up. For horses and cattle give a teaspoonful three times a day; for sheep and pigs give one-half a teaspoonful. It is also used in the same form for killing long round worms and pinworms in horses. For full directions as to how to use it in this case refer to the receipt of worm powders given at the back of this book.

### SULPHATE OF ZINC.

Sulphate of Zinc is got from the mineral kingdom. It is used in the form of a white crystal powder, and resembles Epsom salts in appearance.

**Its Action and Use.**—When used internally it is in the form of an emetic that is to cause vomiting. As to how it is used refer to "dogs poisoned with rat poison, strychnine and arsenic." When used externally it has a healing action. How to use it, refer to the receipt of white lotion at the back of this book. It is also used as a drying wash for clap or gonorrhœa in horses and whites in mares. As to how to mix it refer to those diseases.

### SUGAR OF LEAD.

Sugar of Lead, also called Acetate of Lead, is got from the mineral kingdom.

**Its Action and Use.**—It is not much used internally, but is used outwardly for healing washes, such as white lotion and eye wash, which are fully explained at the back of this book.

### SALICYLIC ACID.

Salicylic Acid is got from a plant. It is used in the form of a white powder.

**Its Action and Use.**—It has a special action in cases of rheumatism. As to how to use it, refer to the receipt given for rheumatism in horses and cattle.

### SALT.

Common Salt is got from the mineral kingdom.

**Its Action and Use.**—It is an essential article of food, and something every animal should have regularly. Horses, cattle, sheep and pigs should have it in front of them all the time for, it should be remembered, stock cannot thrive well without it. Rock salt is the best form in which to have it as the animals can lick it whenever they want it. It is used externally by throwing a handful of salt in a pail of luke warm water, and in this form it makes a very effectual wash for bathing swellings and wounds.

### SWEET SPIRITS OF NITRE.

Sweet Spirits of Nitre, or Nitrous Ether, is chiefly a preparation of alcohol. It is got in the form of a clear liquid which has a sweet taste and smell.

**Its Action and Use.**—In small doses it acts on the kidneys and skin. For this purpose give horses and cattle a quarter of an ounce or one tablespoonful in drinking water once a day. For sheep, pigs and dogs give a teaspoonful in their drinking water or food once a day. In large doses it acts on the bowels and stomach to relieve pain and gases. Thus it is good in the different forms of indigestion and colic.

**Dose.**—For horses, 1 ounce or 4 tablespoonfuls; cattle, 1 ounce or 4 tablespoonfuls; sheep, ½ ounce or 2 tablespoonfuls; pigs and dogs, ¼ ounce or 1 tablespoonful. Mix in a pint of luke warm water and give as a drench. For how often to give it refer to the different diseases it is used in. This is a medicine that is used a great deal in practice, as you will see all through the book, and should be thoroughly understood.

### SPIRITS OF TURPENTINE.

Spirits of Turpentine, also called Oil of Turpentine, is got from a tree. It is used in the form of a clear, oily looking liquid.

**Its Action and Use.**—Internally for horses and cattle it is used in one ounce or four tablespoonful doses mixed with a pint of raw linseed oil in severe cases of acute indigestion and colic to relieve the pain and gases; it is also used in this proportion for killing the long round worms and bots in horses. For further particulars turn to bots and long round worms in horses and other diseases it is used in. Internally for sheep and pigs the dose of turpentine is one-quarter of an ounce or one tablespoonful mixed with half a pint of raw linseed oil. It is used for the same purposes as it is in horses and cattle. Outwardly it is used for making liniments, and for how to use it refer to the receipts for making white linement and acid liniments at the back of this book.

### SALTS.

Salts are used in two forms, Epsom and Glober. Both kinds are got from the mineral kingdom, and are in a white crystal form.

#### EPSOM SALTS.

**Its Action and Use.**—Its chief action is as a physic for cattle, sheep and pigs, and is used to a great extent, as you will notice on reading the treatment of the above mentioned animals. Although salts is a good physic it should not be given to horses, aloes being far the best physic for them.

**Dose.**—Cattle take from one pound to one and one-half pounds dissolved in a quart of luke warm water with a tablespoonful each of ginger and common soda and given as a drench. This physic takes twenty-four hours to operate on the bowels. It is always best after giving a dose to wait twenty-four hours for an action before giving any more. Sheep and pigs take one-quarter of a pound dissolved in a pint of luke warm water with a teaspoonful each of ginger and common soda and given as a drench.

Wait from twelve to sixteen hours for an action on the bowels before giving any more.

GLOBER SALTS.

**Its Action and Use.**—It is not so much used as the Epsom Salts, but is used in horses and cattle for their blood by grinding it up fine and giving a tablespoonful in a hot mash every night.

SULPHUR.

Sulphur, or Brimstone, is got from the mineral kingdom. It is used in a yellow powdered form.

**Its Action and Use.**—It is a great medicine, when given internally, to act on the blood and clear it. It also acts on the skin and helps to kill parasites or germs in the skin, thus it is good in mange and other skin diseases.

**Dose.**—For horses, 1 tablespoonful; cattle, 1 tablespoonful; sheep, 1 teaspoonful; pigs and dogs, 1 teaspoonful. Give every night in a hot mash or soft food; but it is best given combined with gentian root and nitrate of potash or saltpetre as is explained. in the treatment of swelling of the limbs (anasarca) in horses.

VASELINE.

Vaseline is got from the mineral kingdom and is used in the form of a yellow salve.

**Its Action and Use.**—Vaseline has a very healing and soothing action on wounds or irritated parts, and is used in nearly all kinds of healing salves, also for mixing blisters, etc.

VERDIGRIS.

Verdigris is a preparation got from the copper salts, and comes from the mineral kingdom. It is used in the form of a blue, fine, heavy powder.

**Its Action and Use.**—It is not much used now internally, as sulphate of iron and copper take its place. It is used for making up healing salves for wounds, etc. For how to use it see the receipt for green salve.

## CHAPTER II.

# RECEIPTS.

### WHITE LINIMENT.

Proportions to make one quart of the liniment. Use either a quart bottle or a quart self-sealer to mix it in:

FIRST.—Put in one-half pint of hard water.

SECOND.—Put in two ounces or eight tablespoonfuls of spirits of turpentine and shake thoroughly for five minutes.

THIRD.—Beat up one hen egg and put it in and shake thoroughly for five minutes.

FOURTH.—Put in two ounces or eight tablespoonfuls of methylated spirits of alcohol and shake thoroughly for five minutes.

FIFTH.—Put in two ounces or eight tablespoonfuls of strong liquor ammonia and shake thoroughly for five minutes.

SIXTH.—Put in enough hard water to make up a quart, then shake thoroughly and the liniment is ready for use.

This is one of the most effectual remedies known for all kinds of sprains and bruises where the skin is not broken. The longer this liniment stands (if the bottle is kept corked) the stronger and better it gets for using. For making larger or smaller quantities of the liniment add to or take from the proportions given for one quart.

### WHITE LOTION.

Proportions to make up one quart of White Lotion. Use either a quart bottle or a quart self-sealer to mix it in.

Put in one-half ounce of each of the following: Sulphate of zinc, sugar of lead and pulverized alum, add enough water to make a quart and shake thoroughly.

This makes a most effectual lotion for healing all kinds of wounds and bruises where the skin is broken, also where the skin is irritated, such as scratches, grease, etc.

### ACID LINIMENT.

Persons getting this liniment that are not experienced in handling drugs had better get the druggist to mix it for them, as it is a little dangerous mixing the sulphuric acid with other drugs, and it requires to be carefully handled. The following are the proportions for one quart which is best mixed in a quart bottle or a quart self-sealer:

FIRST.—Put in two ounces or eight tablespoonfuls of spirits of turpentine.

SECOND.—Put in one-half ounce of sulphuric acid; pour this very slowly into the bottle by letting it run down the inside of the bottle which is better turned sideways. Take about five minutes to pour it in.

THIRD.—Pour slowly into the bottle two ounces or eight tablespoonfuls of methylated spirits of alcohol.

FOURTH.—Pour in enough cider vinegar to make a quart, then shake well and it is ready for use.

It is a most effectual sweat blister for removing puffy enlargements, such as bog spavin, wind galls, thoroughpins and other puffy swellings around the legs, by applying it every third day. It is also an effectual remedy for sweeny, curbs, etc., where you want to work the horse. This is also an effectual remedy for rheumatism by rubbing the affected joints every third day. It is also used in various other ways as you will notice through the book.

### CREOLIN LOTION.

The following proportions are for one pint of Creolin Lotion:

FIRST.—Pour in one quarter ounce or one tablespoonful of creolin into a pint bottle.

SECOND.—Pour in enough hard water to make a pint, shake well and then it is ready for use.

This makes an excellent healing lotion for wounds, and by making it double strength it makes a most effectual wash for destroying germs, parasites, lice or ticks on all animals. It is used in various other places as you may see in this book.

### EYE WASH.

The following proportions are to make an eight-ounce bottle of eye wash:

| | |
|---|---|
| Sulphate of Zinc | ½ dram. |
| Sugar of Lead | ½ " |
| Fluid Extract of Belladonna | 30 drops. |

Add enough hard water to make eight ounces, shake thoroughly and the wash is ready for use.

This makes a very cheap and effectual wash for sore eyes, or sores around the eyes in all kinds of animals by applying twice a day after bathing with luke warm water or new milk.

### CARBOLIC OIL.

The following proportions are for a four-ounce bottle:

| | |
|---|---|
| Olive or Sweet Oil | 4 ounces. |
| Carbolic Acid | 20 drops. |

Shake well together and this makes a splendid application for healing wounds.

### PHYSIC DRENCH FOR HORSES.

| | |
|---|---|
| Bitter Aloes | 8 drams. |
| Common Soda | 1 teaspoonful. |
| Ginger | 1 " |

Dissolve in a pint of luke warm water and give as a drench, always allowing the horse to stand in the stable a day after giving it.

This is one of the best physics known to clean out a horse's stomach and bowels, and also to purify his blood.

### PHYSIC DRENCH FOR CATTLE.

| | |
|---|---|
| Epsom Salts | 1 pound. |
| Brown Sugar | ½ pound. |
| Common Salt | 2 tablespoonfuls. |
| Ginger | 1 " |
| Common Soda | 1 " |

Dissolve in a quart of luke warm water and give as a drench. This makes a good general physic for a cow that is not thriving well.

### PHYSIC BALL FOR HORSES.

This contains the same ingredients as the physic drench, only it is prepared in a different way, as follows:

First.—Grind up eight drams of bitter aloes good and fine.

Second.—Add a few drops of water to make it sticky when rolled.

Third.—Roll it in a teaspoonful of ginger into the form of a long ball, so it will be about one-half inch in diameter and two or three inches long.

Fourth.—Roll it neatly in a piece of paper, and before giving it to the horse oil the paper, so it will slip down easily; shove it well back into the mouth and hold the horse's head up until he swallows it. The action of this ball is the same as the action of the physic drench.

### BALL TO ACT ON THE LIVER AND WORMS.

First.—Grind up four drams of bitter aloes, moisten it and roll it into the form of a ball.

Second.—Make a hole in the end of the ball and drop in one-half dram of calomel, wrap it in paper, oil the paper and give it as you would a physic ball.

## GREEN SALVE.

The following are the proportions for making green salve:

FIRST.—Take mutton tallow, one-half pound; lard, three-quarters of a pound; beeswax, two ounces; put in a pot and stir over a hot fire until it is melted.

SECOND.—Keep on stirring, and pour in one-half ounce of verdigris. Keep stirring it over the fire for fifteen minutes.

THIRD.—Then take it off the fire and add one ounce of spirits of turpentine and keep stirring it until it is cold, then it is ready for use.

This is one of the best healing salves known, especially when a wound is nearly healed up, for it keeps the wound soft, draws the edges together and allows the hair to grow over better. The way to apply it to a wound is to melt it in a spoon and apply it with a feather.

## MERCURY BLISTER.

| | |
|---|---|
| Biniodid of Mercury or Red Precipitate | 1½ drams. |
| Vaseline or Lard | 1 ounce. |

Mix thoroughly and this is one of the best blisters for blistering bony enlargements, such as splints, spavins, ringbones, sidebones, etc. Rub into the part well, grease it three days after, and in three weeks repeat the blister, and repeat in like manner until you have the required action. In case you want a heavier blister add another half dram of the biniodid of mercury. For further directions look at the separate diseases above mentioned.

## FLY BLISTER.

| | |
|---|---|
| Powdered Cantharides or Spanish Fly | 1½ drams. |
| Vaseline or Lard | 1 ounce. |

Mix thoroughly, and this is one of the best sweat blisters for blistering thickenings or enlargements in any part of the body where they are not on the bone. Use the same directions as are given in the mercury blister.

## POWDERS TO ACT ON THE KIDNEYS AND BLOOD.

| | |
|---|---|
| Nitrate of Potash or Saltpetre | ¼ pound. |
| Sulphur | ¼ " |
| Ground Gentian Root | ¼ " |

Mix thoroughly and give a teaspoonful three times a day in the horse's food or on its tongue with a spoon. If it is a cow, give a tablespoonful twice a day; if a sheep, give a teaspoonful twice a day. This is a good powder for starting an animal to thrive, but before giving it, it is best to give a physic drench. If it is a horse, give aloes; if a cow, give salts.

### WORM POWDERS.

| | |
|---|---|
| Sulphate of Iron | ¼ pound. |
| Ground Gentian Root | ¼ " |

Mix thoroughly and give a teaspoonful three times a day in the animal's food or on its tongue with a spoon. This powder is only intended for bots, round worms and pin worms in horses. It is also a good tonic powder. Before and after using the powder give a physic drench.

### GARGLES.

| | |
|---|---|
| Sulphur | ¼ pound. |
| Nitrate of Potash or Saltpetre | ¼ " |
| Powdered Alum | ⅛ " |

Mix thoroughly and give a teaspoonful three times a day on a horse or cow's tongue with a spoon, if it is a sheep, pig or dog, give half a teaspoonful three times a day on the tongue with a spoon. This is a splendid powder for a gargle in all cases of sore throat, or sore mouth, and also makes a good cough powder.

### GENERAL CONDITION POWDER.

| | |
|---|---|
| Nitrate of Potash or Saltpetre | ¼ pound. |
| Common Soda | ¼ " |
| Ground Aniseed | ¼ " |
| Femigreek | ⅓ " |

Mix thoroughly and give a tablespoonful every night in a bran mash, but if in a case where you want to fatten the animal in a hurry, give a tablespoonful twice a day in a bran mash. In horses that are working hard give a tablespoonful every Saturday night in a bran mash to keep them in good working condition.

### RECEIPT FOR COLIC AND INDIGESTION.

| | |
|---|---|
| Sweet Spirits of Nitre | 1 ounce or 4 tablespoonfuls. |
| Tincture of Laudanum | 1 ounce or 4 tablespoonfuls. |
| Fleming's Tincture of Aconite | 10 drops. |
| Common Soda | 1 teaspoonful. |
| Ginger | 1 " |

Mix in a pint of luke warm water and give as a drench every two hours in cases of colic, indigestion and inflammation. For further particulars refer to diseases of the stomach and bowels.

### HOOF OINTMENT.

| | |
|---|---|
| Raw Linseed Oil | ¼ pint. |
| Crude Petroleum Oil | ¼ " |
| Neatsfoot Oil | ¼ " |
| Pine Tar | ¼ " |

Mix well and apply every night with a brush all over and under the hoof—even a little in the hair above the hoof. Clean out the hoof before applying.

# LIST OF MEDICINES TO BE KEPT ON HAND.

The following is a list of medicines that every stockowner should keep on hand for cases of emergency, that is, where he has an animal take sick with inflammation, colic, indigestion, or any other disease that requires his immediate attention, he can give these medicines, while if he had to go to a drug store for them at the time, he might lose the animal:

| | |
|---|---|
| Sweet Spirits of Nitre | 4 ounces. |
| Tincture of Laudanum | 4 " |
| Fleming's Tincture of Aconite | ½ " |
| Raw Linseed Oil | 1 pint. |
| Epsom Salts | 1 pound. |
| Bitter Aloes | 8 drams. |

It is also advisable to have a bottle of white liniment and a bottle of white lotion on hand for, as they do not cost much when mixed by the quart, they are very handy to have on hand for sprains, sore shoulders, cuts, etc. As for the other medicines in the book they can be got at a drug store any time, as the cases they are used in are not so urgent. The cost of the list of medicines given is trifling, and by having them on hand and able to give it at the time the animal is noticed to be sick will often save the animal's life, whereas, if you had to go to a drug store at the time, the delay would cause the loss of the animal.

NOTE.—In buying fluid medicine always take your bottle to the drug store, for the new bottle often costs you more than the medicine.

## TAKE NOTICE.

All the doses of medicines which are mentioned in the section of the book, dealing with Medicines and Receipts, are intended for an average sized animal, unless otherwise mentioned, so in giving doses to smaller animals you must regulate the dose to suit the animal.

# CONSULTATION AND ADVICE FREE OF CHARGE.

Any person buying a copy of THE VETERINARY SCIENCE from one of our authorized agents will have the privilege of free consultation on any disease they do not understand, or anything in the book they want more information about. In cases where a new disease should break out, by sending the symptoms of the case in full, and causes of the trouble as near as you can tell, we will furnish you by return mail, advice and prescription in full—free of charge. Parties living in Canada should enclose a three cent stamp; if in the United States, five cents in silver, for return postage. Advice given on all kinds of operations, (castrating rigs or original horses a specialty). Address all mail under this heading to J. E. Hodgins, V. S., or to T. H. Haskett, Secretary of the Veterinary Science Company, London, Canada.

# INDEX.

## PART I.

### ANATOMY OF THE HORSE.

# INDEX PART II.

## DISEASES AND TREATMENT OF THE HORSE.

# INDEX PART III.

## ANATOMY, DISEASES AND TREATMENT OF CATTLE.

# INDEX PART IV.

## DISEASES AND TREATMENT OF SHEEP, PIGS, DOGS AND POULTRY.

### SHEEP.

## PIGS.

## DOGS.



## POULTRY.

# INDEX PART V.

## MEDICINES AND RECEIPTS.

## MEDICINES.

## RECEIPTS.



## ILLUSTRATIONS.

www.ingramcontent.com/pod-product-compliance
Lightning Source LLC
Chambersburg PA
CBHW021349210326
41599CB00011B/803

* 9 7 8 3 3 3 7 2 3 6 5 6 4 *